Holt California Geometry

Review for Mastery Workbook

HOLT, RINEHART AND WINSTON
A Harcourt Education Company
Orlando • Austin • New York • San Diego • London

Printed in the United States of America

ISBN 13: 978-0-03-099025-0
ISBN 10: 0-03-099025-4

2 3 4 5 6 7 8 9 862 10 09 08 07

Contents

Chapter 1 1
Chapter 2 15
Chapter 3 29
Chapter 4 41
Chapter 5 57
Chapter 6 73
Chapter 7 85
Chapter 8 97
Chapter 9 109
Chapter 10 121
Chapter 11 137
Chapter 12 151

Name ______________________________________ Date _______________ Class _______________

LESSON **1-1**

Review for Mastery

Understanding Points, Lines, and Planes

A **point** has no size. It is named using a capital letter.
All the figures below contain points.

•P point *P*

Figure	Characteristics	Diagram	Words and Symbols
line	0 endpoints extends forever in two directions	A B	line *AB* or $\overleftrightarrow{AB}$
line segment or segment	2 endpoints has a finite length	X Y	segment *XY* or $\overline{XY}$
ray	1 endpoint extends forever in one direction	Q R	*ray RQ* or $\overrightarrow{RQ}$ *A ray is named starting with its endpoint.*
plane	extends forever in all directions	$\mathcal{V}$ F G H	plane *FGH* or plane $\mathcal{V}$

Draw and label a diagram for each figure.

1. point *W*

2. line *MN*

3. $\overline{JK}$

4. $\overrightarrow{EF}$

Name each figure using words and symbols.

5.

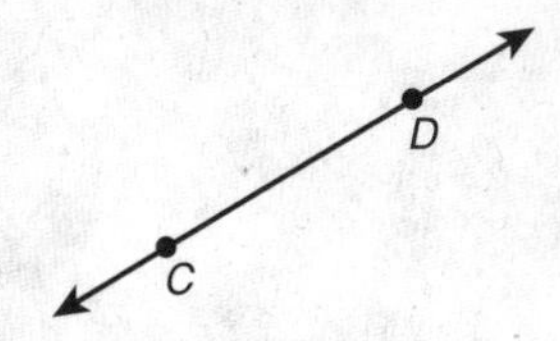

____

6.

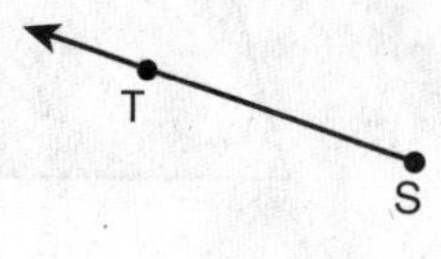

____

7. Name the plane in two different ways.

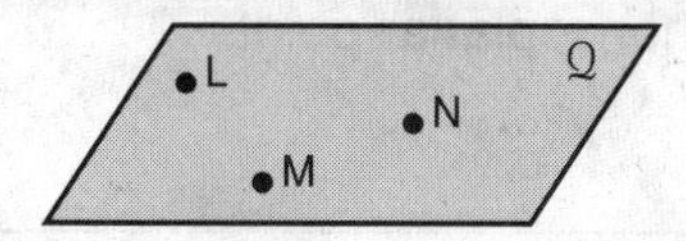

____

8.

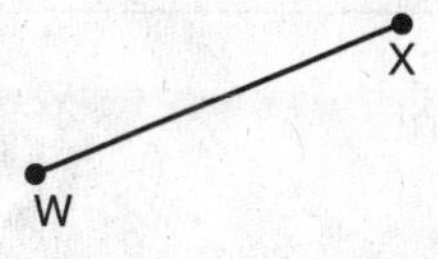

____

Name ______________________________ Date ____________ Class ____________

LESSON **1-1**

Review for Mastery

Understanding Points, Lines, and Planes continued

Term	Meaning	Model
collinear	points that lie on the same line	*F* and *G* are collinear. *F*, *G*, and *H* are noncollinear.
noncollinear	points that do not lie on the same line	
coplanar	points or lines that lie in the same plane	*W*, *X*, and *Y* are coplanar. *W*, *X*, *Y*, and *Z* are noncoplanar.
noncoplanar	points or lines that do not lie in the same plane	

Figures that intersect share a common set of points. In the first model above, $\overrightarrow{FH}$ intersects $\overleftrightarrow{FG}$ at point *F*. In the second model, $\overleftrightarrow{XZ}$ intersects plane *WXY* at point *X*.

Use the figure for Exercises 9–14. Name each of the following.

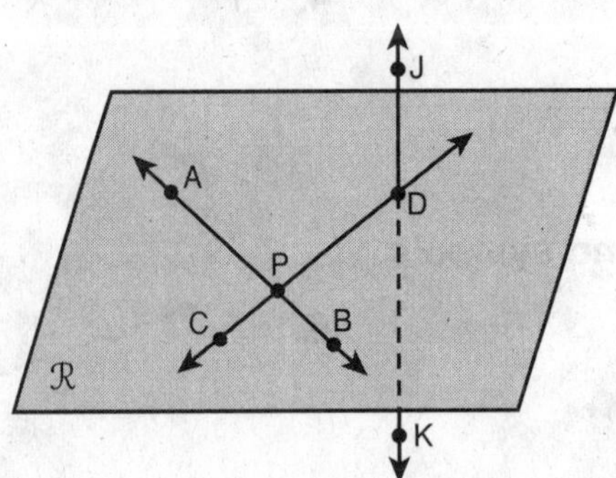

9. three collinear points

____

10. three noncollinear points

____

11. four coplanar points

____

12. four noncoplanar points

____

13. two lines that intersect $\overleftrightarrow{CD}$

____

14. the intersection of $\overleftrightarrow{JK}$ and plane ℛ

____

Name ______________________ Date ____________ Class ____________

LESSON 1-2

Review for Mastery

Measuring and Constructing Segments

The **distance** between any two points is the **length** of the segment that connects them.

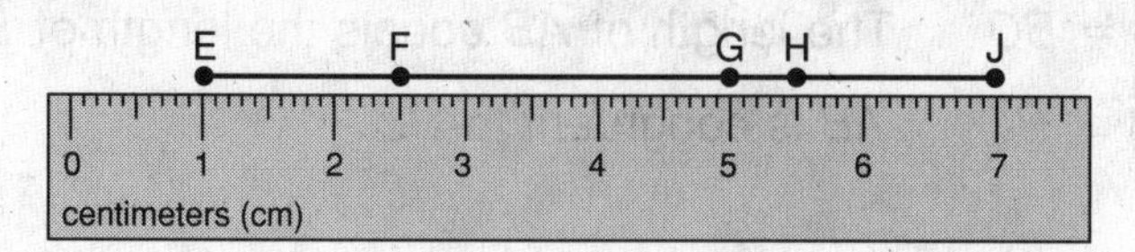

The distance between E and J is EJ, the length of $\overline{EJ}$. To find the distance, subtract the numbers corresponding to the points and then take the absolute value.

$$EJ = |7 - 1|$$
$$= |6|$$
$$= 6 \text{ cm}$$

Use the figure above to find each length.

1. EG ____________

2. EF ____________

3. FH ____________

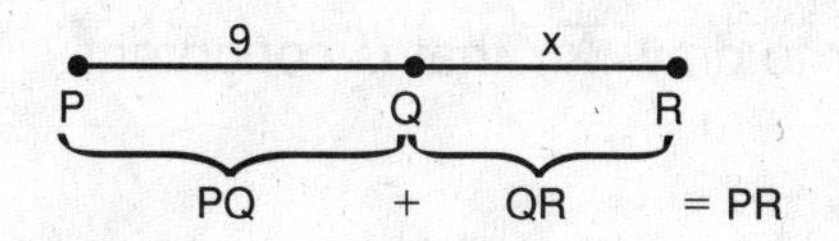

On $\overline{PR}$, Q is between P and R. If $PR = 16$, we can find QR.

$$PQ + QR = PR$$
$$9 + x = 16$$
$$x = 7$$
$$QR = 7$$

4. y, 5; J, K, L; 7

Find JK. ________

5. 8, z; A, B, C; 14

Find BC. ________

6.

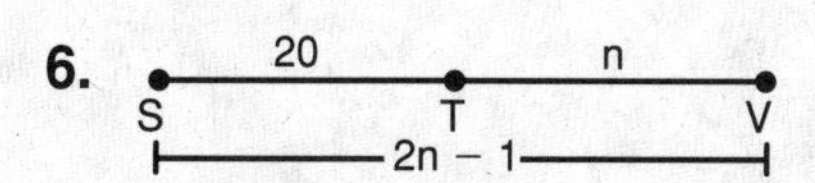

Find SV. ________

7.

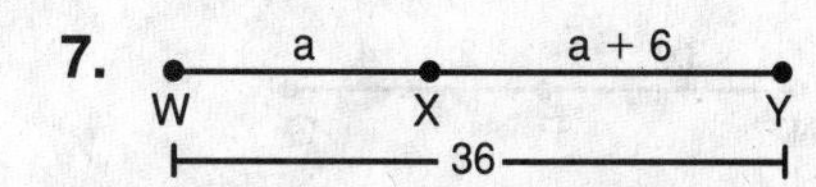

Find XY. ________

8. 90, 45; D, E, F; x

Find DF. ________

9.

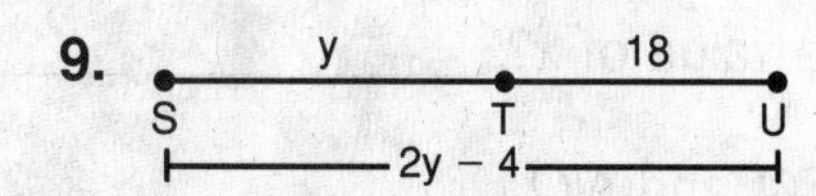

Find ST. ________

Name ______________________ Date ____________ Class ____________

LESSON 1-2

Review for Mastery

Measuring and Constructing Segments continued

Segments are **congruent** if their lengths are equal.

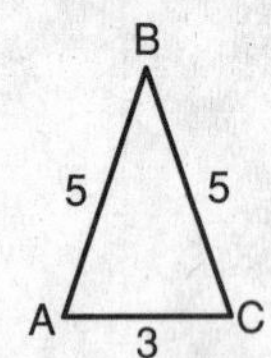

$AB = BC$ The length of $\overline{AB}$ equals the length of $\overline{BC}$.

$\overline{AB} \cong \overline{BC}$ $\overline{AB}$ is congruent to $\overline{BC}$.

Copying a Segment	
Method	**Steps**
sketch using estimation	Estimate the length of the segment. Sketch a segment that is about the same length.
draw with a ruler	Use a ruler to measure the length of the segment. Use the ruler to draw a segment having the same length.
construct with a compass and straightedge	Draw a line and mark a point on it. Open the compass to the length of the original segment. Mark off a segment on your line at the same length.

Refer to triangle *ABC* above for Exercises 10 and 11.

10. Sketch $\overline{LM}$ that is congruent to $\overline{AC}$.

11. Use a ruler to draw $\overline{XY}$ that is congruent to $\overline{BC}$.

12. Use a compass to construct $\overline{ST}$ that is congruent to $\overline{JK}$.

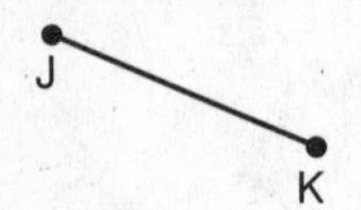

The **midpoint** of a segment separates the segment into two congruent segments. In the figure, P is the midpoint of $\overline{NQ}$.

$3x$ $2x + 4$

N P Q

13. $\overline{PQ}$ is congruent to ______________.

14. What is the value of x? ______________

15. Find NP, PQ, and NQ. ______________

Name ______________________ Date ____________ Class ____________

LESSON 1-3

Review for Mastery

Measuring and Constructing Angles

An **angle** is a figure made up of two rays, or **sides,** that have a common endpoint, called the **vertex** of the angle.

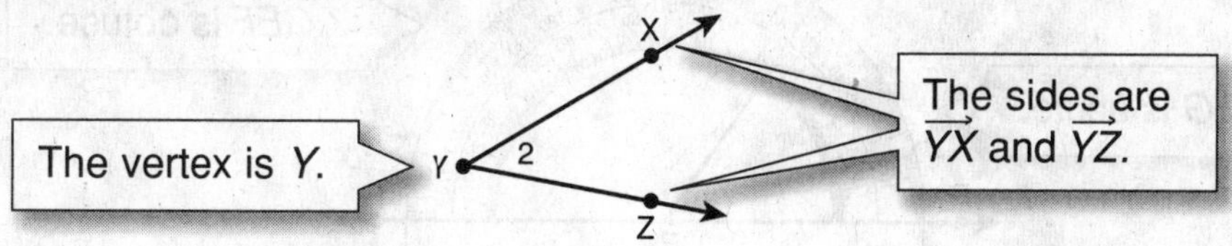

There are four ways to name this angle.

$\angle Y$	*Use the vertex.*
$\angle XYZ$ or $\angle ZYX$	*Use the vertex and a point on each side.*
$\angle 2$	*Use the number.*

Name each angle in three ways.

1.

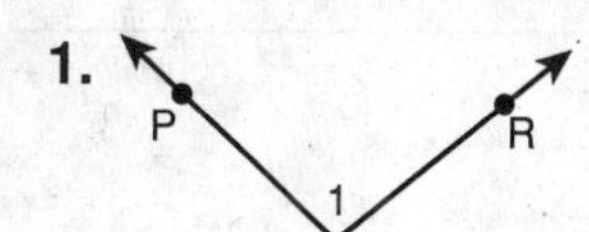

____

2.

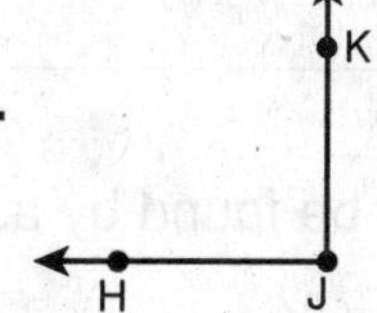

____

3. Name three different angles in the figure.

____

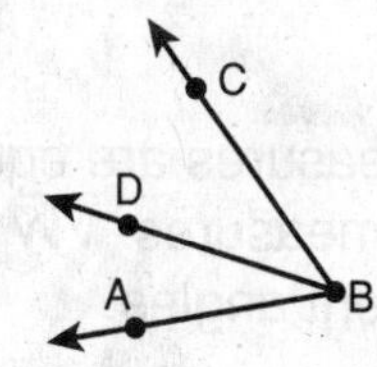

Angle	acute	right	obtuse	straight
Model	a°	a°	a°	a°
Possible Measures	$0° < a° < 90°$	$a° = 90°$	$90° < a° < 180°$	$a° = 180°$

Classify each angle as acute, right, obtuse, or straight.

4. $\angle NMP$

____

5. $\angle QMN$

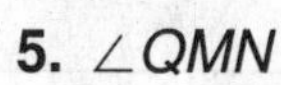

____

6. $\angle PMQ$

____

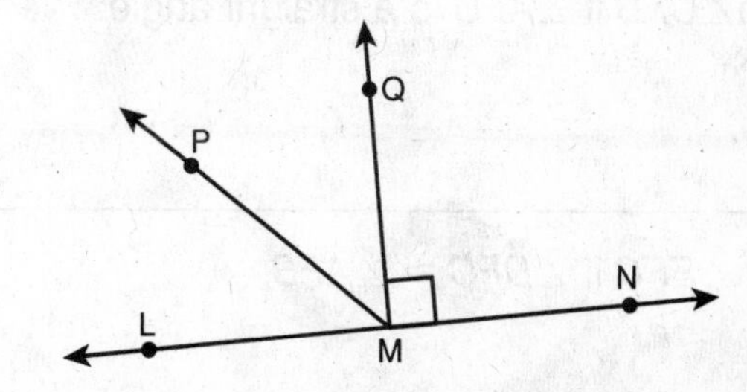

Name ______________________ Date __________ Class __________

LESSON 1-3

Review for Mastery

Measuring and Constructing Angles continued

You can use a protractor to find the measure of an angle.

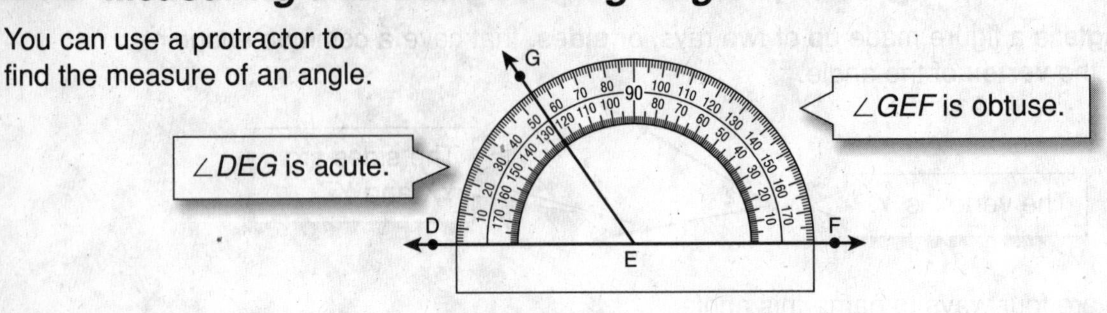

Use the figure above to find the measure of each angle.

7. $\angle DEG$ ______________________

8. $\angle GEF$ ______________________

The measure of $\angle XVU$ can be found by adding.

$m\angle XVU = m\angle XVW + m\angle WVU$

$= 48° + 48°$

$= 96°$

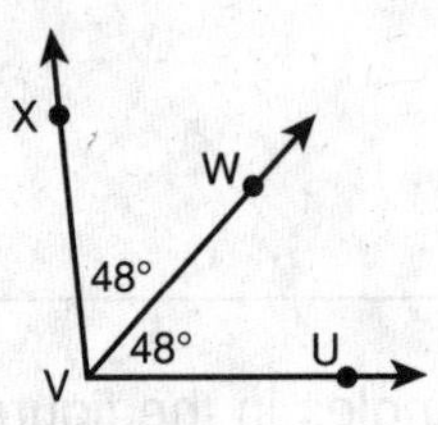

Angles are **congruent** if their measures are equal. In the figure, $\angle XVW \cong \angle WVU$ because the angles have equal measures. $\overrightarrow{VW}$ is an **angle bisector** of $\angle XVU$ because it divides $\angle XVU$ into two congruent angles.

Find each angle measure.

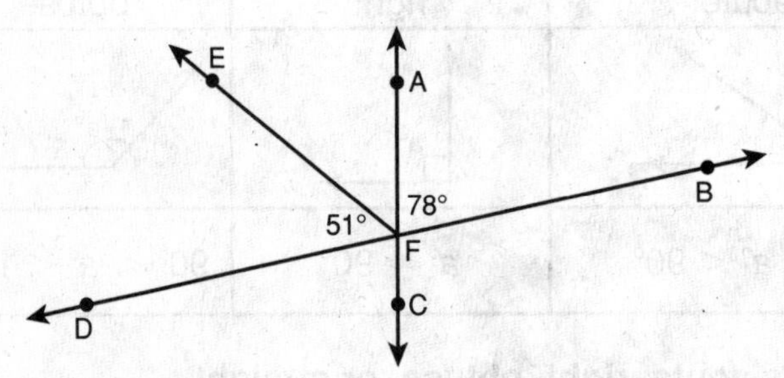

9. $m\angle CFB$ if $\angle AFC$ is a straight angle. ______________________

10. $m\angle EFA$ if the angle is congruent to $\angle DFE$. ______________________

11. $m\angle EFC$ if $\angle DFC \cong \angle AFB$. ______________________

12. $m\angle CFG$ if $\overrightarrow{FG}$ is an angle bisector of $\angle CFB$. ______________________

Name ______________________ Date ____________ Class ____________

LESSON 1-4

Review for Mastery

Pairs of Angles

Angle Pairs		
Adjacent Angles	**Linear Pairs**	**Vertical Angles**
have the same vertex and share a common side	adjacent angles whose noncommon sides are opposite rays	nonadjacent angles formed by two intersecting lines
$\angle 1$ and $\angle 2$ are adjacent.	$\angle 3$ and $\angle 4$ are adjacent and form a linear pair.	$\angle 5$ and $\angle 6$ are vertical angles.

Tell whether $\angle 7$ and $\angle 8$ in each figure are only adjacent, are adjacent and form a linear pair, or are not adjacent.

1.

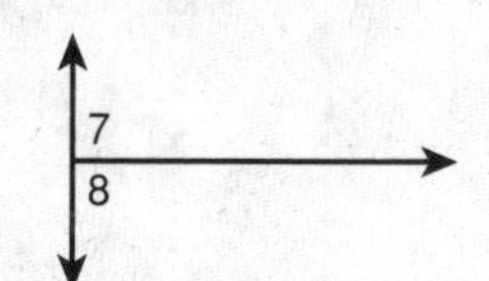

2.

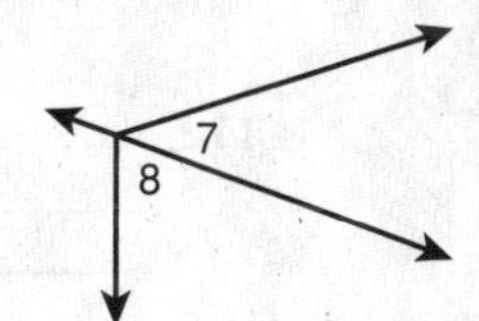

3.

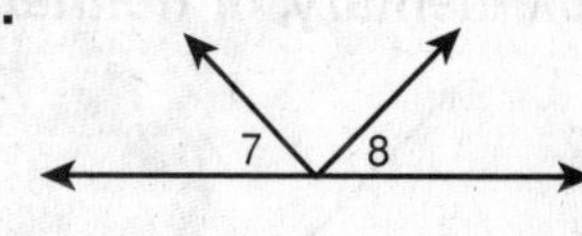

________________ ________________ ________________

Tell whether the indicated angles are only adjacent, are adjacent and form a linear pair, or are not adjacent.

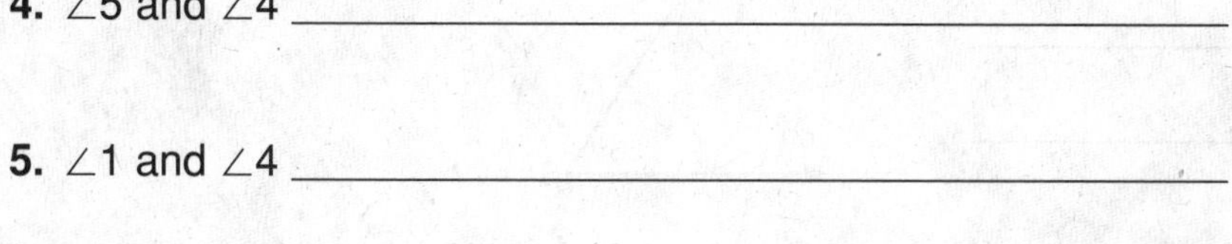

4. $\angle 5$ and $\angle 4$ ________________

5. $\angle 1$ and $\angle 4$ ________________

6. $\angle 2$ and $\angle 3$ ________________

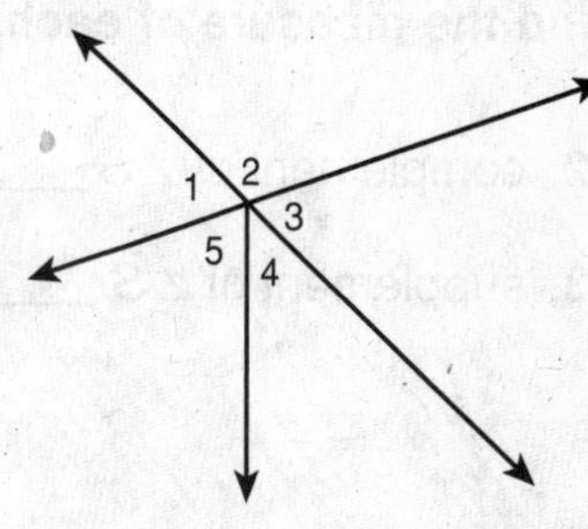

Name each of the following.

7. a pair of vertical angles ________________

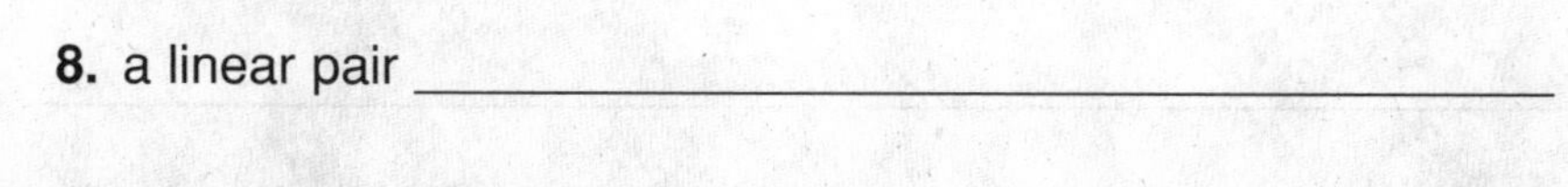

8. a linear pair ________________

9. an angle adjacent to $\angle 4$ ________________

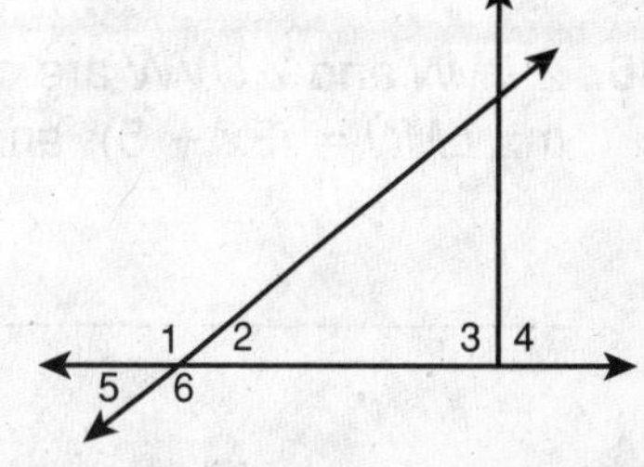

Name ________________________ Date ____________ Class ____________

LESSON 1-4

Review for Mastery

Pairs of Angles continued

Angle Pairs	
Complementary Angles	**Supplementary Angles**
sum of angle measures is 90°	sum of angle measures is 180°
$m\angle 1 + m\angle 2 = 90°$ In each pair, ∠1 and ∠2 are complementary.	$m\angle 3 + m\angle 4 = 180°$ In each pair, ∠3 and ∠4 are supplementary.

Tell whether each pair of labeled angles is complementary, supplementary, or neither.

10.

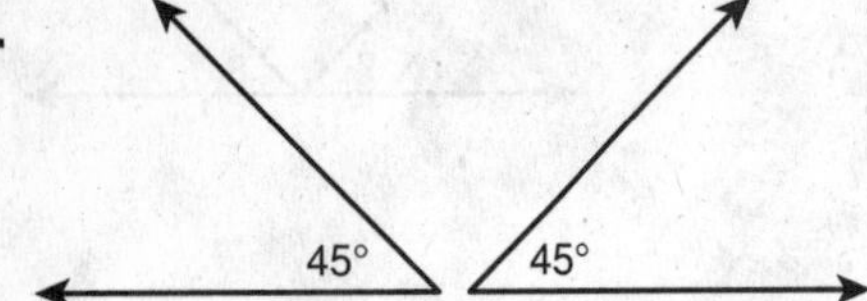

____

11.

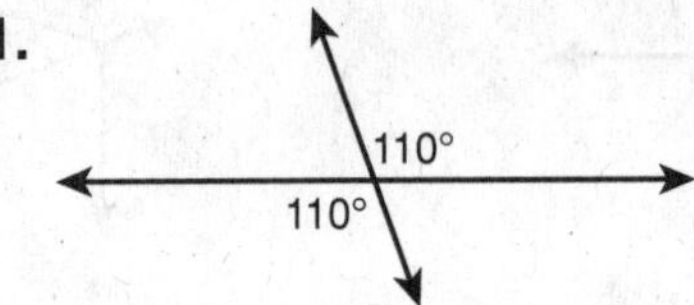

____

Find the measure of each of the following angles.

12. complement of ∠*S* ____________

13. supplement of ∠*S* ____________

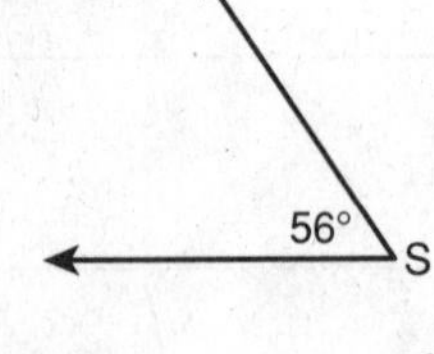

14. complement of ∠*R* ____________

15. supplement of ∠*R* ____________

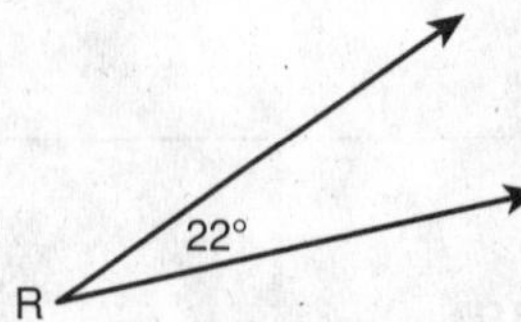

16. ∠*LMN* and ∠*UVW* are complementary. Find the measure of each angle if $m\angle LMN = (3x + 5)°$ and $m\angle UVW = 2x°$.

____

Name ______________________ Date __________ Class __________

LESSON 1-5

Review for Mastery

Using Formulas in Geometry

The **perimeter** of a figure is the sum of the lengths of the sides.
The **area** is the number of square units enclosed by the figure.

Figure	**Rectangle**	**Square**
Model	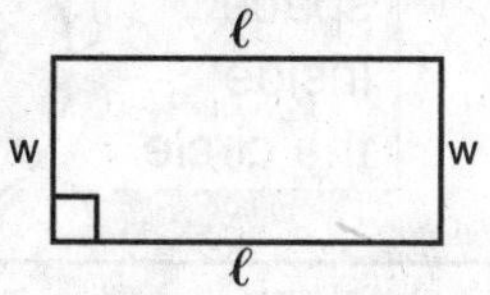	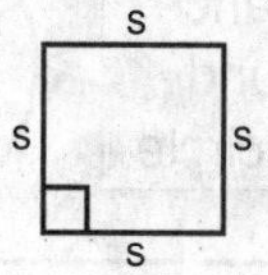
Perimeter	$P = 2\ell + 2w$ or $2(\ell + w)$	$P = 4s$
Area	$A = \ell w$	$A = s^2$

Find the perimeter and area of each figure.

1. rectangle with $\ell = 4$ ft, $w = 1$ ft

____

2. square with $s = 8$ mm

____

3.

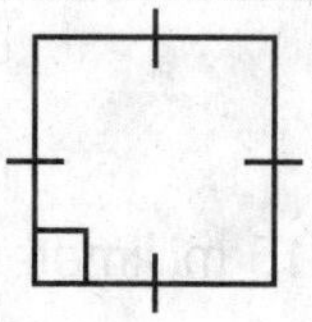

____

4.

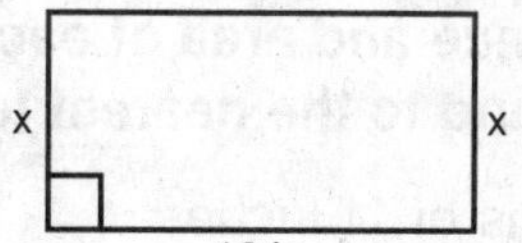

____

The perimeter of a triangle is the sum of its side lengths.
The base and height are used to find the area.

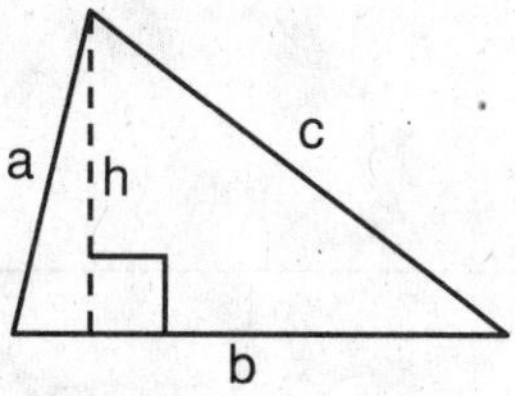

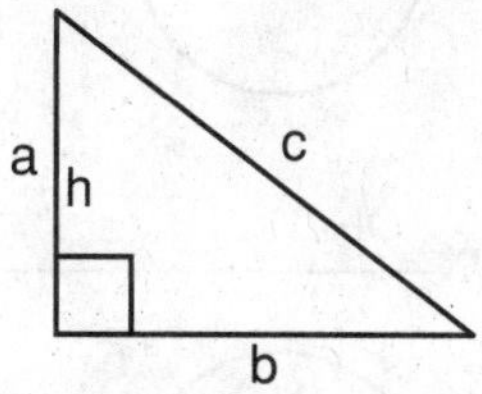

Perimeter

$P = a + b + c$

Area

$A = \frac{1}{2}bh$ or $\frac{bh}{2}$

Find the perimeter and area of each triangle.

5.

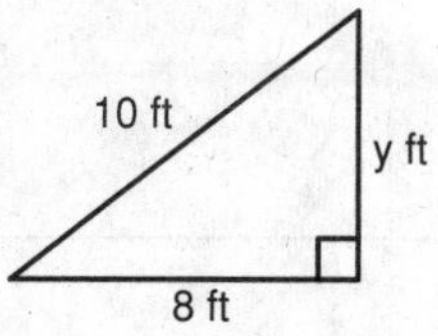

____

6.

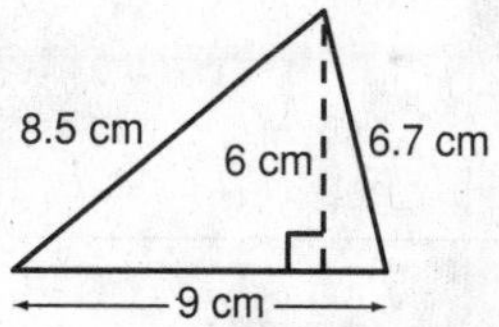

____

Name ______________________ Date ____________ Class ____________

LESSON 1-5

Review for Mastery

Using Formulas in Geometry continued

Circles		
	Circumference	**Area**
Models	distance around the circle (d)	space inside the circle (r)
Words	pi times the diameter or 2 times pi times the radius	pi times the square of the radius
Formulas	$C = \pi d$ or $C = 2\pi r$	$A = \pi r^2$

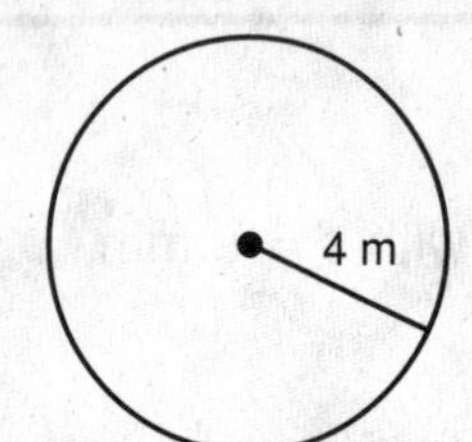

$C = 2\pi r$ $\quad A = \pi r^2$

$C = 2\pi(4)$ $\quad A = \pi(4)^2$

$C = 8\pi$ $\quad A = 16\pi$

$C \approx 25.1$ m $\quad A \approx 50.3$ m^2

Find the circumference and area of each circle. Use the π key on your calculator. Round to the nearest tenth.

7. circle with a radius of 11 inches

____

8. circle with a diameter of 15 millimeters

____

9.

____

10.

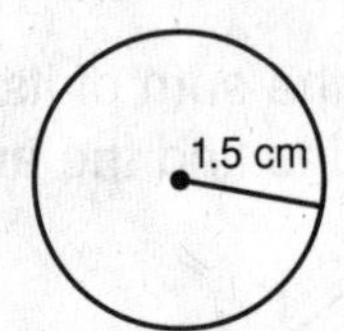

____

11.

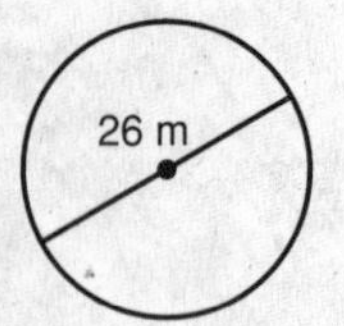

____

12.

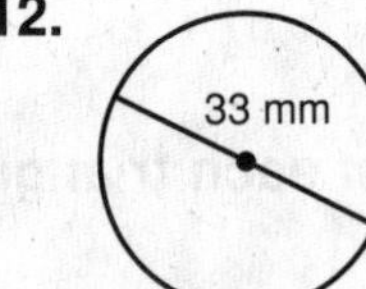

____

Name ______________________ Date ____________ Class ____________

LESSON 1-6

Review for Mastery

Midpoint and Distance in the Coordinate Plane

The **midpoint** of a line segment separates the segment into two halves. You can use the **Midpoint Formula** to find the midpoint of the segment with endpoints $G(1, 2)$ and $H(7, 6)$.

$$M\left(\frac{x_1 + x_2}{2}, \frac{y_1 + y_2}{2}\right) = M\left(\frac{1 + 7}{2}, \frac{2 + 6}{2}\right)$$

$$= M\left(\frac{8}{2}, \frac{8}{2}\right)$$

$$= M(4, 4)$$

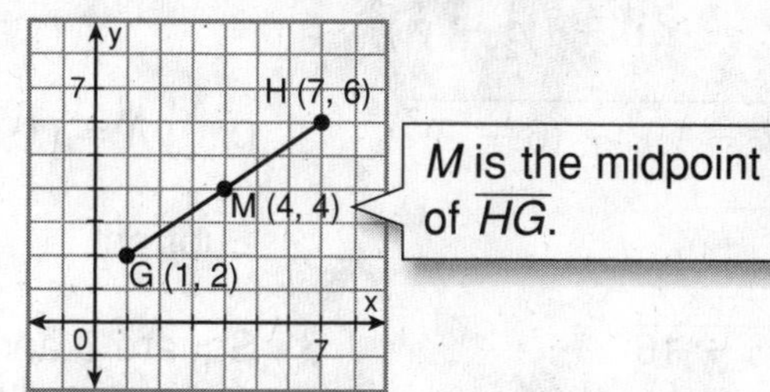

Find the coordinates of the midpoint of each segment.

1. 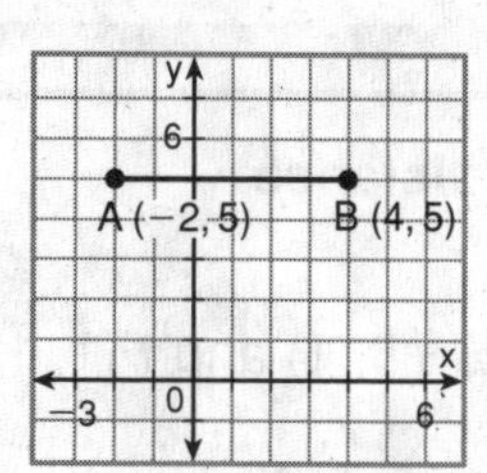

____

2. 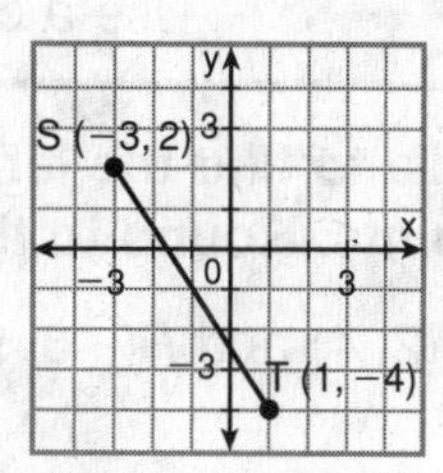

____

3. $\overline{QR}$ with endpoints $Q(0, 5)$ and $R(6, 7)$ ______________

4. $\overline{JK}$ with endpoints $J(1, –4)$ and $K(9, 3)$ ______________

Suppose $M(3, -1)$ is the midpoint of $\overline{CD}$ and C has coordinates (1, 4). You can use the Midpoint Formula to find the coordinates of D.

$$M(3, -1) = M\left(\frac{x_1 + x_2}{2}, \frac{y_1 + y_2}{2}\right)$$

x-coordinate of D		y-coordinate of D
$3 = \frac{x_1 + x_2}{2}$	Set the coordinates equal.	$-1 = \frac{y_1 + y_2}{2}$
$3 = \frac{1 + x_2}{2}$	Replace (x_1, y_1) with (1, 4).	$-1 = \frac{4 + y_2}{2}$
$6 = 1 + x_2$	Multiply both sides by 2.	$-2 = 4 + y_2$
$5 = x_2$	Subtract to solve for x_2 and y_2.	$-6 = y_2$

The coordinates of D are $(5, -6)$.

5. $M(-3, 2)$ is the midpoint of $\overline{RS}$, and R has coordinates (6, 0). What are the coordinates of S? ______________

6. $M(7, 1)$ is the midpoint of $\overline{WX}$, and X has coordinates $(-1, 5)$. What are the coordinates of W? ______________

Name ________________________ Date ____________ Class ____________

LESSON 1-6

Review for Mastery

Midpoint and Distance in the Coordinate Plane continued

The **Distance Formula** can be used to find the distance d between points A and B in the coordinate plane.

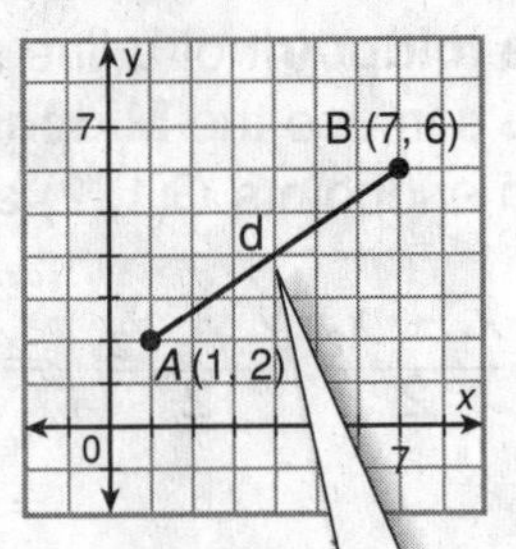

$d = \sqrt{(x_2 - x_1)^2 + (y_2 - y_1)^2}$

$= \sqrt{(7 - 1)^2 + (6 - 2)^2}$ $(x_1, y_1) = (1, 2); (x_2, y_2) = (7, 6)$

$= \sqrt{6^2 + 4^2}$ Subtract.

$= \sqrt{36 + 16}$ Square 6 and 4.

$= \sqrt{52}$ Add.

≈ 7.2 Use a calculator.

Use the Distance Formula to find the length of each segment or the distance between each pair of points. Round to the nearest tenth.

7. $\overline{QR}$ with endpoints $Q(2, 4)$ and $R(-3, 9)$

____

8. $\overline{EF}$ with endpoints $E(-8, 1)$ and $F(1, 1)$

____

9. $T(8, -3)$ and $U(5, 5)$

____

10. $N(4, -2)$ and $P(-7, 1)$

____

You can also use the Pythagorean Theorem to find distances in the coordinate plane. Find the distance between J and K.

$c^2 = a^2 + b^2$ Pythagorean Theorem

$= 5^2 + 6^2$ $a = 5$ units and $b = 6$ units

$= 25 + 36$ Square 5 and 6.

$= 61$ Add.

$c = \sqrt{61}$ or about 7.8 Take the square root.

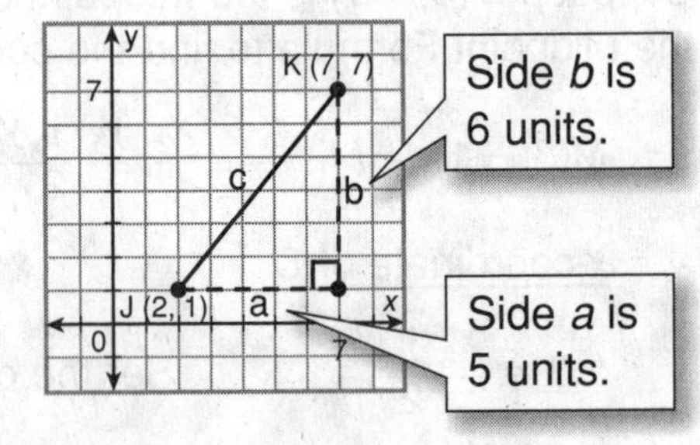

Use the Pythagorean Theorem to find the distance, to the nearest tenth, between each pair of points.

11.

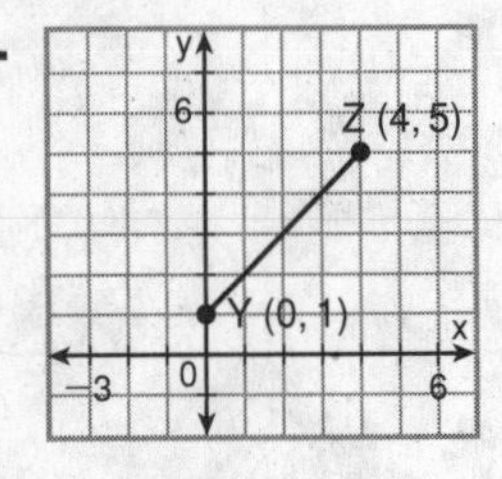

____

12.

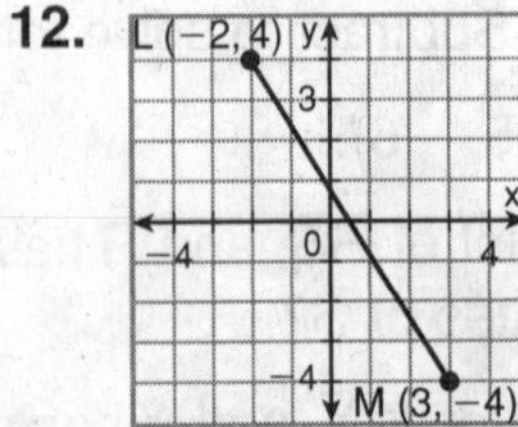

____

Name ______________________ Date __________ Class __________

LESSON 1-7

Review for Mastery

Transformations in the Coordinate Plane

In a transformation, each point of a figure is moved to a new position.

Reflection	Rotation	Translation
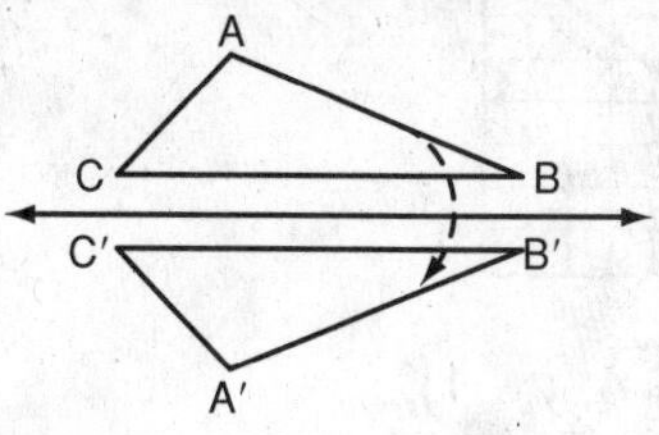 $\triangle ABC \rightarrow \triangle A'B'C'$	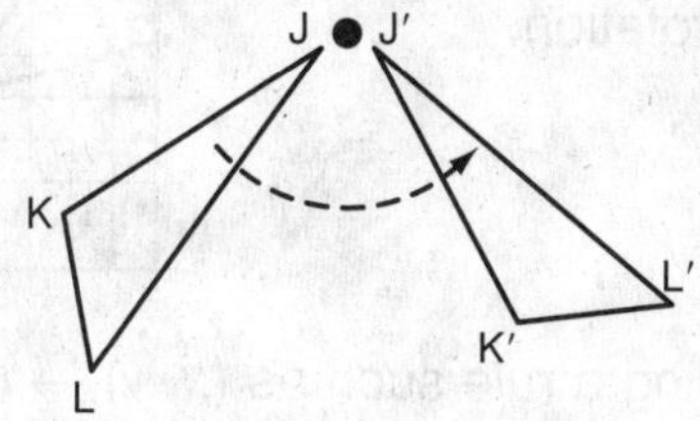 $\triangle JKL \rightarrow \triangle J'K'L'$	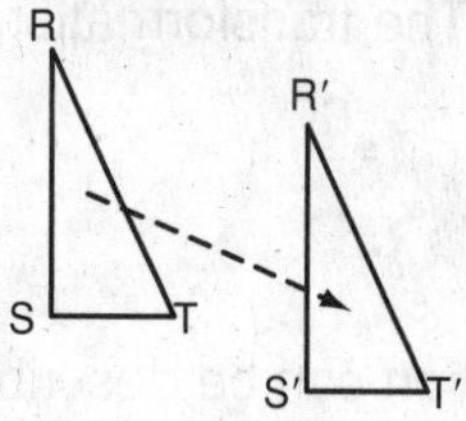 $\triangle RST \rightarrow \triangle R'S'T'$
A figure is flipped over a line.	A figure is turned around a fixed point.	A figure is slid to a new position without turning.

Identify each transformation. Then use arrow notation to describe the transformation.

1.

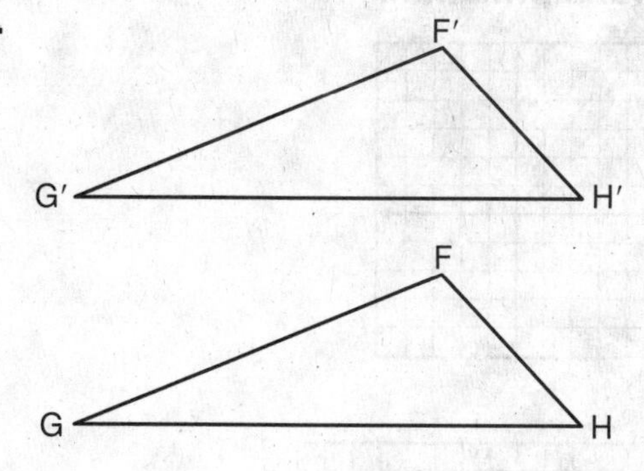

____

2.

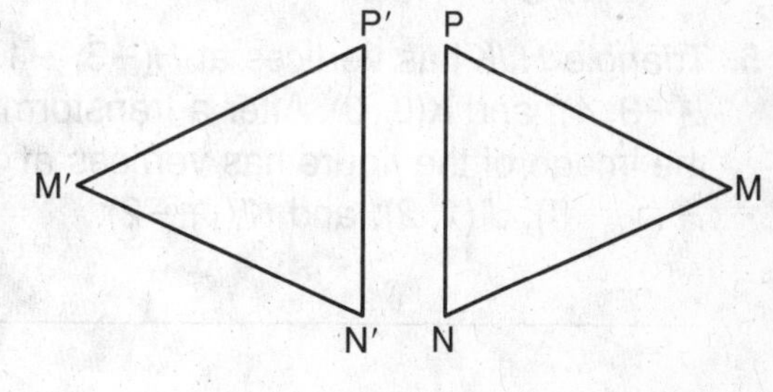

____

3.

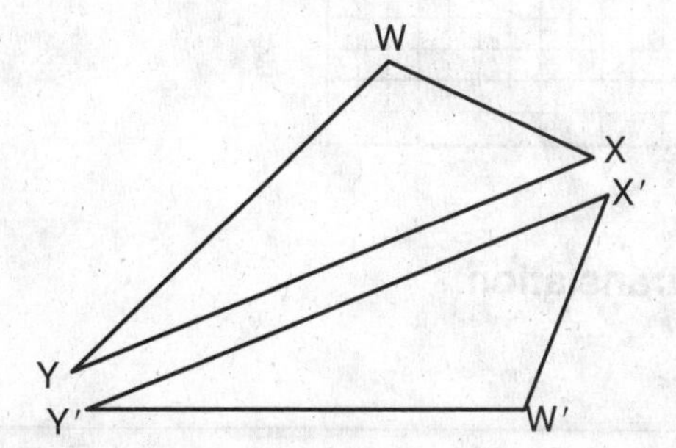

____

4.

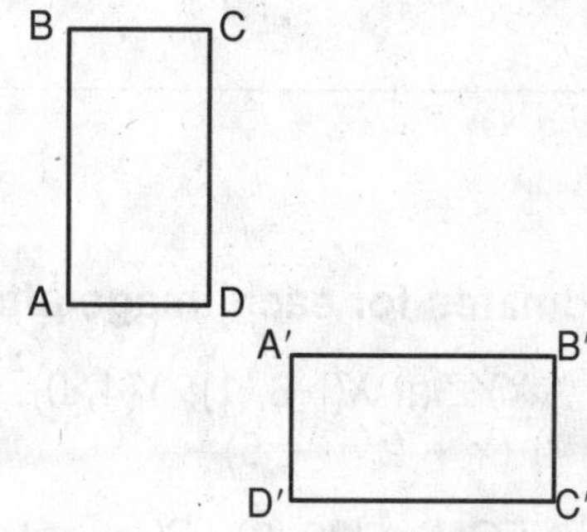

____

Name ______________________ Date __________ Class __________

LESSON 1-7

Review for Mastery

Transformations in the Coordinate Plane continued

Triangle *QRS* has vertices at *Q*(−4, 1), *R*(−3, 4), and *S*(0, 0). After a transformation, the image of the figure has vertices at *Q*′(1, 4), *R*′(4, 3), and *S*′(0, 0). The transformation is a rotation.

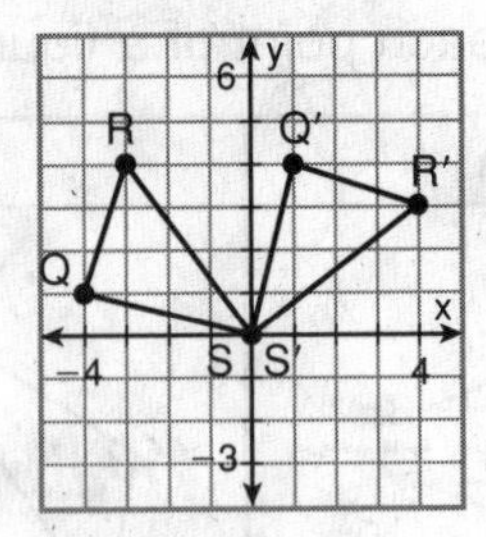

A translation can be described using a rule such as $(x, y) \rightarrow (x + 4, y - 1)$.

Preimage	Apply Rule	Image
R(3, 5)	*R*(3 + 4, 5 − 1)	*R*′(7, 4)
S(0, 1)	*S*(0 + 4, 1 − 1)	*S*′(4, 0)
T(2, −1)	*T*(2 + 4, −1 − 1)	*T*′(6, −2)

Draw each figure and its image. Then identify the transformation.

5. Triangle *HJK* has vertices at *H*(−3, −1), *J*(−3, 4), and *K*(0, 0). After a transformation, the image of the figure has vertices at *H*′(1, −3), *J*′(1, 2), and *K*′(4, −2).

   ____

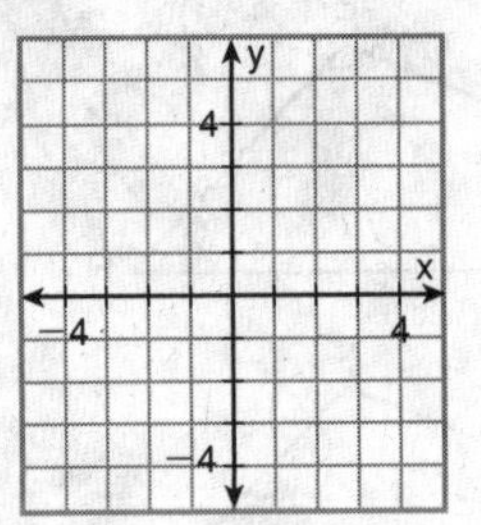

6. Triangle *CDE* has vertices at *C*(−4, 6), *D*(−1, 6), and *E*(−2, 1). After a transformation, the image of the figure has vertices at *C*′(4, 6), *D*′(1, 6), and *E*′(2, 1).

   ____

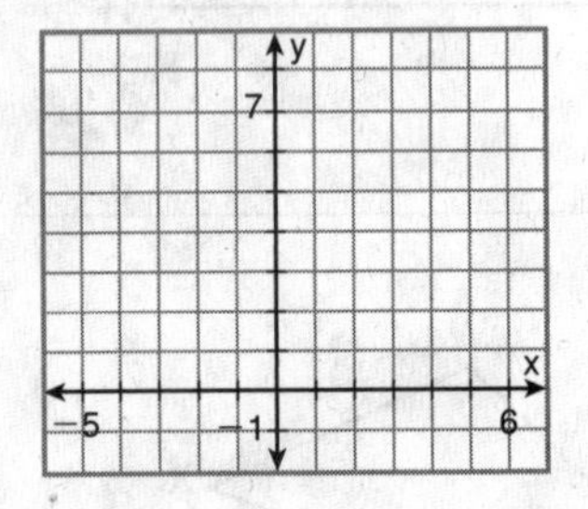

Find the coordinates for each image after the given translation.

7. preimage: △*XYZ* at *X*(−6, 1), *Y*(4, 0), *Z*(1, 3)
   rule: $(x, y) \rightarrow (x + 2, y + 5)$ ______________________

8. preimage: △*FGH* at *F*(9, 8), *G*(−6, 1), *H*(−2, 4)
   rule: $(x, y) \rightarrow (x - 3, y + 1)$ ______________________

9. preimage: △*BCD* at *B*(0, 2), *C*(−7, 1), *D*(1, 5)
   rule: $(x, y) \rightarrow (x + 7, y - 1)$ ______________________

Name ______________________ Date ____________ Class ____________

LESSON 2-1

Review for Mastery

Using Inductive Reasoning to Make Conjectures

When you make a general rule or conclusion based on a pattern, you are using **inductive reasoning.** A conclusion based on a pattern is called a **conjecture.**

Pattern	Conjecture	Next Two Items
$-8, -3, 2, 7, \ldots$	Each term is 5 more than the previous term.	$7 + 5 = 12$ $12 + 5 = 17$
45°	The measure of each angle is half the measure of the previous angle.	22.5° 11.25°

Find the next item in each pattern.

1. $\frac{1}{4}, \frac{1}{2}, \frac{3}{4}, 1, \ldots$

2. 100, 81, 64, 49, . . .

3.

3 6 10

4.

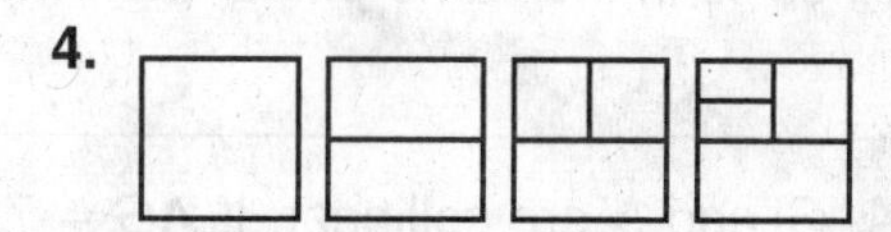

Complete each conjecture.

5. If the side length of a square is doubled, the perimeter of the square is ____________________.

6. The number of nonoverlapping angles formed by *n* lines intersecting in a point is ____________________.

Use the figure to complete the conjecture in Exercise 7.

7. The perimeter of a figure that has *n* of these triangles is ____________________.

1 1 1
P = 3

1 1 1 1
P = 4

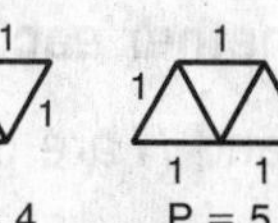

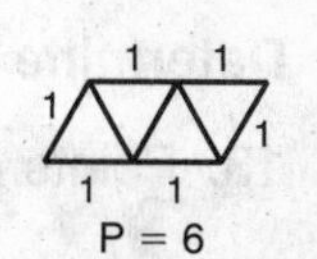

Name ______________________ Date ____________ Class ____________

LESSON 2-1

Review for Mastery

Using Inductive Reasoning to Make Conjectures continued

Since a conjecture is an educated guess, it may be true or false. It takes only one example, or **counterexample,** to prove that a conjecture is false.

Conjecture: For any integer n, $n \leq 4n$.

n	$n \leq 4n$	True or False?
3	$3 \leq 4(3)$ $3 \leq 12$	true
0	$0 \leq 4(0)$ $0 \leq 0$	true
−2	$-2 \leq 4(-2)$ $-2 \leq -8$	false

$n = -2$ is a counterexample, so the conjecture is false.

Show that each conjecture is false by finding a counterexample.

8. If three lines lie in the same plane, then they intersect in at least one point.

____

9. Points A, G, and N are collinear. If $AG = 7$ inches and $GN = 5$ inches, then $AN = 12$ inches.

____

10. For any real numbers x and y, if $x > y$, then $x^2 > y^2$.

____

11. The total number of angles in the figure is 3.

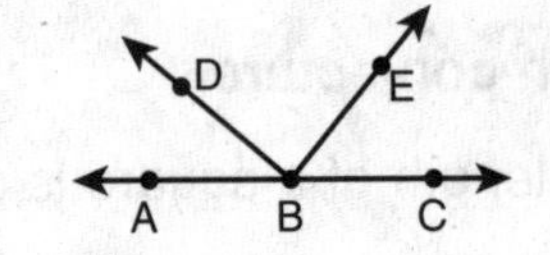

____

____

12. If two angles are acute, then the sum of their measures equals the measure of an obtuse angle.

____

____

Determine whether each conjecture is true. If not, write or draw a counterexample.

13. Points Q and R are collinear.

____

14. If J is between H and K, then $HJ = JK$.

____

Name ______________________ Date __________ Class __________

LESSON 2-2

Review for Mastery

Conditional Statements

A **conditional statement** is a statement that can be written as an if-then statement, "if *p*, then *q*."

The **hypothesis** comes after the word *if*.

The **conclusion** comes after the word *then*.

If you buy this cell phone, then you will receive 10 free ringtone downloads.

Sometimes it is necessary to rewrite a conditional statement so that it is in if-then form.

Conditional: A person who practices putting will improve her golf game.
If-Then Form: If a person practices putting, then she will improve her golf game.

A conditional statement has a false **truth value** *only* if the hypothesis (H) is true and the conclusion (C) is false.

For each conditional, underline the hypothesis and double-underline the conclusion.

1. If x is an even number, then x is divisible by 2.

2. The circumference of a circle is 5π inches if the diameter of the circle is 5 inches.

3. If a line containing the points J, K, and L lies in plane $\mathcal{P}$, then J, K, and L are coplanar.

For Exercises 4–6, write a conditional statement from each given statement.

4. Congruent segments have equal measures.

____

5. On Tuesday, play practice is at 6:00.

____

6. 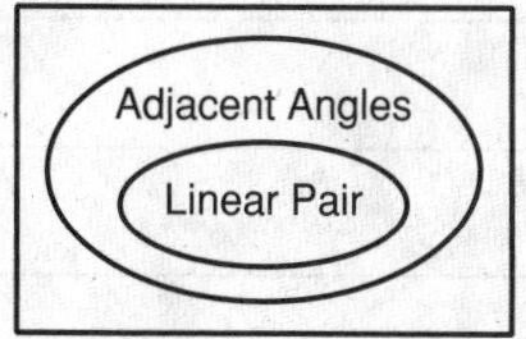

____

Determine whether the following conditional is true. If false, give a counterexample.

7. If two angles are supplementary, then they form a linear pair.

____

Name ______________________ Date ____________ Class ____________

LESSON 2-2

Review for Mastery

Conditional Statements continued

The **negation** of a statement, "not *p*," has the opposite truth value of the original statement.

If *p* is true, then *not p* is false.

If *p* is false, then *not p* is true.

Statement	Example	Truth Value
Conditional	If a figure is a square, then it has four right angles. (H: a figure is a square; C: it has four right angles)	True
Converse: Switch H and C.	If a figure has four right angles, then it is a square.	False
Inverse: Negate H and C.	If a figure is not a square, then it does not have four right angles.	False
Contrapositive: Switch and negate H and C.	If a figure does not have four right angles, then it is not a square.	True

Write the converse, inverse, and contrapositive of each conditional statement. Find the truth value of each.

8. If an animal is an armadillo, then it is nocturnal.

____

____

____

____

9. If $y = 1$, then $y^2 = 1$.

____

____

____

10. If an angle has a measure less than 90°, then it is acute.

____

____

____

____

Name ______________________ Date ____________ Class ____________

LESSON 2-3

Review for Mastery

Using Deductive Reasoning to Verify Conjectures

With inductive reasoning, you use examples to make a conjecture. With **deductive reasoning,** you use facts, definitions, and properties to draw conclusions and prove that conjectures are true.

Given: If two points lie in a plane, then the line containing those points also lies in the plane. *A* and *B* lie in plane $\mathcal{N}$.

$\mathcal{N}$ B A

Conjecture: $\overleftrightarrow{AB}$ lies in plane $\mathcal{N}$.

One valid form of deductive reasoning that lets you draw conclusions from true facts is called the **Law of Detachment.**

Given	If you have \$2, then you can buy a snack. You have \$2.	If you have \$2, then you can buy a snack. You can buy a snack.
Conjecture	You can buy a snack.	You have \$2.
Valid Conjecture?	Yes; the conditional is true and the hypothesis is true.	No; the hypothesis may or may not be true. For example, if you borrowed money, you could also buy a snack.

Tell whether each conclusion uses inductive or deductive reasoning.

1. A sign in the cafeteria says that a car wash is being held on the last Saturday of May. Tomorrow is the last Saturday of May, so Justin concludes that the car wash is tomorrow. ____________

2. So far, at the beginning of every Latin class, the teacher has had students review vocabulary. Latin class is about to start, and Jamilla assumes that they will first review vocabulary. ____________

3. Opposite rays are two rays that have a common endpoint and form a line. $\overrightarrow{YX}$ and $\overrightarrow{YZ}$ are opposite rays. ____________

X Y Z

Determine whether each conjecture is valid by the Law of Detachment.

4. Given: If you ride the Titan roller coaster in Arlington, Texas, then you will drop 255 feet.
   Michael rode the Titan roller coaster.
   Conjecture: Michael dropped 255 feet. ____________

5. Given: A segment that is a diameter of a circle has endpoints on the circle.
   $\overline{GH}$ has endpoints on a circle.
   Conjecture: $\overline{GH}$ is a diameter. ____________

Name ______________________ Date __________ Class __________

LESSON 2-3

Review for Mastery

Using Deductive Reasoning to Verify Conjectures continued

Another valid form of deductive reasoning is the **Law of Syllogism.** It is similar to the Transitive Property of Equality.

Transitive Property of Equality	Law of Syllogism
If $y = 10x$ and $10x = 20$, then $y = 20$.	**Given:** If you have a horse, then you have to feed it. If you have to feed a horse, then you have to get up early every morning. **Conjecture:** If you have a horse, then you have to get up early every morning.

Determine whether each conjecture is valid by the Law of Syllogism.

6. Given: If you buy a car, then you can drive to school. If you can drive to school, then you will not ride the bus.

   Conjecture: If you buy a car, then you will not ride the bus. ____________

7. Given: If $\angle K$ is obtuse, then it does not have a measure of 90°. If an angle does not have a measure of 90°, then it is not a right angle.

   Conjecture: If $\angle K$ is obtuse, then it is not a right angle. ____________

8. Given: If two segments are congruent, then they have the same measure. If two segments each have a measure of 6.5 centimeters, then they are congruent.

   Conjecture: If two segments are congruent, then they each have a measure of 6.5 centimeters. ____________

Draw a conclusion from the given information.

9. If $\triangle LMN$ is translated in the coordinate plane, then it has the same size and shape as its preimage. If an image and preimage have the same size and shape, then the figures have equal perimeters. $\triangle LMN$ is translated in the coordinate plane.

   ____

10. If $\angle R$ and $\angle S$ are complementary to the same angle, then the two angles are congruent. If two angles are congruent, then they are supplementary to the same angle. $\angle R$ and $\angle S$ are complementary to the same angle.

    R S

    ____

Name ______________________ Date ____________ Class ____________

LESSON 2-4

Review for Mastery

Biconditional Statements and Definitions

A **biconditional statement** combines a conditional statement, "if p, then q," with its converse, "if q, then p."

Conditional: If the sides of a triangle are congruent (p), then the angles are congruent (q).

Converse: If the angles of a triangle are congruent (q), then the sides are congruent (p).

Biconditional: The sides of a triangle are congruent (p) if and only if the angles are congruent (q).

Write the conditional statement and converse within each biconditional.

1. Lindsay will take photos for the yearbook if and only if she doesn't play soccer.

____

____

____

2. $m\angle ABC = m\angle CBD$ if and only if $\overrightarrow{BC}$ is an angle bisector of $\angle ABD$.

A C B D

____

____

____

For each conditional, write the converse and a biconditional statement.

3. If you can download 6 songs for \$5.94, then each song costs \$0.99.

____

____

4. If a figure has 10 sides, then it is a decagon.

____

____

Name ______________________ Date __________ Class __________

LESSON 2-4

Review for Mastery

Biconditional Statements and Definitions continued

A biconditional statement is false if either the conditional statement is false or its converse is false.

The midpoint of $\overline{QR}$ is $M(-3, 3)$ if and only if the endpoints are $Q(-6, 1)$ and $R(0, 5)$.

Conditional: If the midpoint of $\overline{QR}$ is $M(-3, 3)$, then the endpoints are $Q(-6, 1)$ and $R(0, 5)$. *false*

Converse: If the endpoints of $\overline{QR}$ are $Q(-6, 1)$ and $R(0, 5)$, then the midpoint of $\overline{QR}$ is $M(-3, 3)$. *true*

R(0, 5)
M(−3, 3)
Q(−6, 1)
y
x
3
−3
0

The conditional is false because the endpoints of $\overline{QR}$ could be $Q(-3, 6)$ and $R(-3, 0)$. So the biconditional statement is false.

Definitions can be written as biconditionals.

Definition: Circumference is the distance around a circle.

Biconditional: A measure is the circumference if and only if it is the distance around a circle.

Determine if each biconditional is true. If false, give a counterexample.

5. Students perform during halftime at the football games if and only if they are in the high school band.

____

6. An angle in a triangle measures 90° if and only if the triangle is a right triangle.

____

7. $a = 4$ and $b = 3$ if and only if $ab = 12$.

____

Write each definition as a biconditional.

8. An isosceles triangle has at least two congruent sides.

____

____

9. Deductive reasoning requires the use of facts, definitions, and properties to draw conclusions.

____

____

Name ______________________________ Date ____________ Class ____________

LESSON 2-5

Review for Mastery

Algebraic Proof

A **proof** is a logical argument that shows a conclusion is true. An algebraic proof uses algebraic properties, including the Distributive Property and the properties of equality.

Properties of Equality	Symbols	Examples
Addition	If $a = b$, then $a + c = b + c$.	If $x = -4$, then $x + 4 = -4 + 4$.
Subtraction	If $a = b$, then $a - c = b - c$.	If $r + 1 = 7$, then $r + 1 - 1 = 7 - 1$.
Multiplication	If $a = b$, then $ac = bc$.	If $\frac{k}{2} = 8$, then $\frac{k}{2}(2) = 8(2)$.
Division	If $a = 2$ and $c \neq 0$, then $\frac{a}{c} = \frac{b}{c}$.	If $6 = 3t$, then $\frac{6}{3} = \frac{3t}{3}$.
Reflexive	$a = a$	$15 = 15$
Symmetric	If $a = b$, then $b = a$.	If $n = 2$, then $2 = n$.
Transitive	If $a = b$ and $b = c$, then $a = c$.	If $y = 3^2$ and $3^2 = 9$, then $y = 9$.
Substitution	If $a = b$, then b can be substituted for a in any expression.	If $x = 7$, then $2x = 2(7)$.

When solving an algebraic equation, justify each step by using a definition, property, or piece of given information.

$2(a + 1) = -6$	Given equation
$2a + 2 = -6$	Distributive Property
$\underline{-2} \quad \underline{-2}$	Subtraction Property of Equality
$2a = -8$	Simplify.
$\frac{2a}{2} = \frac{-8}{2}$	Division Property of Equality
$a = -4$	Simplify.

Solve each equation. Write a justification for each step.

1. $\frac{n}{6} - 3 = 10$

2. $5 + x = 2x$

3. $\frac{y + 4}{7} = 3$

4. $4(t - 3) = -20$

Name ______________________ Date ____________ Class ____________

LESSON 2-5

Review for Mastery

Algebraic Proof continued

When writing algebraic proofs in geometry, you can also use definitions, postulates, properties, and pieces of given information to justify the steps.

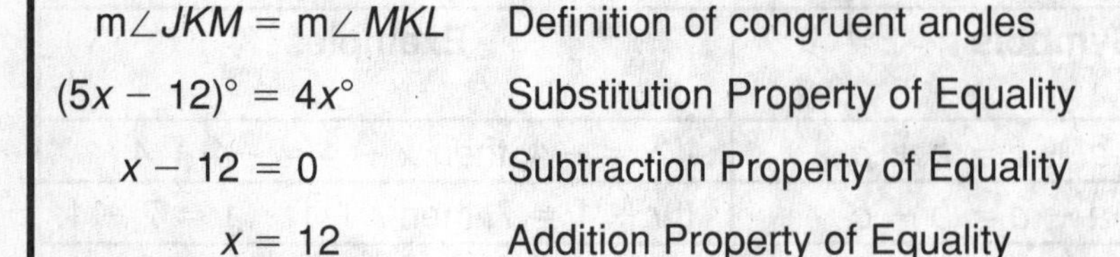

$m\angle JKM = m\angle MKL$	Definition of congruent angles
$(5x - 12)° = 4x°$	Substitution Property of Equality
$x - 12 = 0$	Subtraction Property of Equality
$x = 12$	Addition Property of Equality

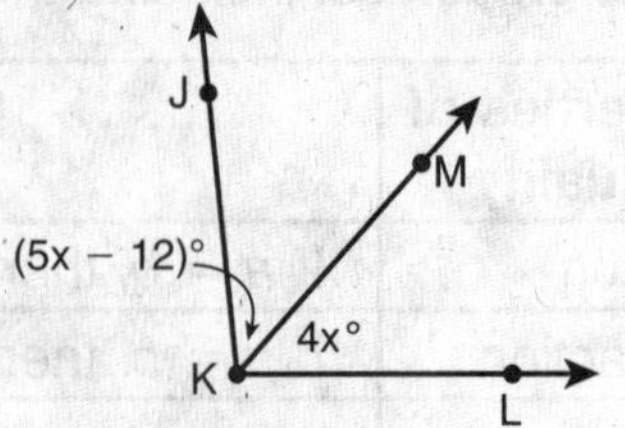

Properties of Congruence	Symbols	Examples
Reflexive	figure $A \cong$ figure A	$\angle CDE \cong \angle CDE$
Symmetric	If figure $A \cong$ figure B, then figure $B \cong$ figure A.	If $\overline{JK} \cong \overline{LM}$, then $\overline{LM} \cong \overline{JK}$.
Transitive	If figure $A \cong$ figure B and figure $B \cong$ figure C, then figure $A \cong$ figure C.	If $\angle N \cong \angle P$ and $\angle P \cong \angle Q$, then $\angle N \cong \angle Q$.

Write a justification for each step.

5. $CE = CD + DE$ ______________________

$6x = 8 + (3x + 7)$ ______________________

$6x = 15 + 3x$ ______________________

$3x = 15$ ______________________

$x = 5$ ______________________

6x; C 8 D 3x + 7 E

6. $m\angle PQR = m\angle PQS + m\angle SQR$ ______________________

$90° = 2x° + (4x - 12)°$ ______________________

$90 = 6x - 12$ ______________________

$102 = 6x$ ______________________

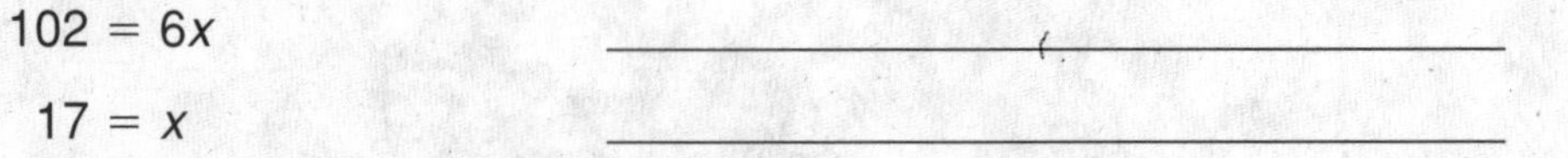

$17 = x$ ______________________

P, S, 2x°, (4x − 12)°, Q, R

Identify the property that justifies each statement.

7. If $\angle ABC \cong \angle DEF$, then $\angle DEF \cong \angle ABC$.

____

8. $\angle 1 \cong \angle 2$ and $\angle 2 \cong \angle 3$, so $\angle 1 \cong \angle 3$.

____

9. If $FG = HJ$, then $HJ = FG$.

____

10. $\overline{WX} \cong \overline{WX}$

____

Name ______________________ Date ____________ Class ____________

LESSON 2-6

Review for Mastery

Geometric Proof

To write a geometric proof, start with the hypothesis of a conditional.

Apply deductive reasoning.

Prove that the conclusion of the conditional is true.

Hypothesis

↓

Deductive Reasoning
- Definitions
- Properties
- Postulates
- Theorems

↓

Conclusion

Conditional: If $\overrightarrow{BD}$ is the angle bisector of $\angle ABC$, and $\angle ABD \cong \angle 1$, then $\angle DBC \cong \angle 1$.

Given: $\overrightarrow{BD}$ is the angle bisector of $\angle ABC$, and $\angle ABD \cong \angle 1$.

Prove: $\angle DBC \cong \angle 1$

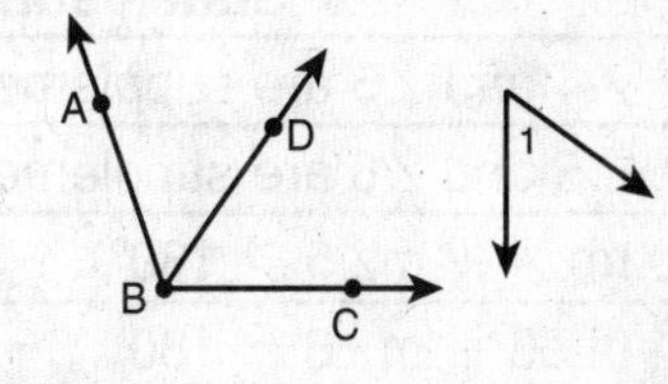

Proof:

1. $\overrightarrow{BD}$ is the angle bisector of $\angle ABC$. — 1. Given
2. $\angle ABD \cong \angle DBC$ — 2. Def. of $\angle$ bisector
3. $\angle ABD \cong \angle 1$ — 3. Given
4. $\angle DBC \cong \angle 1$ — 4. Transitive Prop. of $\cong$

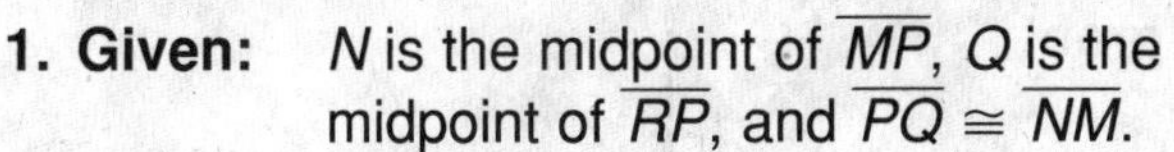

1. Given: N is the midpoint of $\overline{MP}$, Q is the midpoint of $\overline{RP}$, and $\overline{PQ} \cong \overline{NM}$.

Prove: $\overline{PN} \cong \overline{QR}$

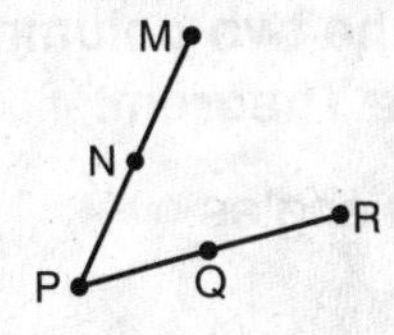

Write a justification for each step.

Proof:

1. N is the midpoint of $\overline{MP}$. — 1. ______________________
2. Q is the midpoint of $\overline{RP}$. — 2. ______________________
3. $\overline{PN} \cong \overline{NM}$ — 3. ______________________
4. $\overline{PQ} \cong \overline{NM}$ — 4. ______________________
5. $\overline{PN} \cong \overline{PQ}$ — 5. ______________________
6. $\overline{PQ} \cong \overline{QR}$ — 6. ______________________
7. $\overline{PN} \cong \overline{QR}$ — 7. ______________________

Name ____________________ Date __________ Class __________

LESSON 2-6

Review for Mastery
Geometric Proof continued

A **theorem** is any statement that you can prove. You can use **two-column proofs** and deductive reasoning to prove theorems.

Congruent Supplements Theorem	If two angles are supplementary to the same angle (or to two congruent angles), then the two angles are congruent.
Right Angle Congruence Theorem	All right angles are congruent.

Here is a two-column proof of one case of the Congruent Supplements Theorem.

Given: $\angle 4$ and $\angle 5$ are supplementary and $\angle 5$ and $\angle 6$ are supplementary.

Prove: $\angle 4 \cong \angle 6$

Proof:

Statements	Reasons
1. $\angle 4$ and $\angle 5$ are supplementary.	1. Given
2. $\angle 5$ and $\angle 6$ are supplementary.	2. Given
3. $m\angle 4 + m\angle 5 = 180°$	3. Definition of supplementary angles
4. $m\angle 5 + m\angle 6 = 180°$	4. Definition of supplementary angles
5. $m\angle 4 + m\angle 5 = m\angle 5 + m\angle 6$	5. Substitution Property of Equality
6. $m\angle 4 = m\angle 6$	6. Subtraction Property of Equality
7. $\angle 4 \cong \angle 6$	7. Definition of congruent angles

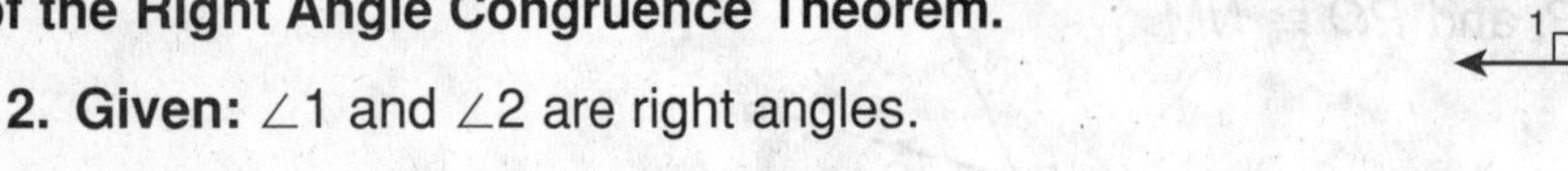

Fill in the blanks to complete the two-column proof of the Right Angle Congruence Theorem.

2. Given: $\angle 1$ and $\angle 2$ are right angles.

Prove: $\angle 1 \cong \angle 2$

Proof:

Statements	Reasons
1. **a.** ____________________	1. Given
2. $m\angle 1 = 90°$	2. **b.** ____________________
3. **c.** ____________________	3. Definition of right angle
4. $m\angle 1 = m\angle 2$	4. **d.** ____________________
5. **e.** ____________________	5. Definition of congruent angles

Name ______________________________ Date ____________ Class ____________

LESSON 2-7

Review for Mastery

Flowchart and Paragraph Proofs

In addition to the two-column proof, there are other types of proofs that you can use to prove conjectures are true.

Flowchart Proof	• Uses boxes and arrows. • Steps go left to right or top to bottom, as shown by arrows. • The justification for each step is written below the box.

You can write a flowchart proof of the Right Angle Congruence Theorem.

Given: $\angle 1$ and $\angle 2$ are right angles.

Prove: $\angle 1 \cong \angle 2$

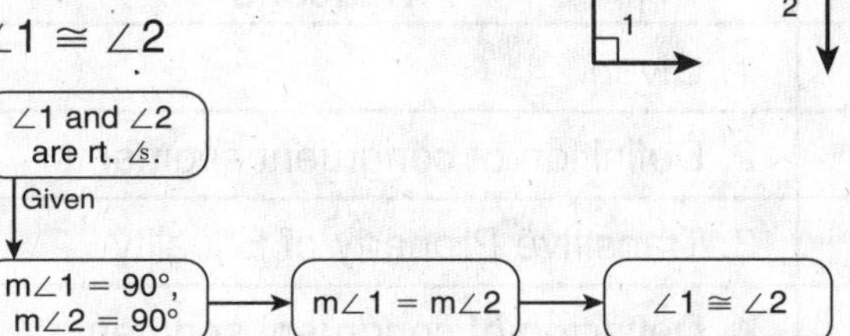

1. Use the given two-column proof to write a flowchart proof.

Given: V is the midpoint of $\overline{SW}$, and W is the midpoint of $\overline{VT}$.

Prove: $\overline{SV} \cong \overline{WT}$

Two-Column Proof:

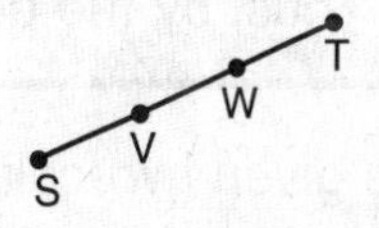

Statements	Reasons
1. V is the midpoint of $\overline{SW}$.	1. Given
2. W is the midpoint of $\overline{VT}$.	2. Given
3. $\overline{SV} \cong \overline{VW}$, $\overline{VW} \cong \overline{WT}$	3. Definition of midpoint
4. $\overline{SV} \cong \overline{WT}$	4. Transitive Property of Equality

Name ______________________ Date ____________ Class ____________

LESSON 2-7

Review for Mastery

Flowchart and Paragraph Proofs continued

To write a paragraph proof, use sentences to write a paragraph that presents the statements and reasons.

You can use the given two-column proof to write a paragraph proof.

Given: $\overline{AB} \cong \overline{BC}$ and $\overline{BC} \cong \overline{DE}$

Prove: $\overline{AB} \cong \overline{DE}$

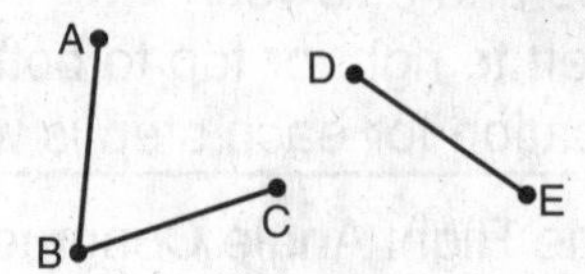

Two-Column Proof:

Statements	Reasons
1. $\overline{AB} \cong \overline{BC}$, $\overline{BC} \cong \overline{DE}$	1. Given
2. $AB = BC$, $BC = DE$	2. Definition of congruent segments
3. $AB = DE$	3. Transitive Property of Equality
4. $\overline{AB} \cong \overline{DE}$	4. Definition of congruent segments

Paragraph Proof: It is given that $\overline{AB} \cong \overline{BC}$ and $\overline{BC} \cong \overline{DE}$, so $AB = BC$ and $BC = DE$ by the definition of congruent segments. By the Transitive Property of Equality, $AB = DE$. Thus, by the definition of congruent segments, $\overline{AB} \cong \overline{DE}$.

2. Use the given two-column proof to write a paragraph proof.

Given: $\angle JKL$ is a right angle.

Prove: $\angle 1$ and $\angle 2$ are complementary angles.

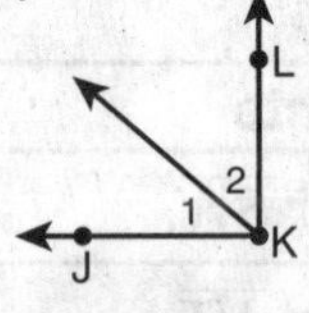

Two-Column Proof:

Statements	Reasons
1. $\angle JKL$ is a right angle.	1. Given
2. $m\angle JKL = 90°$	2. Definition of right angle
3. $m\angle JKL = m\angle 1 + m\angle 2$	3. Angle Addition Postulate
4. $90° = m\angle 1 + m\angle 2$	4. Substitution
5. $\angle 1$ and $\angle 2$ are complementary angles.	5. Definition of complementary angles

Paragraph Proof: ____

____

____

____

____

Name ______________________ Date ______________ Class ______________

LESSON 3-1

Review for Mastery

Lines and Angles

Lines	Description	Examples
parallel	lines that lie in the same plane and do not intersect **symbol:** $\parallel$	$\ell \parallel m$
perpendicular	lines that form 90° angles **symbol:** $\perp$	k and m are skew.
skew	lines that do not lie in the same plane and do not intersect	$k \perp \ell$

Parallel planes are planes that do not intersect. For example, the top and bottom of a cube represent parallel planes.

Use the figure for Exercises 1–3. Identify each of the following.

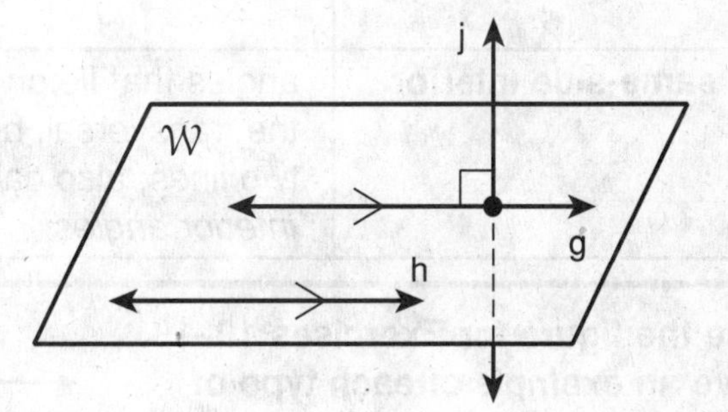

1. a pair of parallel lines

____

2. a pair of skew lines

____

3. a pair of perpendicular lines

____

Use the figure f or Exercises 4–9. Identify each of the following.

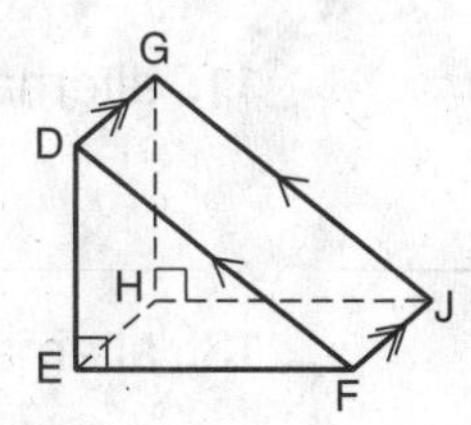

4. a segment that is parallel to $\overline{DG}$

____

5. a segment that is perpendicular to $\overline{GH}$

____

6. a segment that is skew to $\overline{JF}$

____

7. one pair of parallel planes

____

8. one pair of perpendicular segments, not including $\overline{GH}$

____

9. one pair of skew segments, not including $\overline{JF}$

____

Name ______________________ Date ____________ Class ____________

LESSON 3-1

Review for Mastery

Lines and Angles continued

A transversal is a line that intersects two lines in a plane at different points. Eight angles are formed. Line *t* is a transversal of lines *a* and *b*.

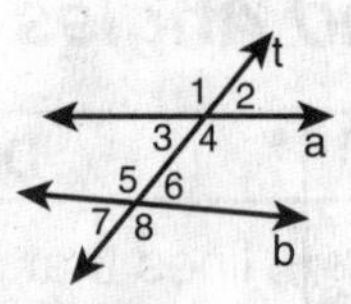

Angle Pairs Formed by a Transversal		
Angles	**Description**	**Examples**
corresponding	angles that lie on the same side of the transversal and on the same sides of the other two lines	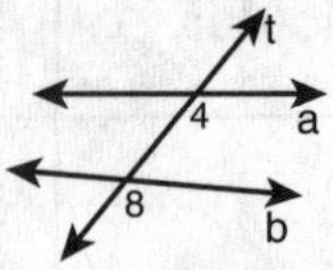
alternate interior	angles that lie on opposite sides of the transversal, between the other two lines	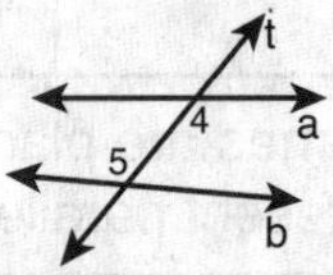
alternate exterior	angles that lie on opposite sides of the transversal, outside the other two lines	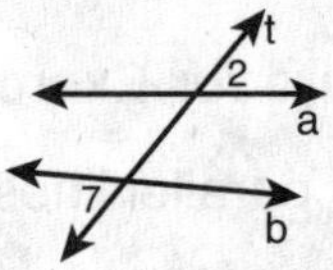
same-side interior	angles that lie on the same side of the transversal, between the other two lines; also called *consecutive interior angles*	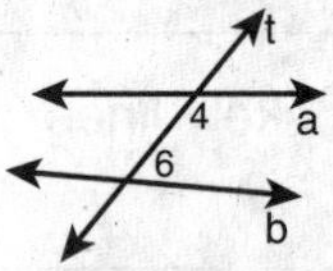

Use the figure for Exercises 10–13. Give an example of each type of angle pair.

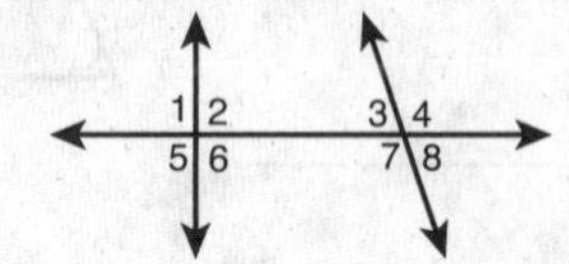

10. corresponding angles

____

11. alternate exterior angles

____

12. same-side interior angles

____

13. alternate interior angles

____

Use the figure for Exercises 14–16. Identify the transversal and classify each angle pair.

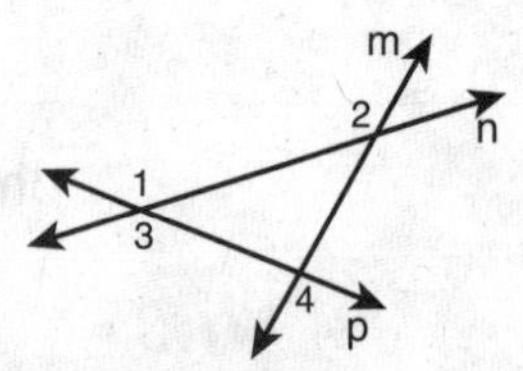

14. ∠1 and ∠2

____

15. ∠2 and ∠4

____

16. ∠3 and ∠4

____

Name ______________________ Date ___________ Class ___________

LESSON 3-2

Review for Mastery

Angles Formed by Parallel Lines and Transversals

According to the **Corresponding Angles Postulate,** if two parallel lines are cut by a transversal, then the pairs of corresponding angles are congruent.

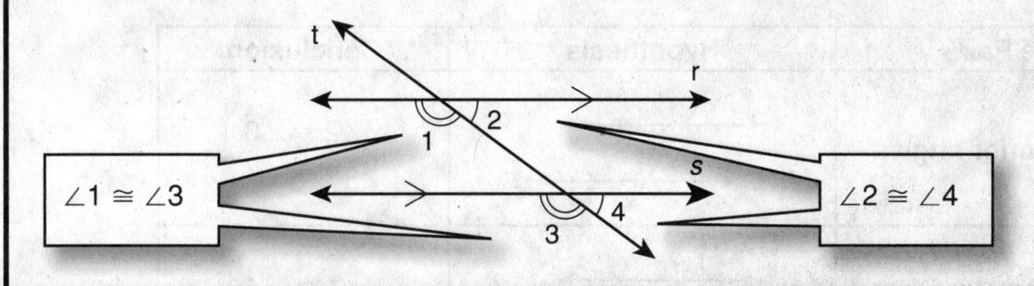

Determine whether each pair of angles is congruent according to the Corresponding Angles Postulate.

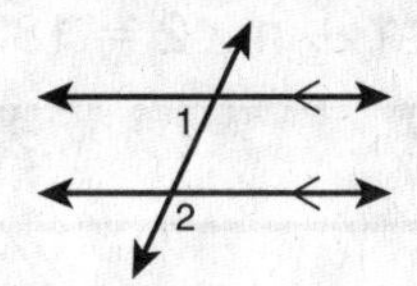

1. $\angle 1$ and $\angle 2$

____

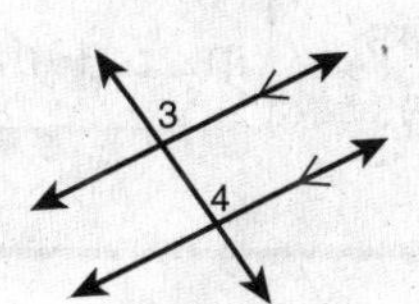

2. $\angle 3$ and $\angle 4$

____

Find each angle measure.

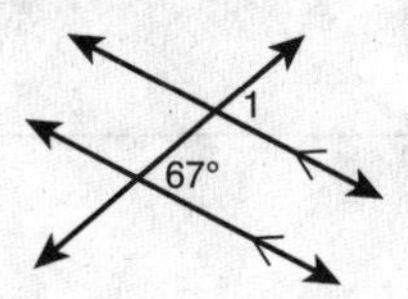

3. $m\angle 1$

____

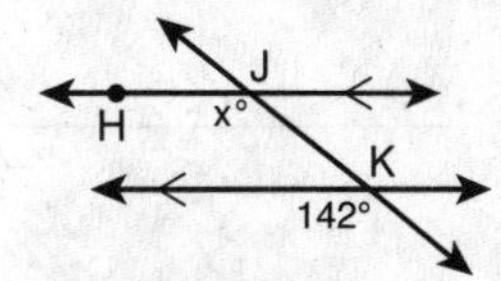

4. $m\angle HJK$

____

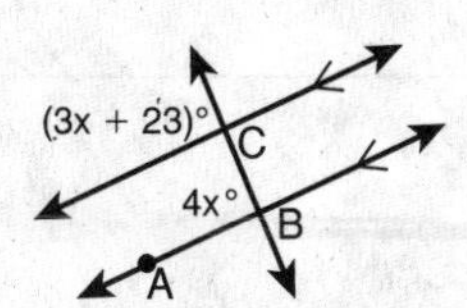

5. $m\angle ABC$

____

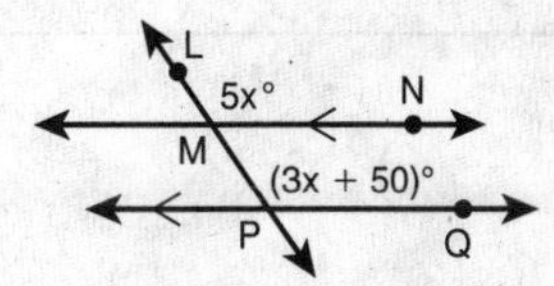

6. $m\angle MPQ$

____

Name ______________________ Date ____________ Class ____________

LESSON 3-2

Review for Mastery

Angles Formed by Parallel Lines and Transversals continued

If two parallel lines are cut by a transversal, then the following pairs of angles are also congruent.

Angle Pairs	Hypothesis	Conclusion
alternate interior angles	t, c, d; 6, 2, 3, 7	$\angle 2 \cong \angle 3$ $\angle 6 \cong \angle 7$
alternate exterior angles	t, q, r; 5, 1, 4, 8	$\angle 1 \cong \angle 4$ $\angle 5 \cong \angle 8$

If two parallel lines are cut by a transversal, then the pairs of same-side interior angles are supplementary.

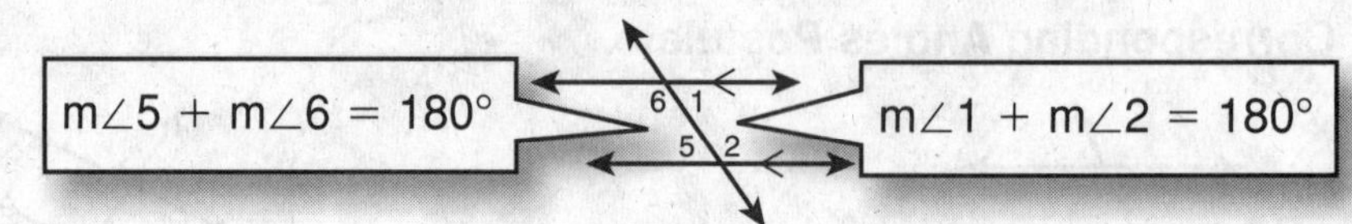

Find each angle measure.

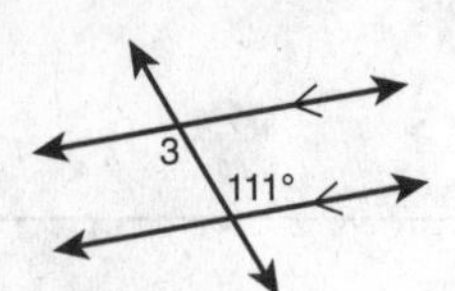

7. $m\angle 3$

____

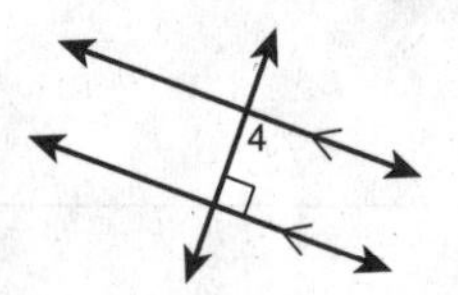

8. $m\angle 4$

____

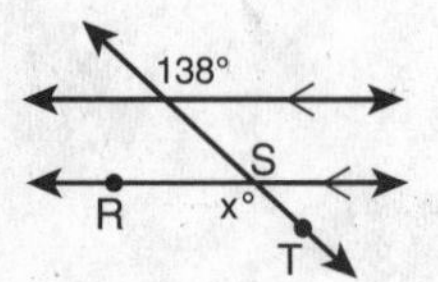

9. $m\angle RST$

____

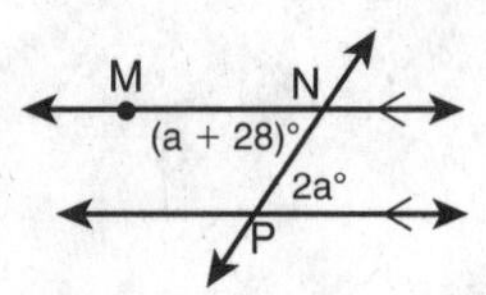

10. $m\angle MNP$

____

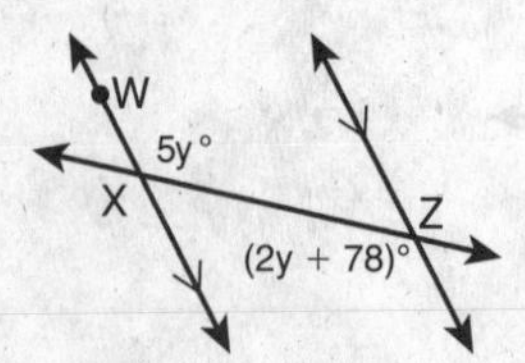

11. $m\angle WXZ$

____

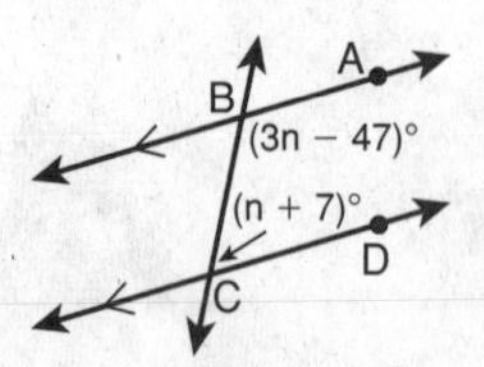

12. $m\angle ABC$

____

Name ______________________ Date __________ Class __________

LESSON 3-3

Review for Mastery

Proving Lines Parallel

Converse of the Corresponding Angles Postulate	If two coplanar lines are cut by a transversal so that a pair of corresponding angles are congruent, then the two lines are parallel.

You can use the Converse of the Corresponding Angles Postulate to show that two lines are parallel.

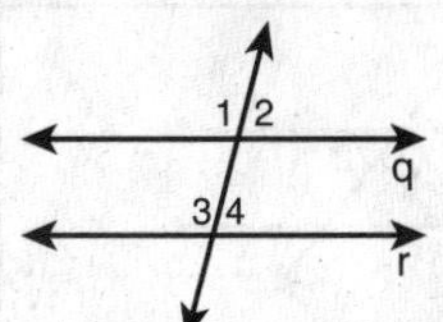

Given: $\angle 1 \cong \angle 3$

$\angle 1 \cong \angle 3$	$\angle 1 \cong \angle 3$ are corresponding angles.
$q \parallel r$	Converse of the Corresponding Angles Postulate

Given: $m\angle 2 = 3x°$, $m\angle 4 = (x + 50)°$, $x = 25$

$m\angle 2 = 3(25)° = 75°$	Substitute 25 for *x*.
$m\angle 4 = (25 + 50)° = 75°$	Substitute 25 for *x*.
$m\angle 2 = m\angle 4$	Transitive Property of Equality
$\angle 2 \cong \angle 4$	Definition of congruent angles
$q \parallel r$	Converse of the Corresponding Angles Postulate

For Exercises 1 and 2, use the Converse of the Corresponding Angles Postulate and the given information to show that $c \parallel d$.

1. Given: $\angle 2 \cong \angle 4$

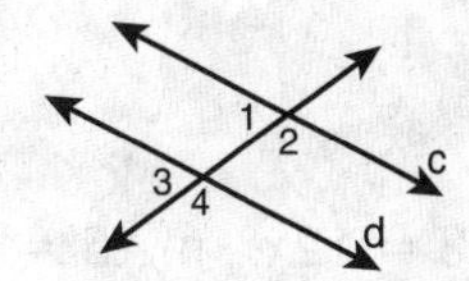

2. Given: $m\angle 1 = 2x°$, $m\angle 3 = (3x - 31)°$, $x = 31$

Name ______________________________ Date ___________ Class ___________

LESSON 3-3

Review for Mastery

Proving Lines Parallel continued

You can also prove that two lines are parallel by using the converse of any of the other theorems that you learned in Lesson 3-2.

Theorem	Hypothesis	Conclusion
Converse of the Alternate Interior Angles Theorem	$\angle 2 \cong \angle 3$	$a \parallel b$
Converse of the Alternate Exterior Angles Theorem	$\angle 1 \cong \angle 4$	$f \parallel g$
Converse of the Same-Side Interior Angles Theorem	$m\angle 1 + m\angle 2 = 180°$	$s \parallel t$

For Exercises 3–5, use the theorems and the given information to show that $j \parallel k$.

3. Given: $\angle 4 \cong \angle 5$

4. Given: $m\angle 3 = 12x°$, $m\angle 5 = 18x°$, $x = 6$

5. Given: $m\angle 2 = 8x°$, $m\angle 7 = (7x + 9)°$, $x = 9$

Name ______________________ Date __________ Class __________

LESSON 3-4

Review for Mastery

Perpendicular Lines

The **perpendicular bisector** of a segment is a line perpendicular to the segment at the segment's midpoint.

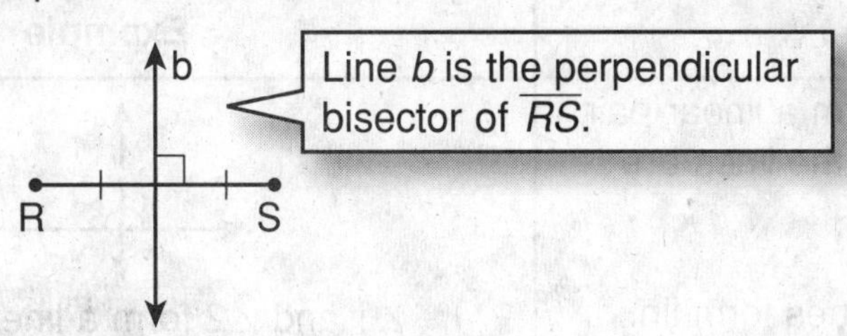

The **distance from a point to a line** is the length of the shortest segment from the point to the line. It is the length of the perpendicular segment that joins them.

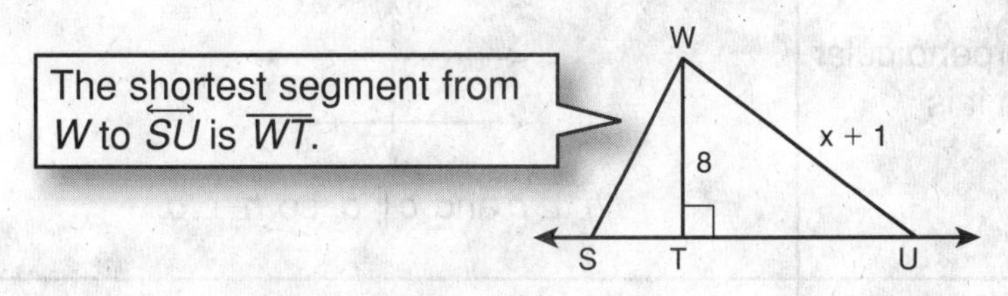

You can write and solve an inequality for *x*.

$WU > WT$	$\overline{WT}$ is the shortest segment.
$x + 1 > 8$	Substitute $x + 1$ for WU and 8 for WT.
$-1 \quad -1$	Subtract 1 from both sides of the equality.
$x > 7$	

Use the figure for Exercises 1 and 2.

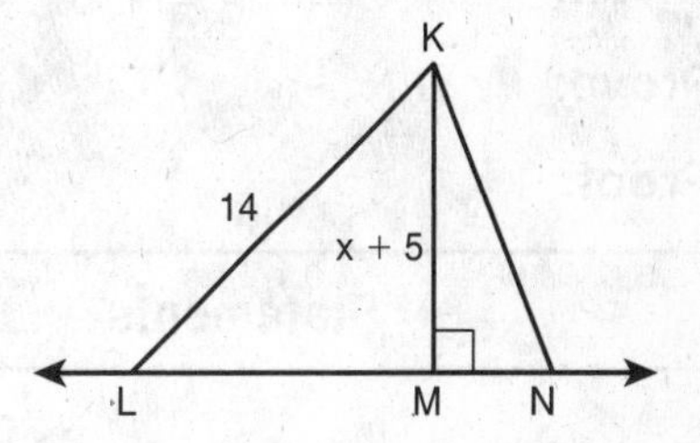

1. Name the shortest segment from point *K* to $\overleftrightarrow{LN}$.

____

2. Write and solve an inequality for *x*.

____

Use the figure for Exercises 3 and 4.

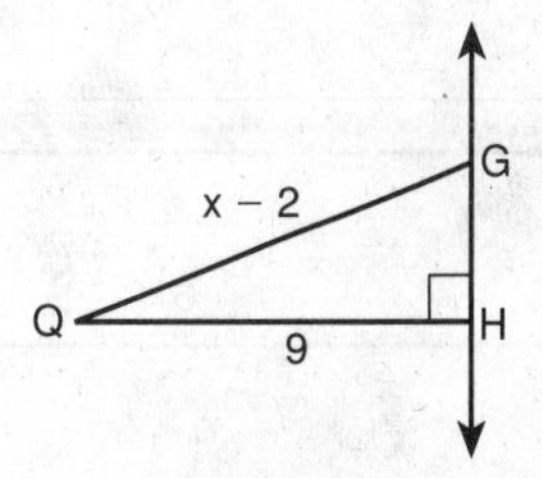

3. Name the shortest segment from point *Q* to $\overleftrightarrow{GH}$.

____

4. Write and solve an inequality for *x*.

____

Name ______________________________ Date ____________ Class ____________

LESSON 3-4

Review for Mastery

Perpendicular Lines continued

You can use the following theorems about perpendicular lines in your proofs.

Theorem	Example
If two intersecting lines form a linear pair of congruent angles, then the lines are perpendicular. **Symbols:** 2 intersecting lines form lin. pair of $\cong$ ∠s $\rightarrow$ lines $\perp$.	$\angle 1$ and $\angle 2$ form a linear pair and $\angle 1 \cong \angle 2$, so $a \perp b$.
Perpendicular Transversal Theorem In a plane, if a transversal is perpendicular to one of two parallel lines, then it is perpendicular to the other line. **Symbols:** $\perp$ Transv. Thm.	$h \perp c$ and $c \parallel d$, so $h \perp d$.
If two coplanar lines are perpendicular to the same line, then the two lines are parallel to each other. **Symbols:** 2 lines $\perp$ to same line $\rightarrow$ 2 lines $\parallel$.	$j \perp \ell$ and $k \perp \ell$, so $j \parallel k$.

5. Complete the two-column proof.

Given: $\angle 1 \cong \angle 2$, $s \perp t$

Prove: $r \perp t$

Proof:

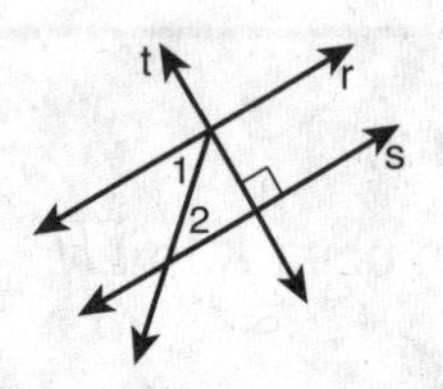

Statements	Reasons
1. $\angle 1 \cong \angle 2$	1. Given
2. **a.** ______________	2. Conv. of Alt. Int. ∠s Thm.
3. $s \perp t$	3. **b.** ______________
4. $r \perp t$	4. **c.** ______________

Name ______________________ Date ____________ Class ____________

LESSON 3-5

Review for Mastery

Slopes of Lines

The **slope** of a line describes how steep the line is. You can find the slope by writing the ratio of the **rise** to the **run.**

slope $= \frac{\text{rise}}{\text{run}} = \frac{3}{6} = \frac{1}{2}$

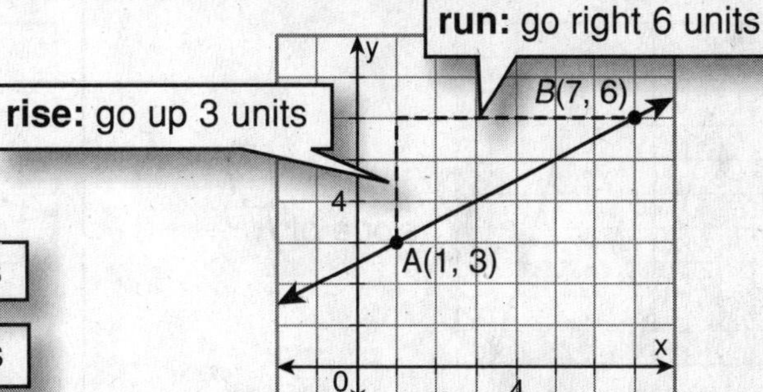

You can use a formula to calculate the slope m of the line through points (x_1, y_1) and (x_2, y_2).

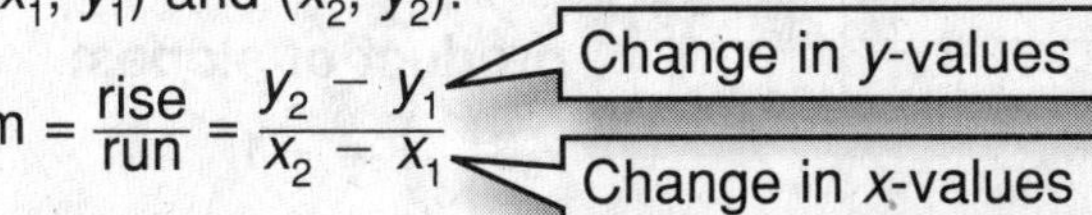

$$m = \frac{\text{rise}}{\text{run}} = \frac{y_2 - y_1}{x_2 - x_1}$$

To find the slope of $\overleftrightarrow{AB}$ using the formula, substitute (1, 3) for (x_1, y_1) and (7, 6) for (x_2, y_2).

$m = \frac{y_2 - y_1}{x_2 - x_1}$	Slope formula
$= \frac{6 - 3}{7 - 1}$	Substitution
$= \frac{3}{6}$	Simplify.
$= \frac{1}{2}$	Simplify.

Use the slope formula to determine the slope of each line.

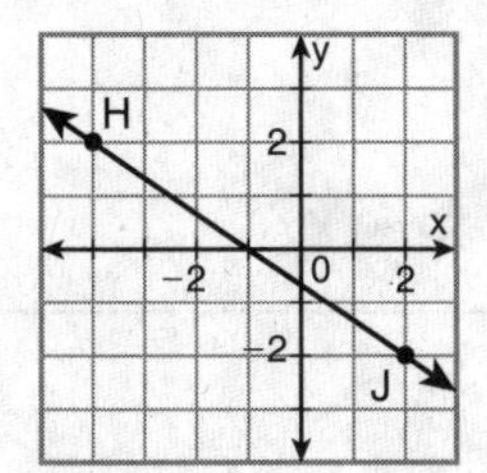

1. $\overleftrightarrow{HJ}$

____

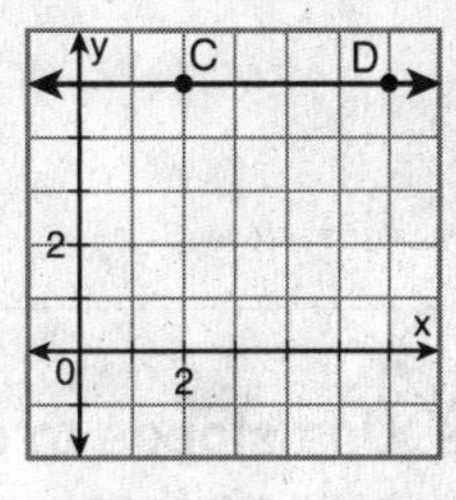

2. $\overleftrightarrow{CD}$

____

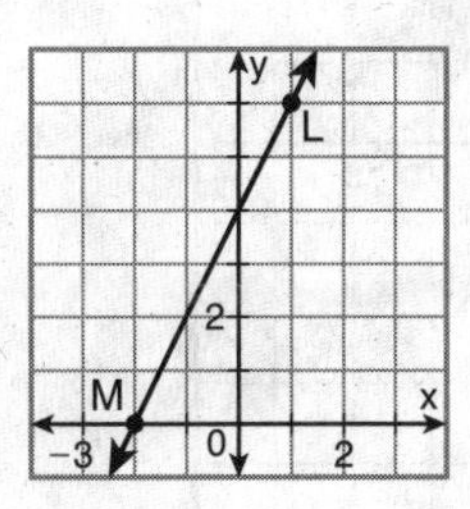

3. $\overleftrightarrow{LM}$

____

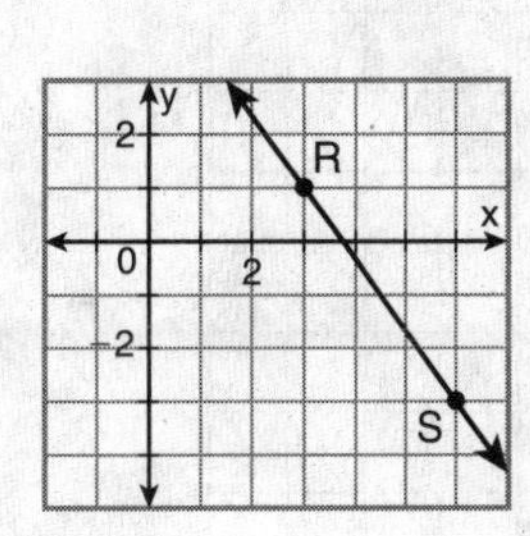

4. $\overleftrightarrow{RS}$

____

Name ______________________ Date ____________ Class ____________

LESSON 3-5

Review for Mastery

Slopes of Lines continued

Slopes of Parallel and Perpendicular Lines

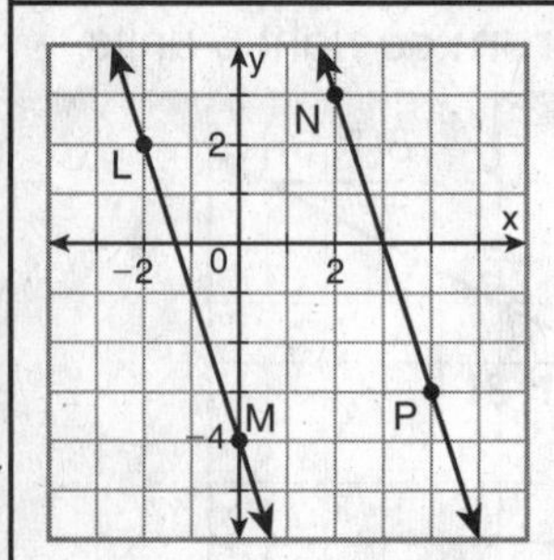

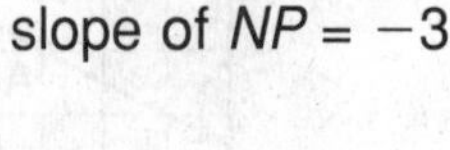

slope of $\overleftrightarrow{LM} = -3$

slope of $\overleftrightarrow{NP} = -3$

Parallel lines have the same slope.

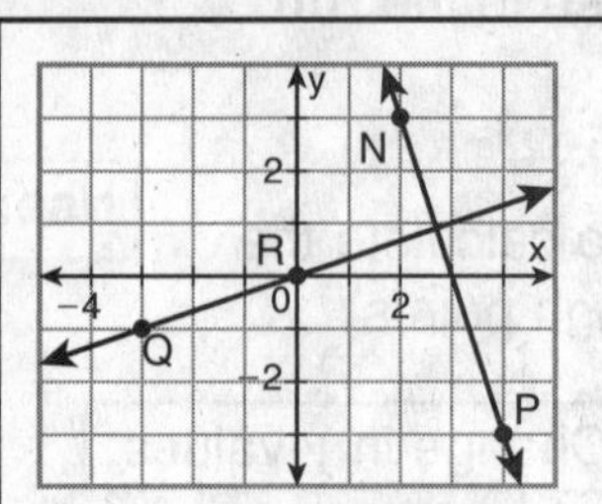

slope of $\overleftrightarrow{NP} = -3$

slope of $\overleftrightarrow{QR} = \frac{1}{3}$

product of slopes:

$-3\left(\frac{1}{3}\right) = -1$

Perpendicular lines have slopes that are *opposite reciprocals*. The product of the slopes is -1.

Use slopes to determine whether each pair of distinct lines is parallel, perpendicular, or neither.

5. slope of $\overleftrightarrow{PQ} = 5$

slope of $\overleftrightarrow{JK} = -\frac{1}{5}$

____

6. slope of $\overleftrightarrow{EF} = -\frac{3}{4}$

slope of $\overleftrightarrow{CD} = -\frac{3}{4}$

____

7. slope of $\overleftrightarrow{BC} = -\frac{5}{3}$

slope of $\overleftrightarrow{ST} = \frac{3}{5}$

____

8. slope of $\overleftrightarrow{WX} = \frac{1}{2}$

slope of $\overleftrightarrow{YZ} = -\frac{1}{2}$

____

Graph each pair of lines. Use slopes to determine whether the lines are parallel, perpendicular, or neither.

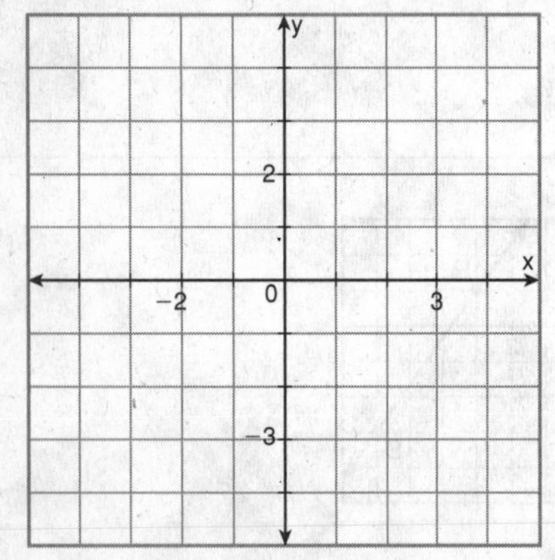

9. $\overleftrightarrow{FG}$ and $\overleftrightarrow{HJ}$ for $F(-1, 2)$, $G(3, -4)$, $H(-2, -3)$, and $J(4, 1)$

____

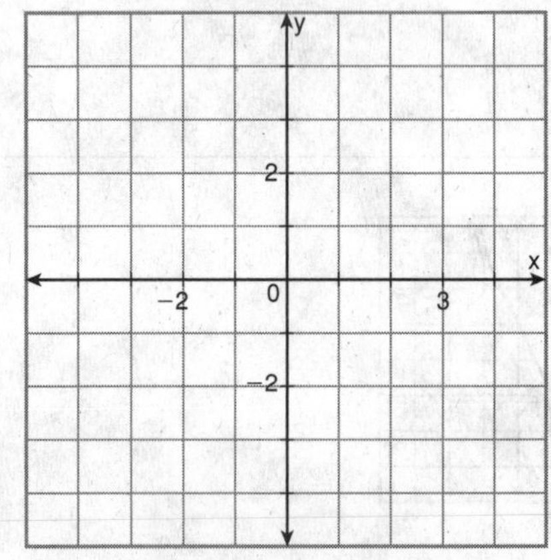

10. $\overleftrightarrow{RS}$ and $\overleftrightarrow{TU}$ for $R(-2, 3)$, $S(3, 3)$, $T(-3, 1)$, and $U(3, -1)$

____

Name ______________________ Date ____________ Class ____________

LESSON 3-6

Review for Mastery

Lines in the Coordinate Plane

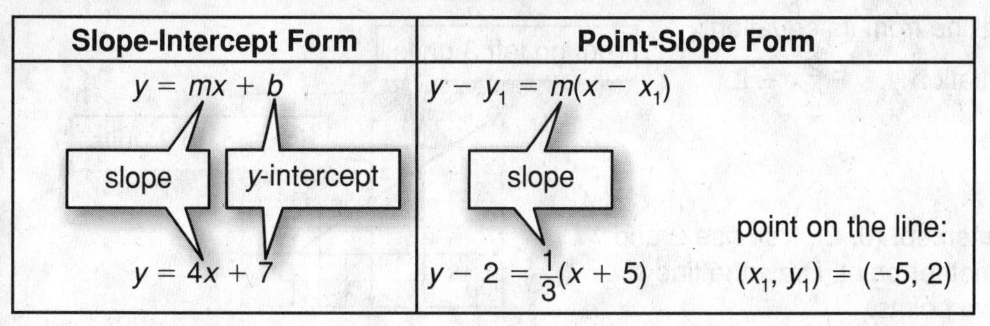

Write the equation of the line through (0, 1) and (2, 7) in slope-intercept form.

Step 1: Find the slope.

$m = \frac{y_2 - y_1}{x_2 - x_1}$ Formula for slope

$= \frac{7 - 1}{2 - 0} = \frac{6}{2} = 3$

Step 2: Find the *y*-intercept.

$y = mx + b$ Slope-intercept form

$1 = 3(0) + b$ Substitute 3 for *m*, 0 for *x*, and 1 for *y*.

$1 = b$ Simplify.

Step 3: Write the equation.

$y = mx + b$ Slope-intercept form

$y = 3x + 1$ Substitute 3 for *m* and 1 for *b*.

Write the equation of each line in the given form.

1. the line through (4, 2) and (8, 5) in slope-intercept form

____

2. the line through (4, 6) with slope $\frac{1}{2}$ in point-slope form

____

3. the line through (−5, 1) with slope 2 in point-slope form

____

4. the line with *x*-intercept −5 and *y*-intercept 3 in slope-intercept form

____

5. the line through (8, 0) with slope $-\frac{3}{4}$ in slope-intercept form

____

6. the line through (1, 7) and (−6, 7) in point-slope form

____

Name ______________________________ Date ___________ Class ___________

LESSON 3-6

Review for Mastery

Lines in the Coordinate Plane continued

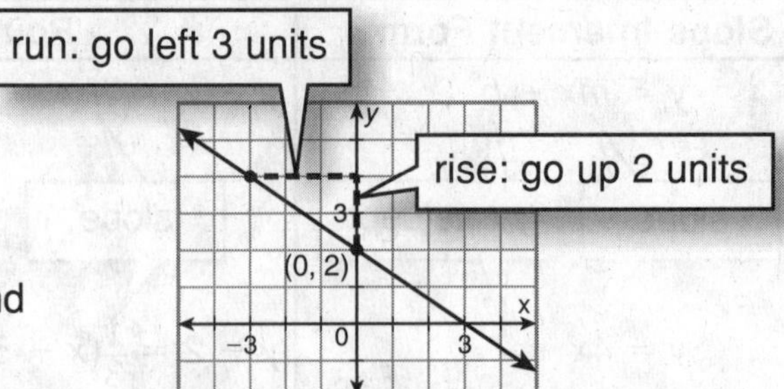

You can graph a line from its equation.

Consider the equation $y = -\frac{2}{3}x + 2$.

y-intercept $= 2$

slope $= -\frac{2}{3}$

First plot the y-intercept (0, 2). Use rise 2 and run −3 to find another point. Draw the line containing the two points.

Parallel Lines	Intersecting Lines	Coinciding Lines
$y = \frac{1}{3}x + 2$ $y = \frac{1}{3}x$ same slope different y-intercepts	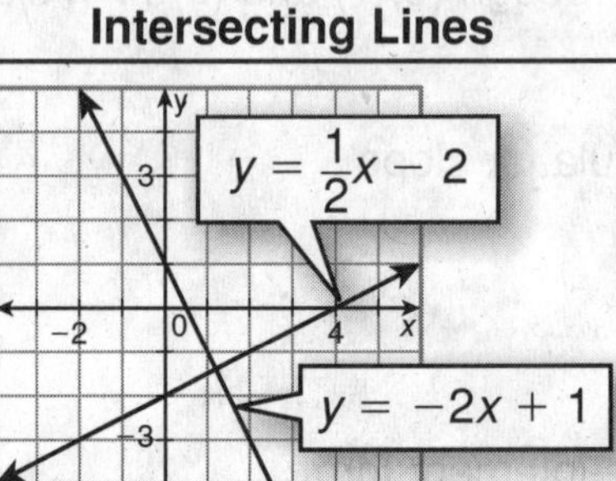 different slopes	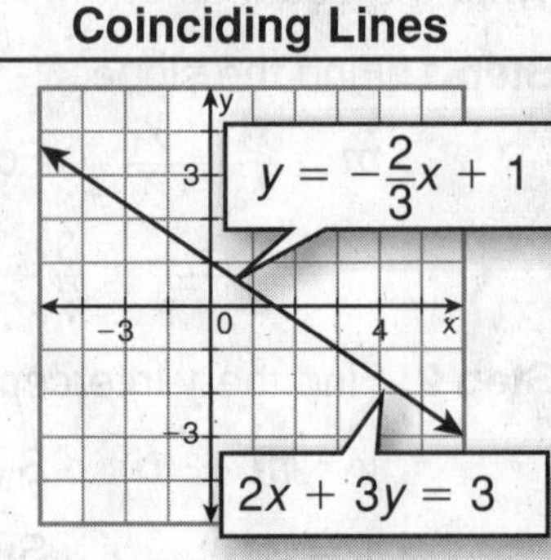 same slope same y-intercept

Graph each line.

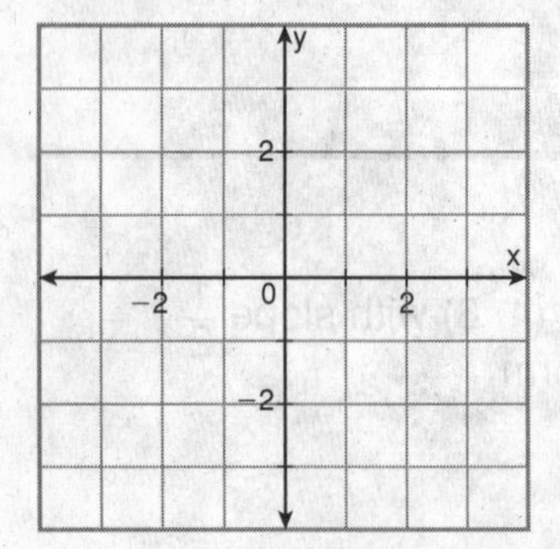

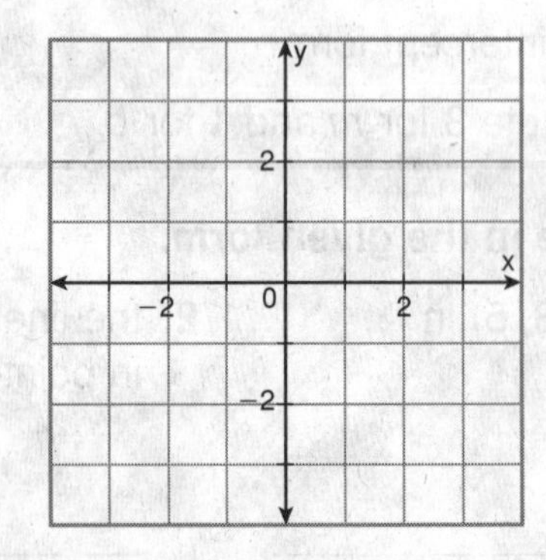

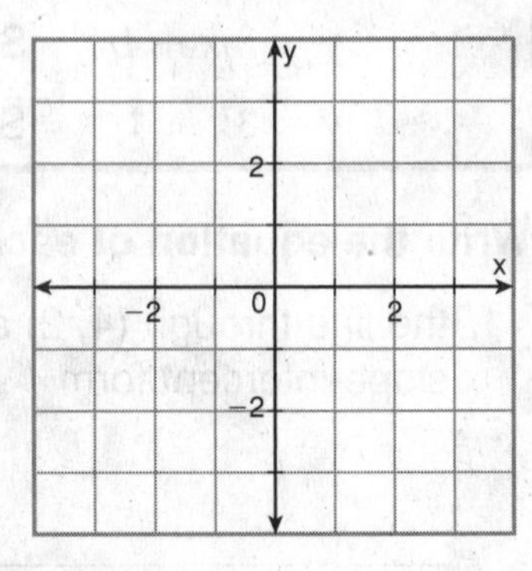

7. $y = x - 2$

8. $y = -\frac{1}{3}x + 3$

9. $y - 2 = \frac{1}{4}(x + 1)$

Determine whether the lines are parallel, intersect, or coincide.

10. $y = 2x + 5$
$y = 2x - 1$

____

11. $y = \frac{1}{3}x + 4$
$x - 3y = -12$

____

12. $y = 5x - 2$
$x + 4y = 8$

____

13. $5y + 2x = 1$
$y = -\frac{2}{5}x + 3$

____

Name ______________________ Date ____________ Class ____________

LESSON 4-1

Review for Mastery

Classifying Triangles

You can classify triangles by their angle measures. An **equiangular triangle,** for example, is a triangle with three congruent angles.

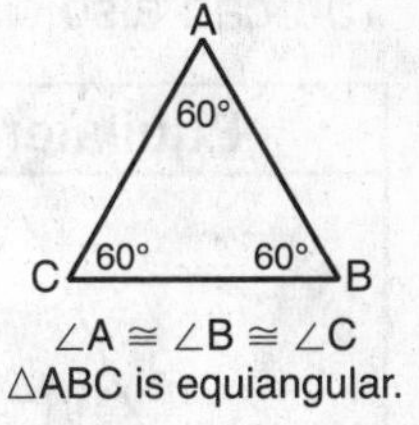

$\angle A \cong \angle B \cong \angle C$
$\triangle ABC$ is equiangular.

Examples of three other triangle classifications are shown in the table.

Acute Triangle	Right Triangle	Obtuse Triangle
72°, 50°, 58°	53°, 37°	45°, 31°, 104°
all acute angles	one right angle	one obtuse angle

You can use angle measures to classify $\triangle JML$ at right.

$\angle JKL$ is obtuse so $\triangle JLK$ is an obtuse triangle.

$\angle JLM$ and $\angle JLK$ form a linear pair, so they are supplementary.

$m\angle JLM + m\angle JLK = 180°$	Def. of supp. ∠s
$m\angle JLM + 120° = 180°$	Substitution
$m\angle JLM = 60°$	Subtract.

Since all the angles in $\triangle JLM$ are congruent, $\triangle JLM$ is an equiangular triangle.

Classify each triangle by its angle measures.

1.

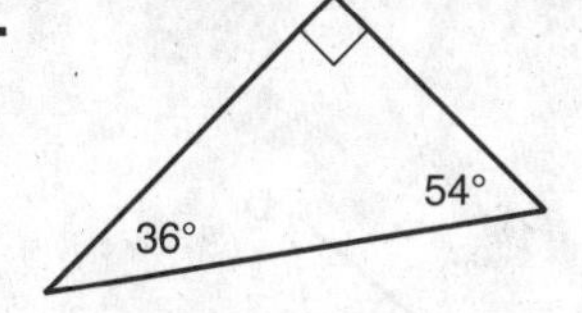

____

2.

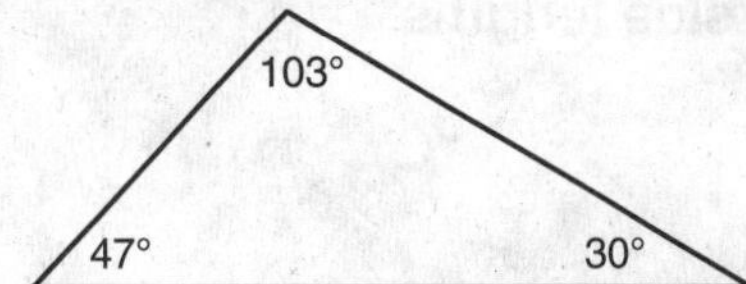

____

3.

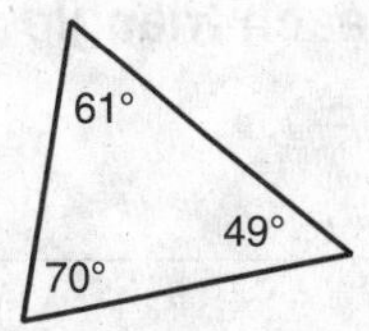

____

Use the figure to classify each triangle by its angle measures.

4. $\triangle DFG$

____

5. $\triangle DEG$

____

6. $\triangle EFG$

____

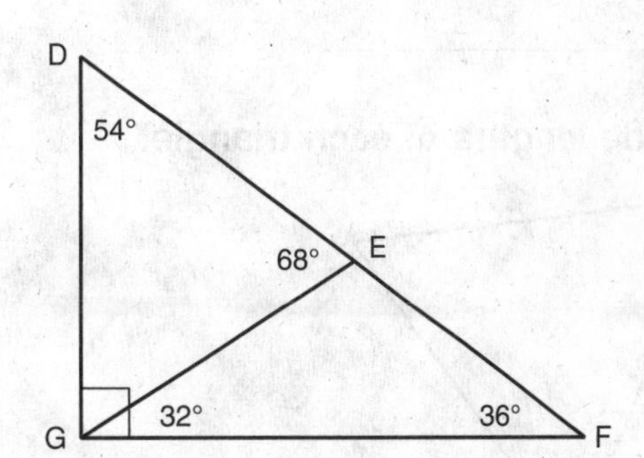

Name ______________________________ Date ___________ Class ___________

LESSON 4-1

Review for Mastery

Classifying Triangles continued

You can also classify triangles by their side lengths.

Equilateral Triangle	Isosceles Triangle	Scalene Triangle
4, 4, 4 all sides congruent	8, 8 at least two sides congruent	11, 7, 9 no sides congruent

You can use triangle classification to find the side lengths of a triangle.

Step 1 Find the value of x.

$QR = RS$	Def. of $\cong$ segs.
$4x = 3x + 5$	Substitution
$x = 5$	Simplify.

Step 2 Use substitution to find the length of a side.

$4x = 4(5)$	Substitute 5 for x.
$= 20$	Simplify.

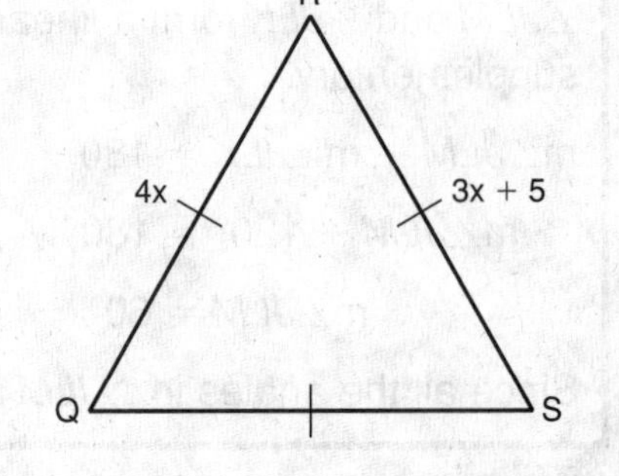

Each side length of $\triangle QRS$ is 20.

Classify each triangle by its side lengths.

7. $\triangle EGF$

____

8. $\triangle DEF$

____

9. $\triangle DFG$

____

D, 14, E, 13, G, F, 20

Find the side lengths of each triangle.

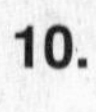

10.

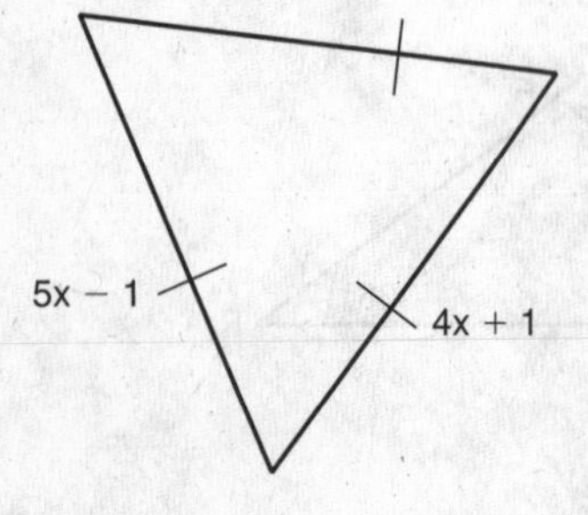

____

11.

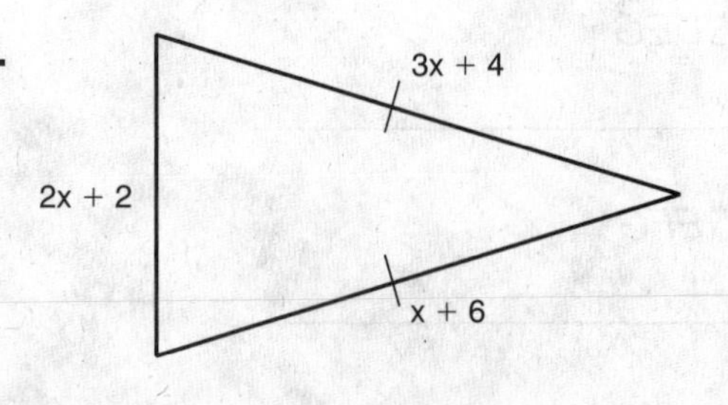

____

Name ______________________________ Date ____________ Class ____________

LESSON 4-2

Review for Mastery

Angle Relationships in Triangles

According to the **Triangle Sum Theorem,** the sum of the angle measures of a triangle is 180°.

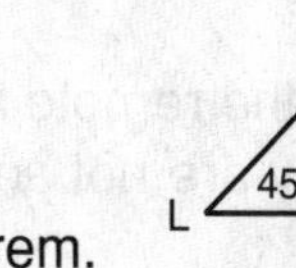

$m\angle J + m\angle K + m\angle L = 62 + 73 + 45$

$= 180°$

The **corollary** below follows directly from the Triangle Sum Theorem.

Corollary	Example
The acute angles of a right triangle are complementary.	$m\angle C = 90 - 39$ $= 51°$ $m\angle C + m\angle E = 90°$

Use the figure for Exercises 1 and 2.

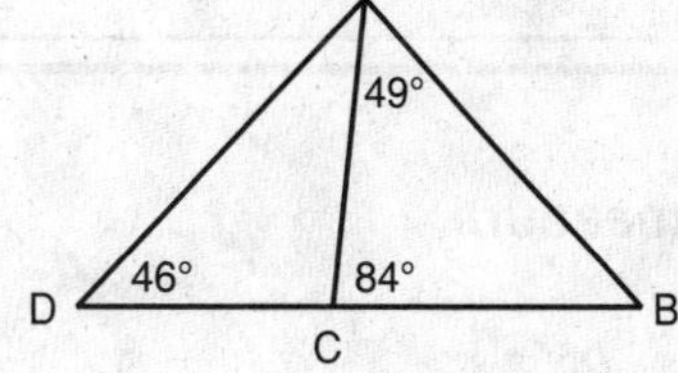

1. Find $m\angle ABC$.

____

2. Find $m\angle CAD$.

____

Use $\triangle RST$ for Exercises 3 and 4.

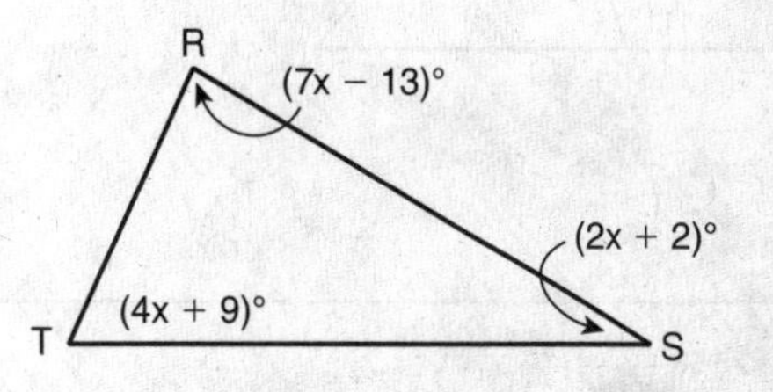

3. What is the value of x?

____

4. What is the measure of each angle?

____

What is the measure of each angle?

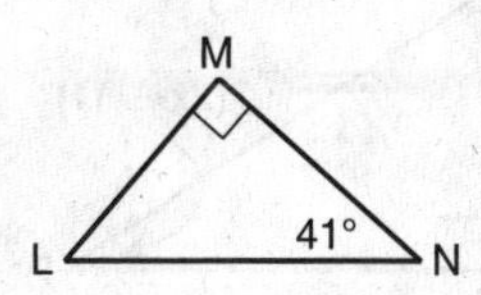

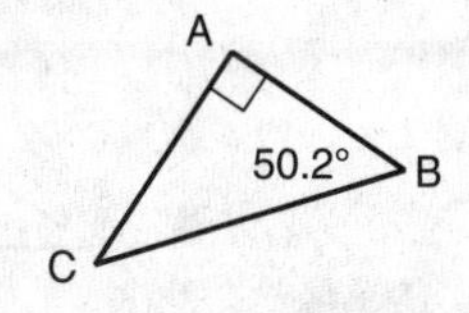

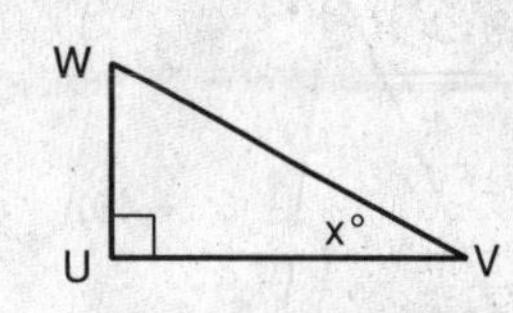

5. $\angle L$ ________________

6. $\angle C$ ________________

7. $\angle W$ ________________

Name ______________________________ Date ______________ Class ______________

LESSON 4-2

Review for Mastery

Angle Relationships in Triangles continued

An **exterior angle** of a triangle is formed by one side of the triangle and the extension of an adjacent side.

$\angle 1$ and $\angle 2$ are the remote interior angles of $\angle 4$ because they are not adjacent to $\angle 4$.

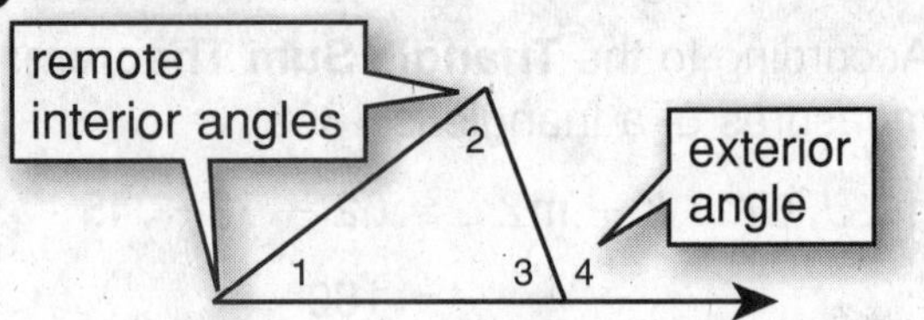

Exterior Angle Theorem
The measure of an exterior angle of a triangle is equal to the sum of the measures of its remote interior angles. 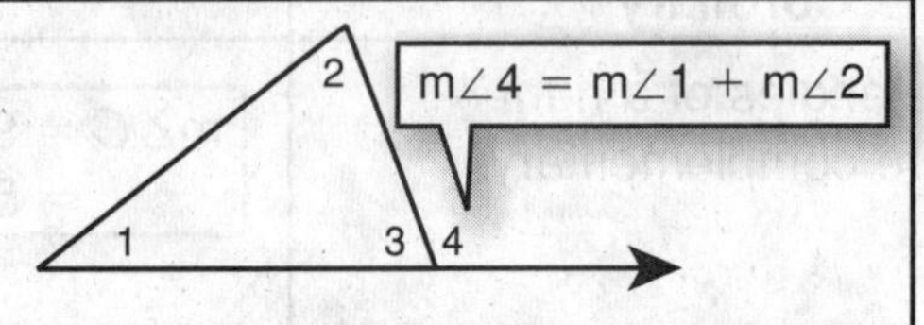

Third Angles Theorem
If two angles of one triangle are congruent to two angles of another triangle, then the third pair of angles are congruent. 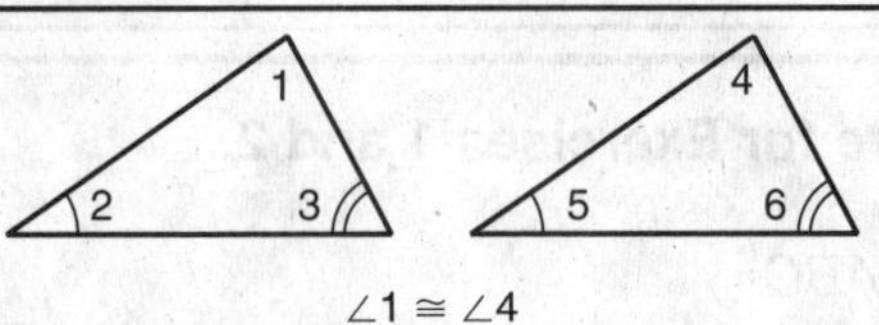

Find each angle measure.

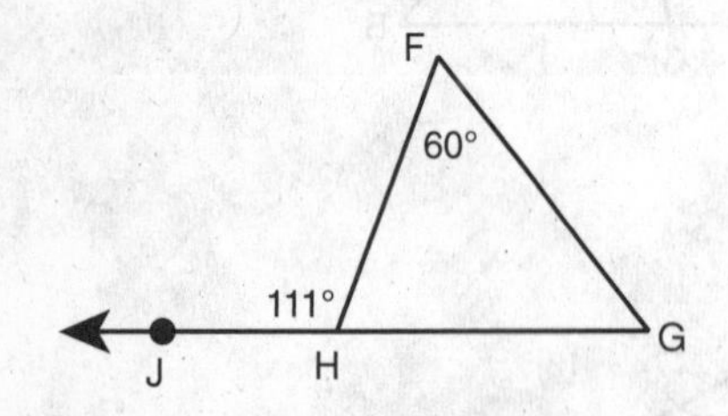

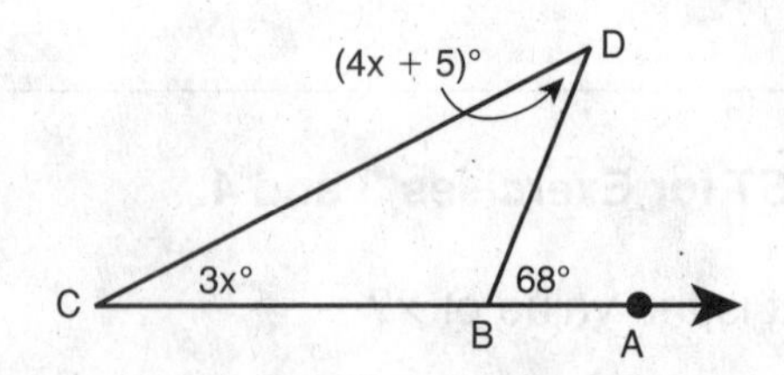

8. $m\angle G$ ______________________

9. $m\angle D$ ______________________

Find each angle measure.

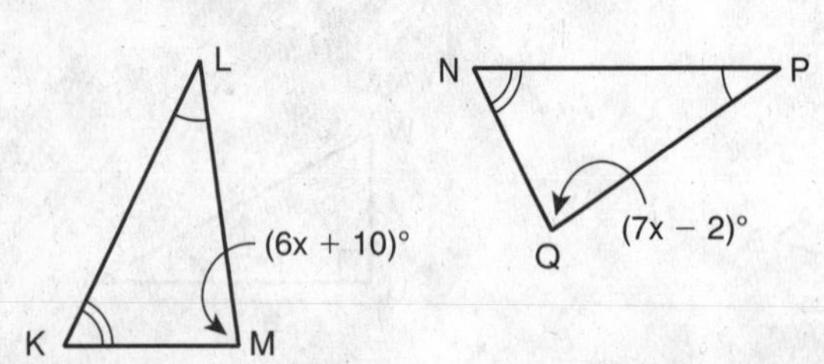

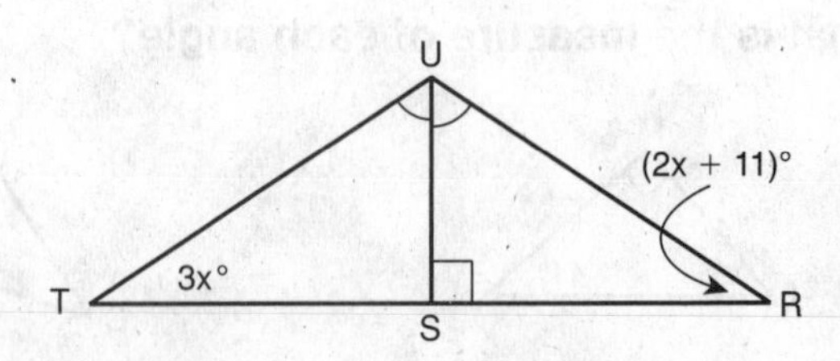

10. $m\angle M$ and $m\angle Q$ ______________________

11. $m\angle T$ and $m\angle R$ ______________________

Name ______________________ Date __________ Class __________

LESSON 4-3

Review for Mastery

Congruent Triangles

Triangles are **congruent** if they have the same size and shape. Their **corresponding parts,** the angles and sides that are in the same positions, are congruent.

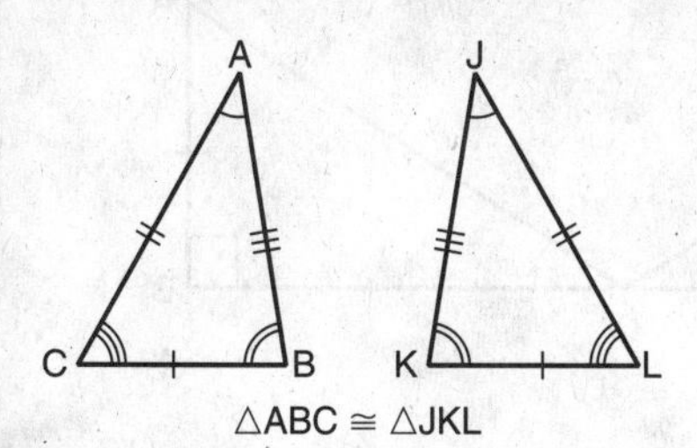

Corresponding Parts	
Congruent Angles	**Congruent Sides**
$\angle A \cong \angle J$	$\overline{AB} \cong \overline{JK}$
$\angle B \cong \angle K$	$\overline{BC} \cong \overline{KL}$
$\angle C \cong \angle L$	$\overline{CA} \cong \overline{LJ}$

To identify corresponding parts of congruent triangles, look at the order of the vertices in the congruence statement such as $\triangle ABC \cong \triangle JKL$.

Given: $\triangle XYZ \cong \triangle NPQ$. Identify the congruent corresponding parts.

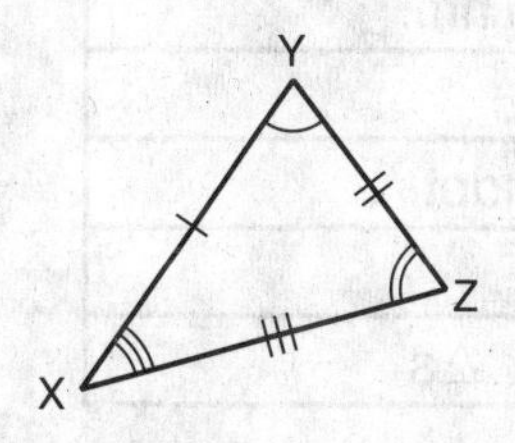

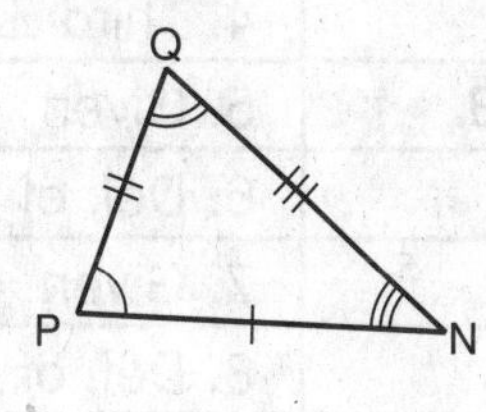

1. $\angle Z \cong$ ______________

2. $\overline{YZ} \cong$ ______________

3. $\angle P \cong$ ______________

4. $\angle X \cong$ ______________

5. $\overline{NQ} \cong$ ______________

6. $\overline{PN} \cong$ ______________

Given: $\triangle EFG \cong \triangle RST$. Find each value below.

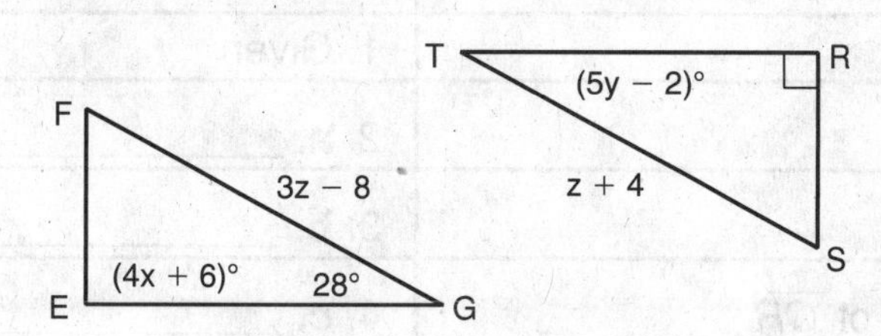

7. $x =$ ______________

8. $y =$ ______________

9. $m\angle F =$ ______________

10. $ST =$ ______________

Name ______________________________ Date __________ Class __________

LESSON 4-3

Review for Mastery

Congruent Triangles continued

You can prove triangles congruent by using the definition of congruence.

Given: $\angle D$ and $\angle B$ are right angles.

$\angle DCE \cong \angle BCA$

C is the midpoint of $\overline{DB}$.

$\overline{ED} \cong \overline{AB}$, $\overline{EC} \cong \overline{AC}$

Prove: $\triangle EDC \cong \triangle ABC$

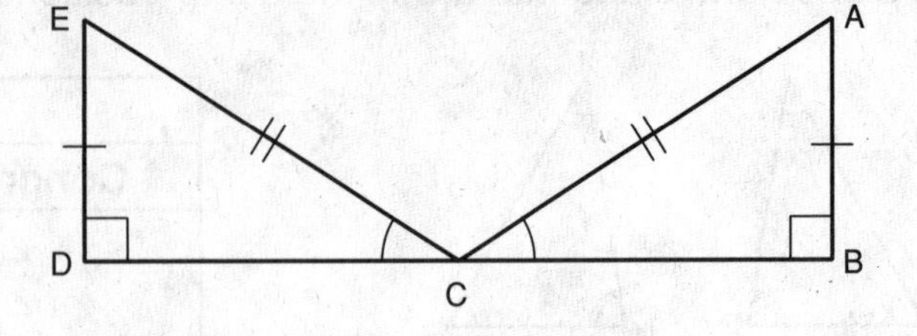

Proof:

Statements	Reasons
1. $\angle D$ and $\angle B$ are rt. $\angle$s.	1. Given
2. $\angle D \cong \angle B$	2. Rt. $\angle \cong$ Thm.
3. $\angle DCE \cong \angle BCA$	3. Given
4. $\angle E \cong \angle A$	4. Third $\angle$s Thm.
5. C is the midpoint of $\overline{DB}$.	5. Given
6. $\overline{DC} \cong \overline{BC}$	6. Def. of mdpt.
7. $\overline{ED} \cong \overline{AB}$, $\overline{EC} \cong \overline{AC}$	7. Given
8. $\triangle EDC \cong \triangle ABC$	8. Def. of $\cong$ $\triangle$s

11. Complete the proof.

Given: $\angle Q \cong \angle R$

P is the midpoint of $\overline{QR}$.

$\overline{NQ} \cong \overline{SR}$, $\overline{NP} \cong \overline{SP}$

Prove: $\triangle NPQ \cong \triangle SPR$

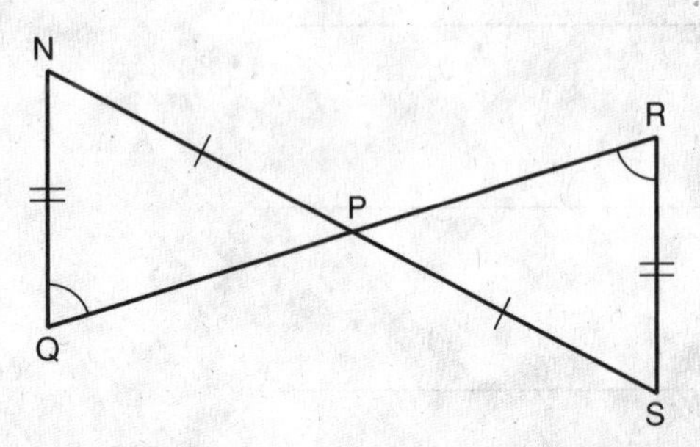

Proof:

Statements	Reasons
1. $\angle Q \cong \angle R$	1. Given
2. $\angle NPQ \cong \angle SPR$	2. **a.** ____________
3. $\angle N \cong \angle S$	3. **b.** ____________
4. P is the midpoint of $\overline{QR}$.	4. **c.** ____________
5. **d.** ____________	5. Def. of mdpt.
6. $\overline{NQ} \cong \overline{SR}$, $\overline{NP} \cong \overline{SP}$	6. **e.** ____________
7. $\triangle NPQ \cong \triangle SPR$	7. **f.** ____________

Name ______________________ Date __________ Class __________

LESSON 4-4

Review for Mastery

Triangle Congruence: SSS and SAS

Side-Side-Side (SSS) Congruence Postulate

If three sides of one triangle are congruent to three sides of another triangle, then the triangles are congruent.

$\overline{QR} \cong \overline{TU}$, $\overline{RP} \cong \overline{US}$, and $\overline{PQ} \cong \overline{ST}$, so $\triangle PQR \cong \triangle STU$.

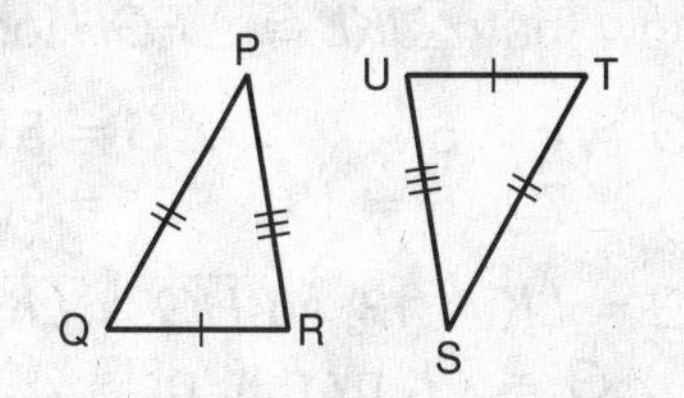

You can use SSS to explain why $\triangle FJH \cong \triangle FGH$.
It is given that $\overline{FJ} \cong \overline{FG}$ and that $\overline{JH} \cong \overline{GH}$. By the Reflex. Prop. of $\cong$, $\overline{FH} \cong \overline{FH}$. So $\triangle FJH \cong \triangle FGH$ by SSS.

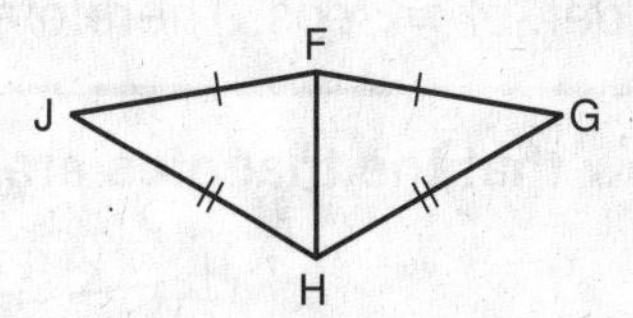

Side-Angle-Side (SAS) Congruence Postulate

If two sides and the included angle of one triangle are congruent to two sides and the included angle of another triangle, then the triangles are congruent.

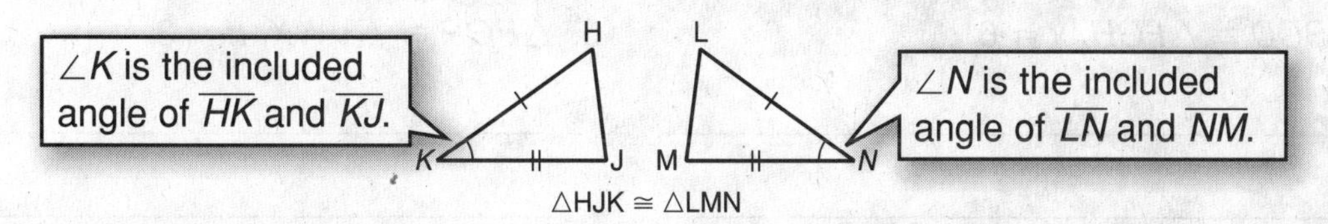

Use SSS to explain why the triangles in each pair are congruent.

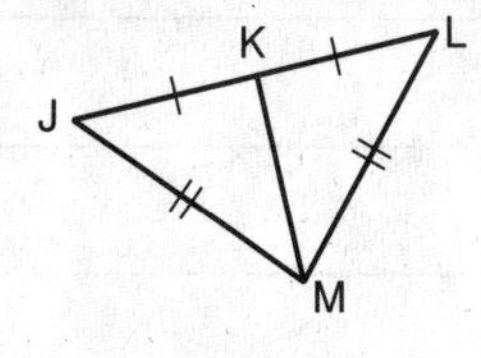

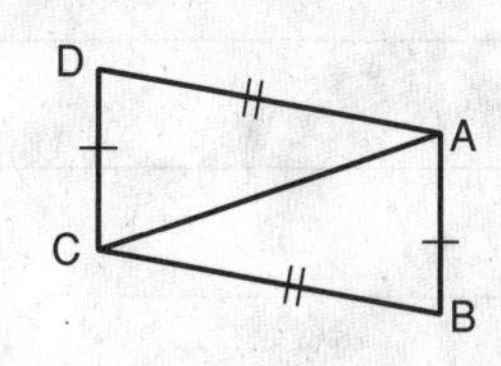

1. $\triangle JKM \cong \triangle LKM$

____

____

____

____

2. $\triangle ABC \cong \triangle CDA$

____

____

____

____

3. Use SAS to explain why $\triangle WXY \cong \triangle WZY$.

____

____

____

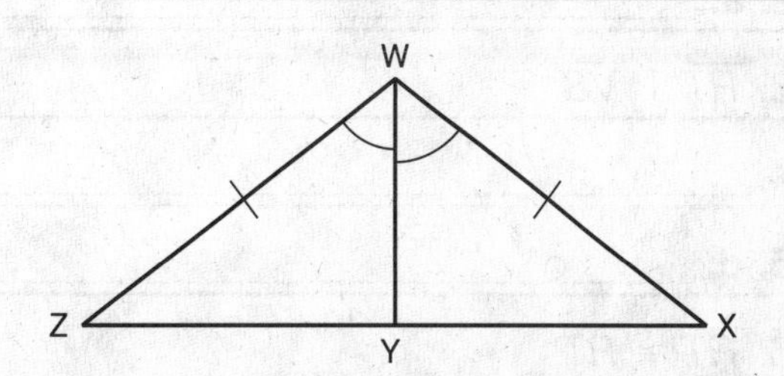

Name ______________________ Date ____________ Class ____________

LESSON **4-4**

Review for Mastery

Triangle Congruence: SSS and SAS continued

You can show that two triangles are congruent by using SSS and SAS.

Show that $\triangle JKL \cong \triangle FGH$ for $y = 7$.

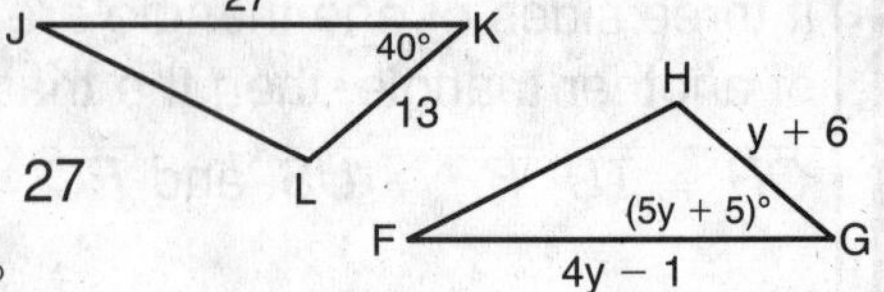

$HG = y + 6$
$= 7 + 6 = 13$

$m\angle G = 5y + 5$
$= 5(7) + 5 = 40°$

$FG = 4y - 1$
$= 4(7) - 1 = 27$

$HG = LK = 13$, so $\overline{HG} \cong \overline{LK}$ by def. of $\cong$ segs. $m\angle G = 40°$, so $\angle G \cong \angle K$ by def. of $\cong$ ∠s. $FG = JK = 27$, so $\overline{FG} \cong \overline{JK}$ by def. of $\cong$ segs. Therefore $\triangle JKL \cong \triangle FGH$ by SAS.

Show that the triangles are congruent for the given value of the variable.

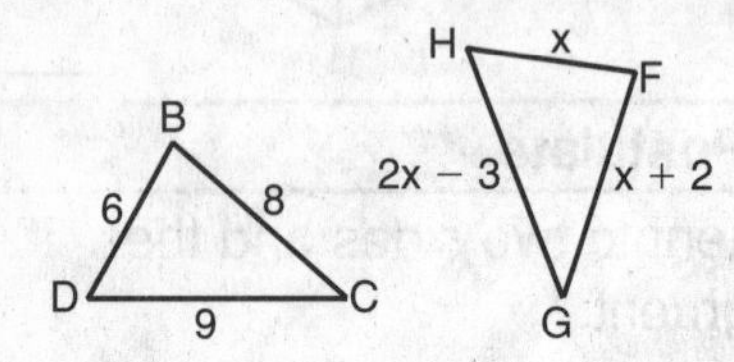

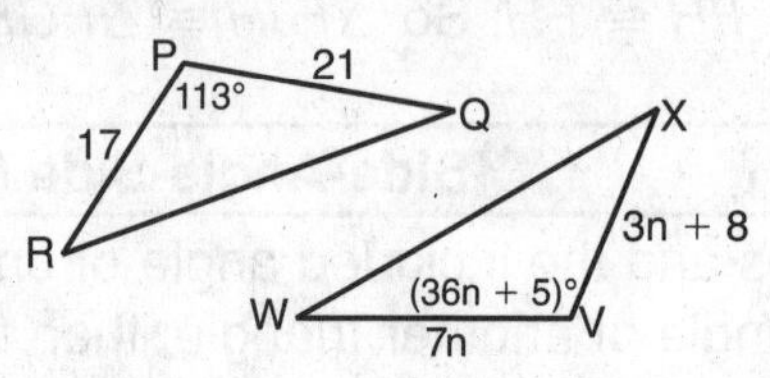

4. $\triangle BCD \cong \triangle FGH$, $x = 6$

____
____
____
____
____
____

5. $\triangle PQR \cong \triangle VWX$, $n = 3$

____
____
____
____
____
____

6. Complete the proof.

Given: T is the midpoint of $\overline{VS}$.
$\overline{RT} \perp \overline{VS}$

Prove: $\triangle RST \cong \triangle RVT$

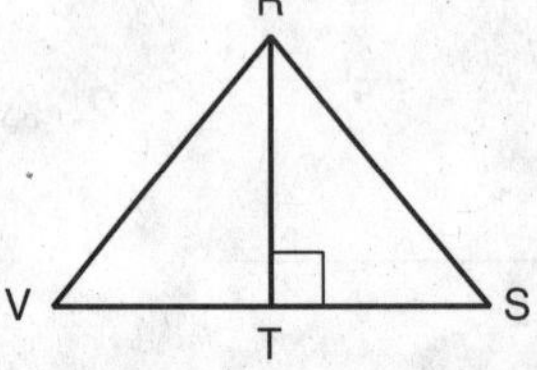

Statements	Reasons
1. T is the midpoint of $\overline{VS}$.	1. Given
2. **a.** ______________	2. Def. of mdpt.
3. $\overline{RT} \perp \overline{VS}$	3. **b.** ______________
4. ______________	4. **c.** ______________
5. **d.** ______________	5. Rt. $\angle \cong$ Thm.
6. $\overline{RT} \cong \overline{RT}$	6. **e.** ______________
7. $\triangle RST \cong \triangle RVT$	7. **f.** ______________

Name ______________________ Date __________ Class __________

LESSON 4-5

Review for Mastery

Triangle Congruence: ASA, AAS, and HL

Angle-Side-Angle (ASA) Congruence Postulate

If two angles and the included side of one triangle are congruent to two angles and the included side of another triangle, then the triangles are congruent.

$\overline{AC}$ is the included side of $\angle A$ and $\angle C$.

$\overline{DF}$ is the included side of $\angle D$ and $\angle F$.

$\triangle ABC \cong \triangle DEF$

Determine whether you can use ASA to prove the triangles congruent. Explain.

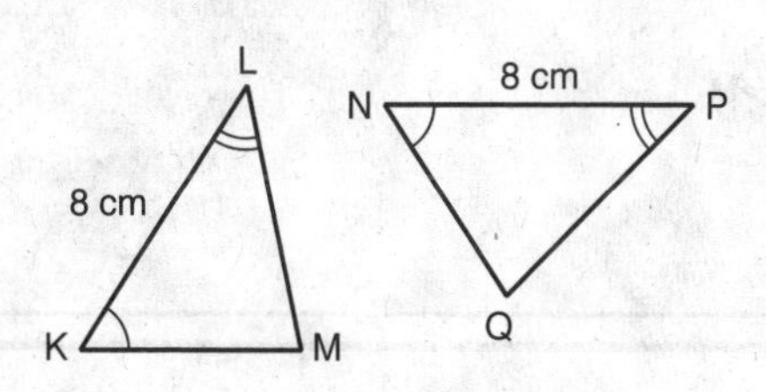

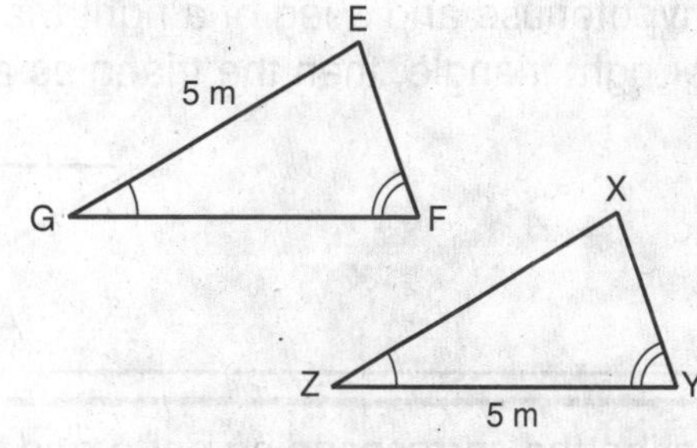

1. $\triangle KLM$ and $\triangle NPQ$

____

____

2. $\triangle EFG$ and $\triangle XYZ$

____

____

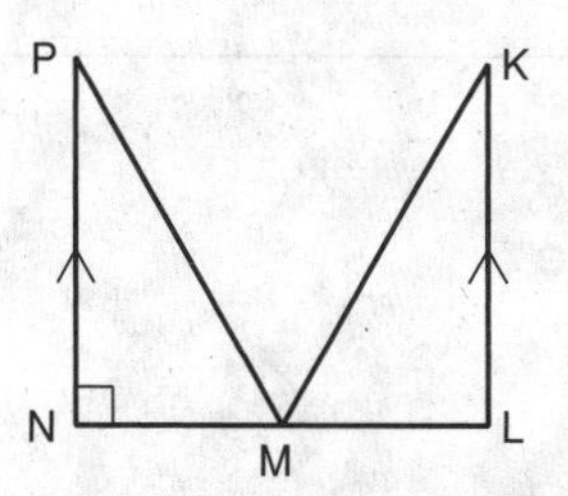

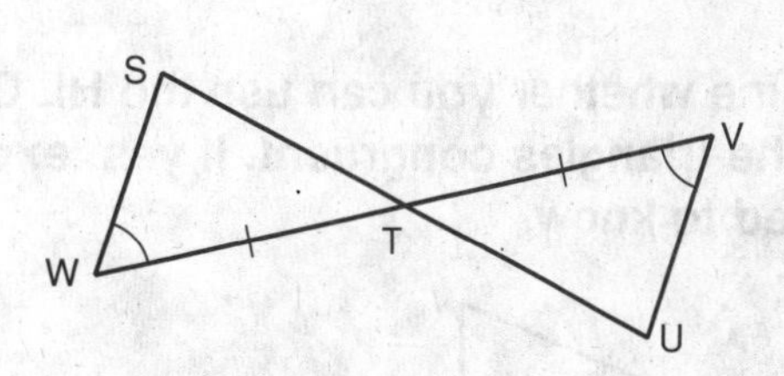

3. $\triangle KLM$ and $\triangle PNM$, given that M is the midpoint of $\overline{NL}$

____

____

____

4. $\triangle STW$ and $\triangle UTV$

____

____

____

Name ______________________ Date __________ Class __________

LESSON 4-5

Review for Mastery

Triangle Congruence: ASA, AAS, and HL continued

Angle-Angle-Side (AAS) Congruence Theorem

If two angles and a nonincluded side of one triangle are congruent to the corresponding angles and nonincluded side of another triangle, then the triangles are congruent.

$\overline{FH}$ is a nonincluded side of $\angle F$ and $\angle G$.

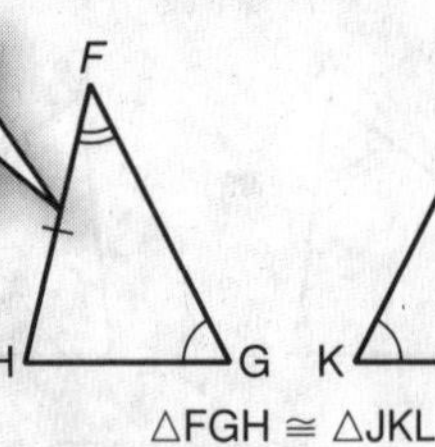

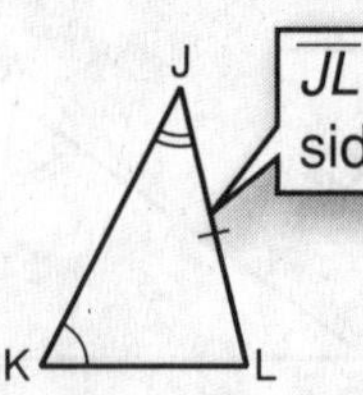

$\overline{JL}$ is a nonincluded side of $\angle J$ and $\angle K$.

$\triangle FGH \cong \triangle JKL$

Special theorems can be used to prove right triangles congruent.

Hypotenuse-Leg (HL) Congruence Theorem

If the hypotenuse and a leg of a right triangle are congruent to the hypotenuse and a leg of another right triangle, then the triangles are congruent.

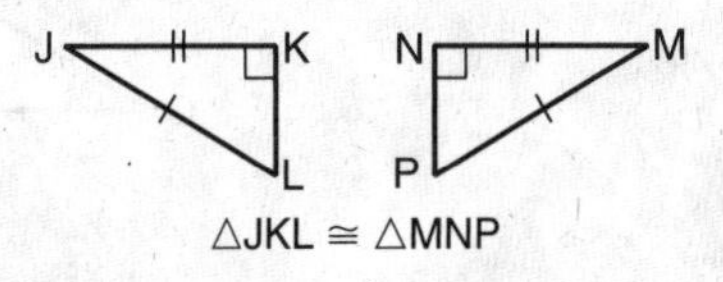

$\triangle JKL \cong \triangle MNP$

5. Describe the corresponding parts and the justifications for using them to prove the triangles congruent by AAS.

Given: $\overline{BD}$ is the angle bisector of $\angle ADC$.

Prove: $\triangle ABD \cong \triangle CBD$

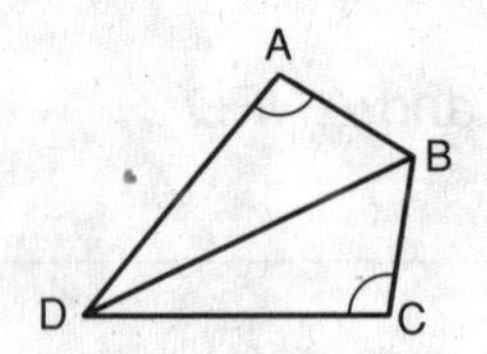

____

____

Determine whether you can use the HL Congruence Theorem to prove the triangles congruent. If yes, explain. If not, tell what else you need to know.

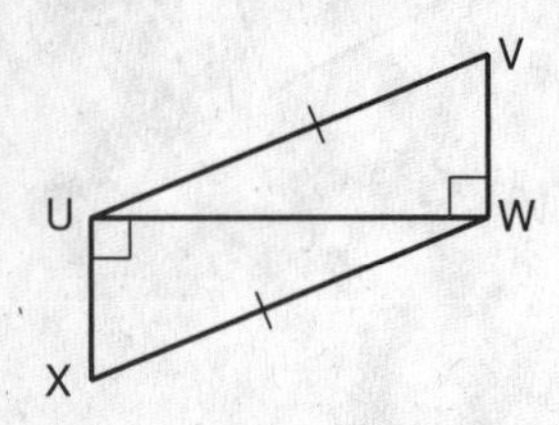

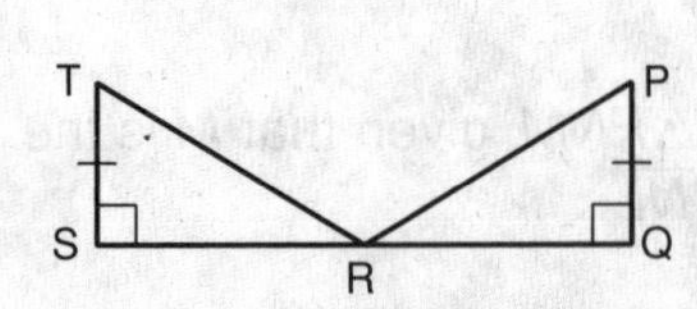

6. $\triangle UVW \cong \triangle WXU$

____

____

7. $\triangle TSR \cong \triangle PQR$

____

____

Name ______________________________ Date ____________ Class ____________

LESSON 4-6

Review for Mastery

Triangle Congruence: CPCTC

Corresponding Parts of Congruent Triangles are Congruent **(CPCTC)** is useful in proofs. If you prove that two triangles are congruent, then you can use CPCTC as a justification for proving corresponding parts congruent.

Given: $\overline{AD} \cong \overline{CD}$, $\overline{AB} \cong \overline{CB}$

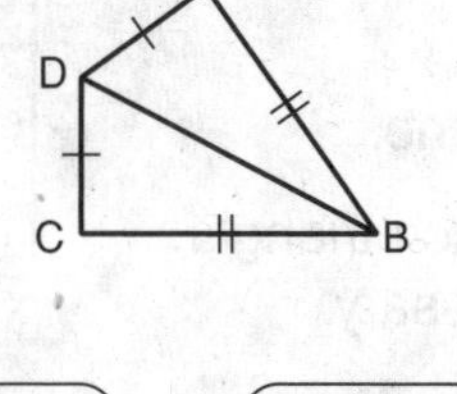

Prove: $\angle A \cong \angle C$

Proof:

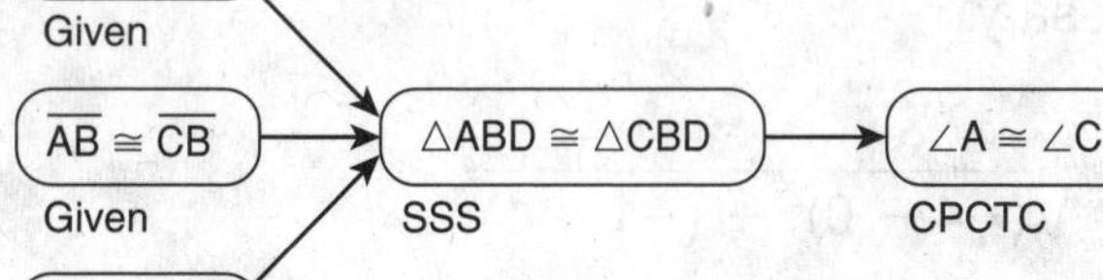

Complete each proof.

1. Given: $\angle PNQ \cong \angle LNM$, $\overline{PN} \cong \overline{LN}$, N is the midpoint of $\overline{QM}$.

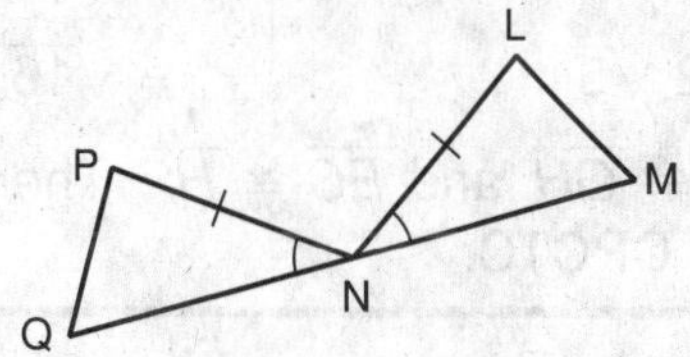

Prove: $\overline{PQ} \cong \overline{LM}$

Proof:

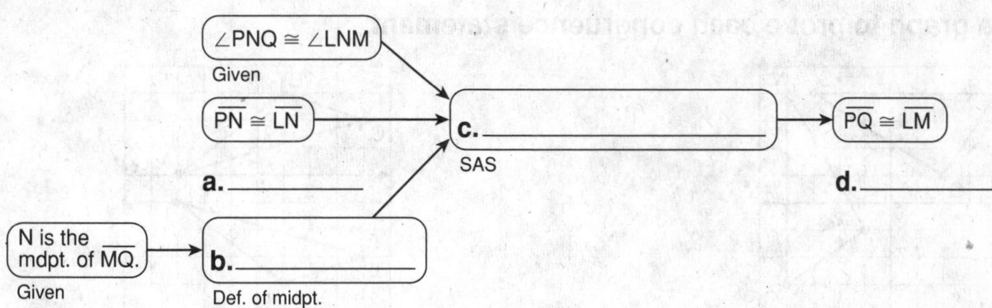

2. Given: $\triangle UXW$ and $\triangle UVW$ are right $\triangle$s. $\overline{UX} \cong \overline{UV}$

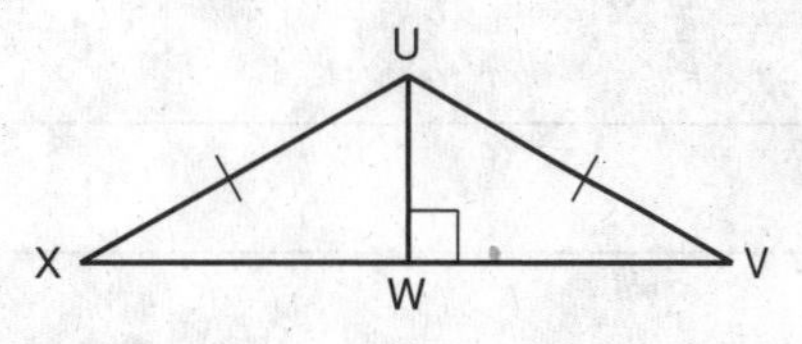

Prove: $\angle X \cong \angle V$

Proof:

Statements	**Reasons**
1. $\triangle UXW$ and $\triangle UVW$ are rt. $\triangle$s.	1. Given
2. $\overline{UX} \cong \overline{UV}$	2. **a.** ____________
3. $\overline{UW} \cong \overline{UW}$	3. **b.** ____________
4. **c.** ____________	4. **d.** ____________
5. $\angle X \cong \angle V$	5. **e.** ____________

Name ______________________ Date ___________ Class ___________

LESSON 4-6

Review for Mastery

Triangle Congruence: CPCTC continued

You can also use CPCTC when triangles are on the coordinate plane.

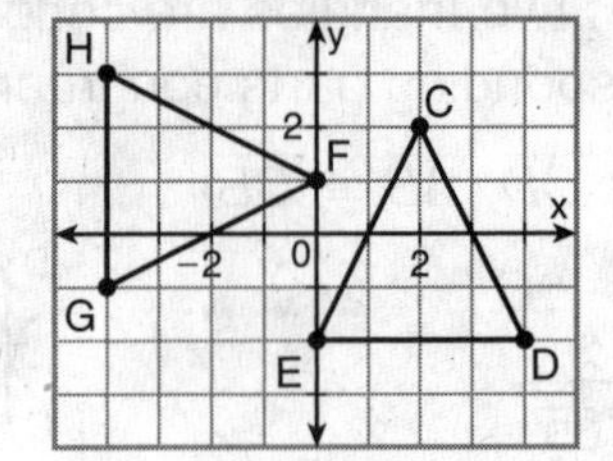

Given: $C(2, 2)$, $D(4, -2)$, $E(0, -2)$,
$F(0, 1)$, $G(-4, -1)$, $H(-4, 3)$

Prove: $\angle CED \cong \angle FHG$

Step 1 Plot the points on a coordinate plane.

Step 2 Find the lengths of the sides of each triangle.
Use the Distance Formula if necessary.

$d = \sqrt{(x_2 - x_1)^2 + (y_2 - y_1)^2}$

$CD = \sqrt{(4 - 2)^2 + (-2 - 2)^2}$ $\quad$ $FG = \sqrt{(-4 - 0)^2 + (-1 - 1)^2}$

$= \sqrt{4 + 16} = 2\sqrt{5}$ $\quad$ $= \sqrt{16 + 4} = 2\sqrt{5}$

$DE = 4$ $\quad$ $GH = 4$

$EC = \sqrt{(2 - 0)^2 + [2 - (-2)]^2}$ $\quad$ $HF = \sqrt{[0 - (-4)]^2 + (1 - 3)^2}$

$= \sqrt{4 + 16} = 2\sqrt{5}$ $\quad$ $= \sqrt{16 + 4} = 2\sqrt{5}$

So, $\overline{CD} \cong \overline{FG}$, $\overline{DE} \cong \overline{GH}$, and $\overline{EC} \cong \overline{HF}$. Therefore $\triangle CDE \cong \triangle FGH$ by SSS, and $\angle CED \cong \angle FHG$ by CPCTC.

Use the graph to prove each congruence statement.

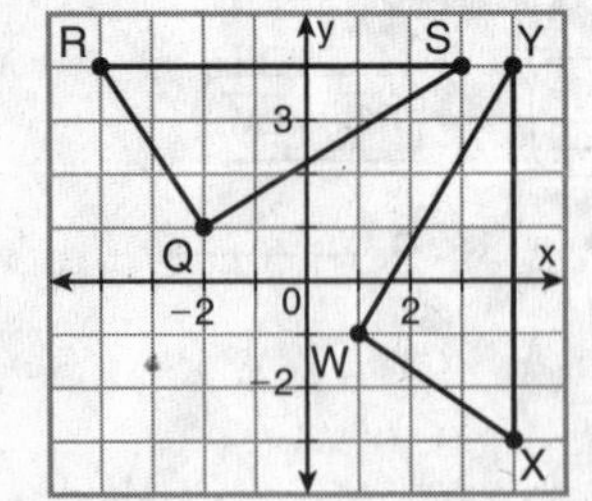

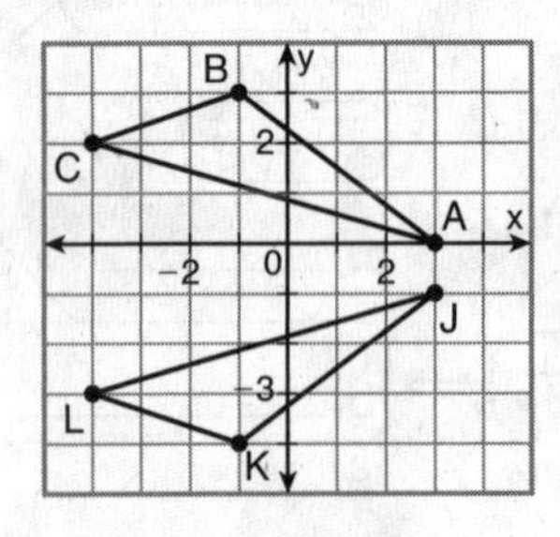

3. $\angle RSQ \cong \angle XYW$

____

____

____

____

4. $\angle CAB \cong \angle LJK$

____

____

____

____

5. Use the given set of points to prove $\angle PMN \cong \angle VTU$.

$M(-2, 4)$, $N(1, -2)$, $P(-3, -4)$, $T(-4, 1)$, $U(2, 4)$, $V(4, 0)$

____

____

Name ______________________________ Date ____________ Class ____________

LESSON 4-7

Review for Mastery

Introduction to Coordinate Proof

A **coordinate proof** is a proof that uses coordinate geometry and algebra. In a coordinate proof, the first step is to position a figure in a plane. There are several ways you can do this to make your proof easier.

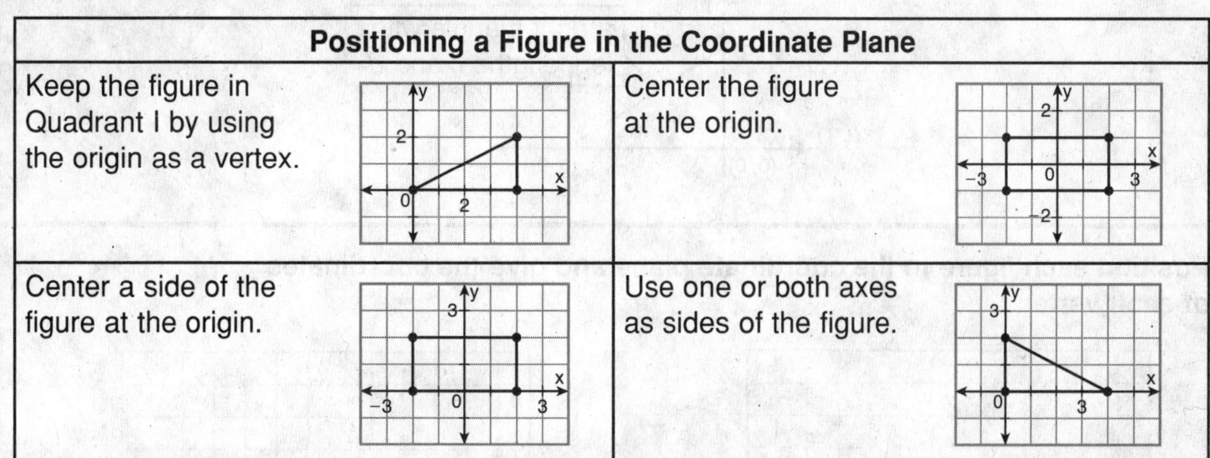

Positioning a Figure in the Coordinate Plane	
Keep the figure in Quadrant I by using the origin as a vertex.	Center the figure at the origin.
Center a side of the figure at the origin.	Use one or both axes as sides of the figure.

Position each figure in the coordinate plane and give the coordinates of each vertex.

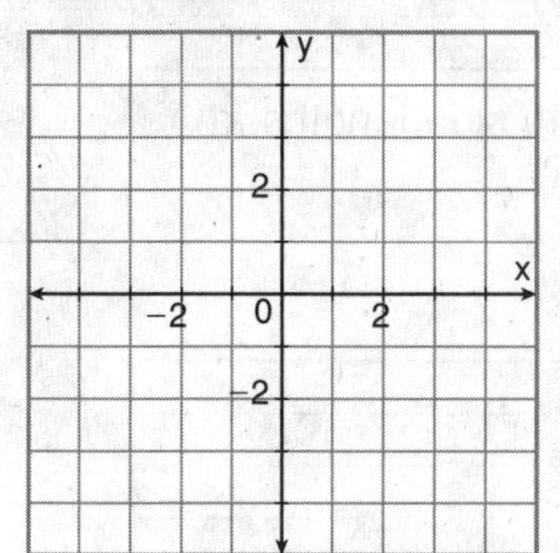

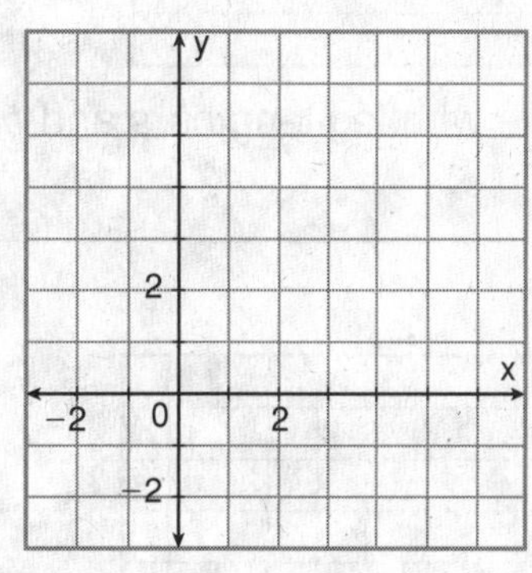

1. a square with side lengths of 6 units

____

2. a right triangle with leg lengths of 3 units and 4 units

____

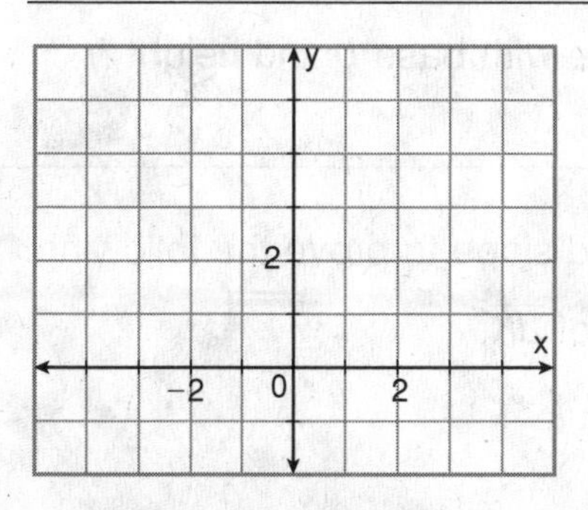

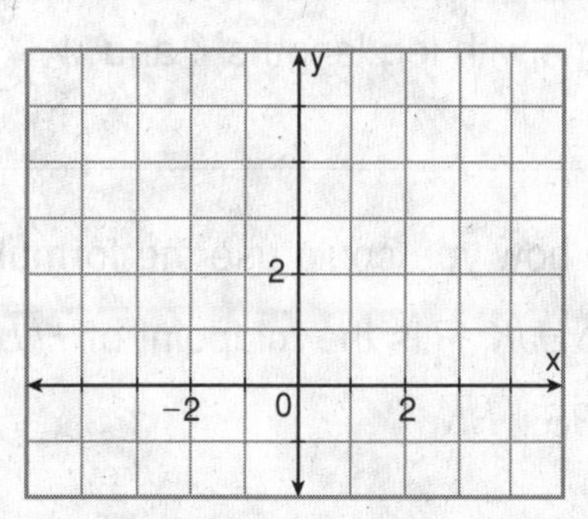

3. a triangle with a base of 8 units and a height of 2 units

____

4. a rectangle with a length of 6 units and a width of 3 units

____

Name ______________________ Date __________ Class __________

LESSON 4-7

Review for Mastery

Introduction to Coordinate Proof continued

You can prove that a statement about a figure is true without knowing the side lengths. To do this, assign variables as the coordinates of the vertices.

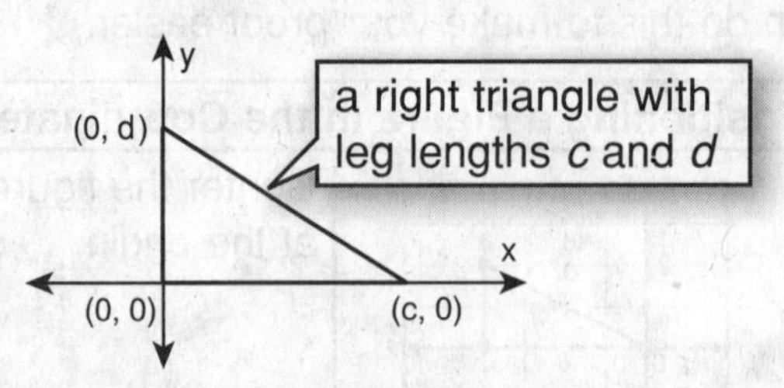

Position each figure in the coordinate plane and give the coordinates of each vertex.

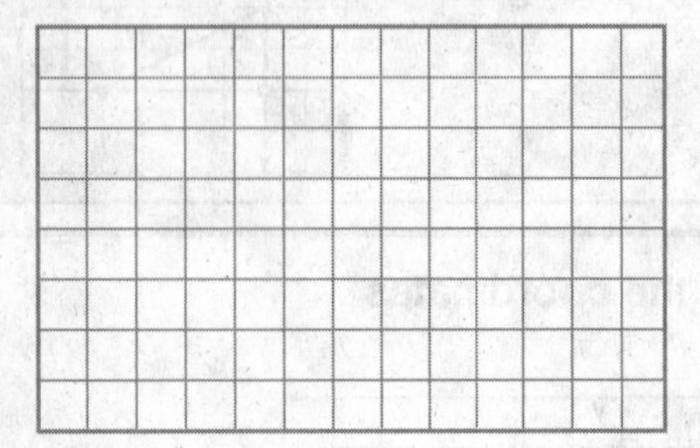

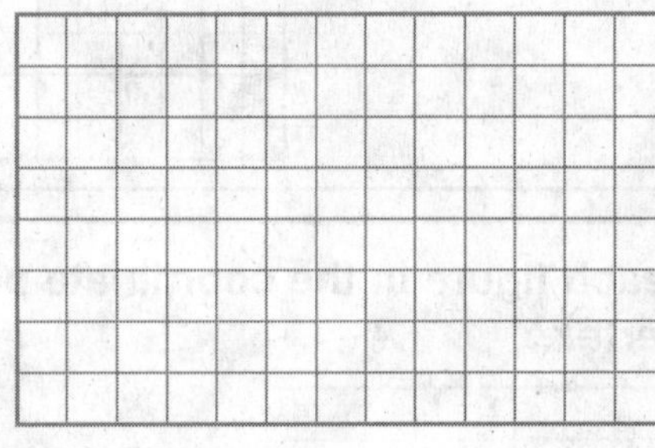

5. a right triangle with leg lengths s and t

____

6. a square with side lengths k

____

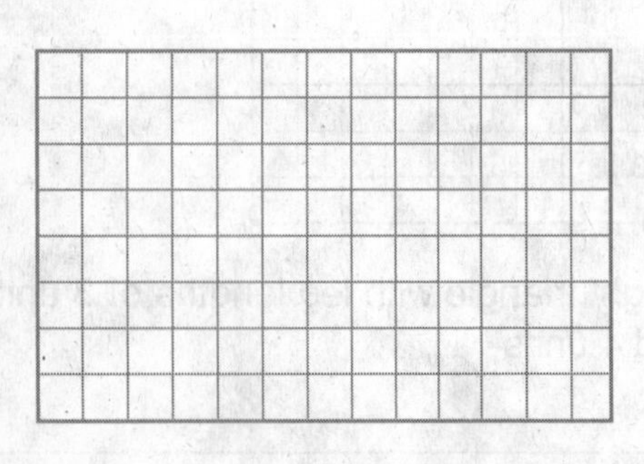

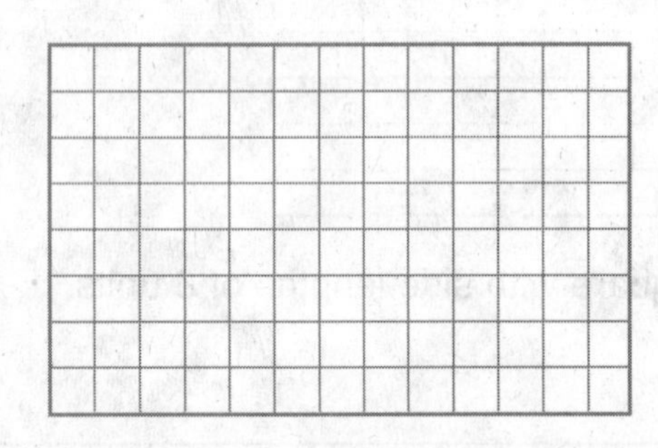

7. a rectangle with leg lengths ℓ and w

____

8. a triangle with base b and height h

____

9. Describe how you could use the formulas for midpoint and slope to prove the following.

Given: $\triangle HJK$, R is the midpoint of $\overline{HJ}$, S is the midpoint of $\overline{JK}$.

Prove: $\overline{RS} \parallel \overline{HK}$

____

____

____

Name ______________________ Date __________ Class __________

LESSON 4-8

Review for Mastery

Isosceles and Equilateral Triangles

Theorem	Examples
Isosceles Triangle Theorem If two sides of a triangle are congruent, then the angles opposite the sides are congruent.	If $\overline{RT} \cong \overline{RS}$, then $\angle T \cong \angle S$.
Converse of Isosceles Triangle Theorem If two angles of a triangle are congruent, then the sides opposite those angles are congruent.	If $\angle N \cong \angle M$, then $\overline{LN} \cong \overline{LM}$.

You can use these theorems to find angle measures in isosceles triangles.

Find m∠*E* in △*DEF*.

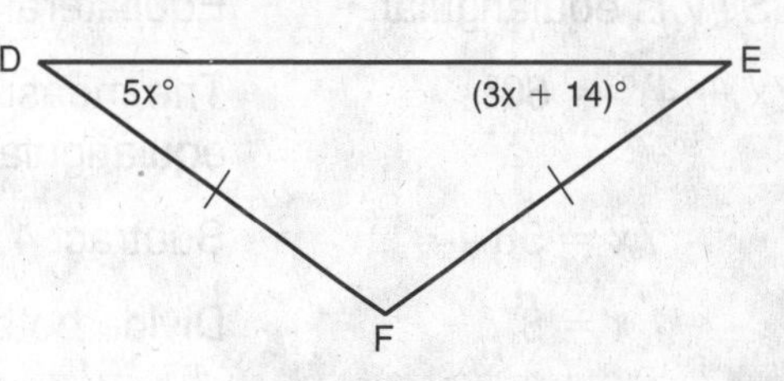

$m\angle D = m\angle E$	Isosc. △ Thm.
$5x° = (3x + 14)°$	Substitute the given values.
$2x = 14$	Subtract 3*x* from both sides.
$x = 7$	Divide both sides by 2.

Thus $m\angle E = 3(7) + 14 = 35°$.

Find each angle measure.

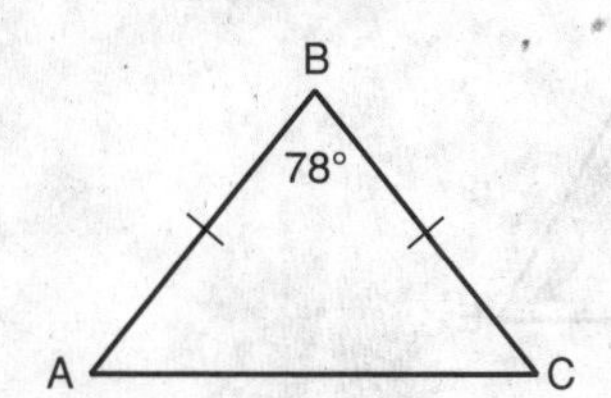

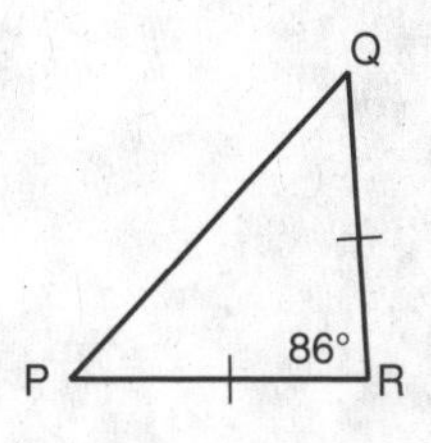

1. $m\angle C =$ ______________

2. $m\angle Q =$ ______________

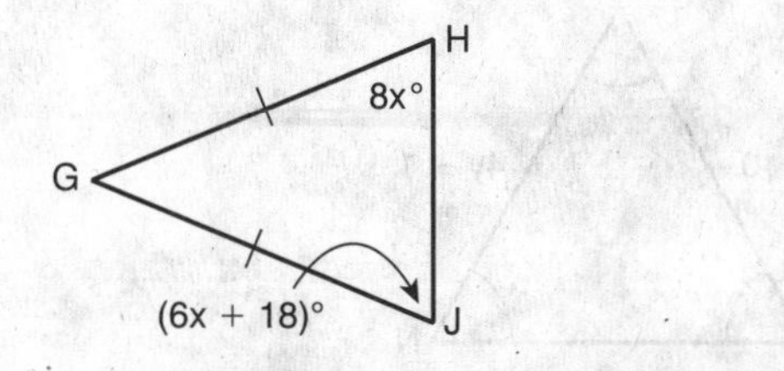

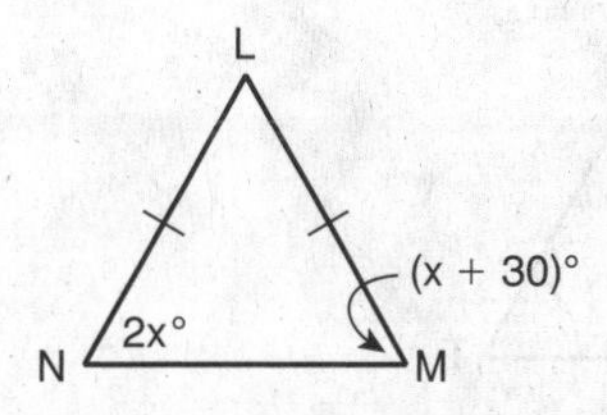

3. $m\angle H =$ ______________

4. $m\angle M =$ ______________

Name ______________________ Date __________ Class __________

LESSON 4-8

Review for Mastery

Isosceles and Equilateral Triangles continued

Equilateral Triangle Corollary
If a triangle is equilateral, then it is equiangular.

(equilateral $\triangle \rightarrow$ equiangular $\triangle$)

Equiangular Triangle Corollary
If a triangle is equiangular, then it is equilateral.

(equiangular $\triangle \rightarrow$ equilateral $\triangle$)

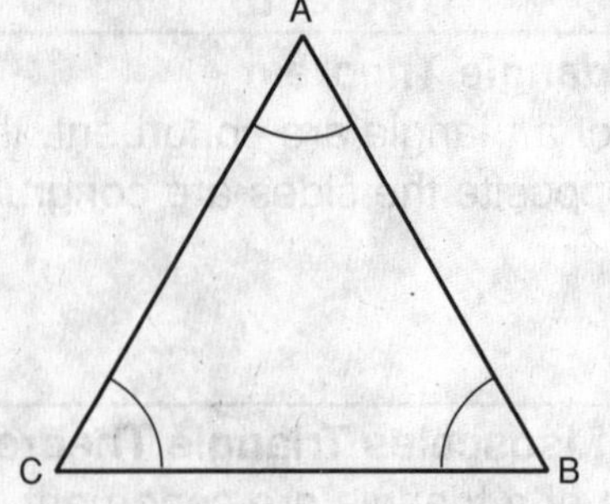

If $\angle A \cong \angle B \cong \angle C$, then $\overline{AB} \cong \overline{BC} \cong \overline{CA}$.

You can use these theorems to find values in equilateral triangles.

Find *x* in $\triangle STV$.

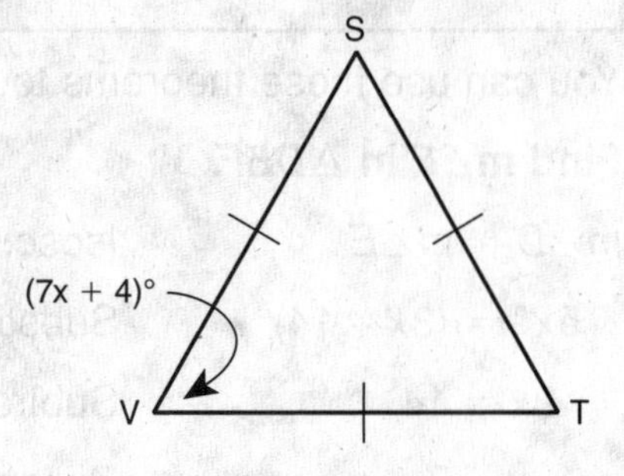

$\triangle STV$ is equiangular.	Equilateral $\triangle \rightarrow$ equiangular $\triangle$
$(7x + 4)^\circ = 60^\circ$	The measure of each $\angle$ of an equiangular $\triangle$ is 60°.
$7x = 56$	Subtract 4 from both sides.
$x = 8$	Divide both sides by 7.

Find each value.

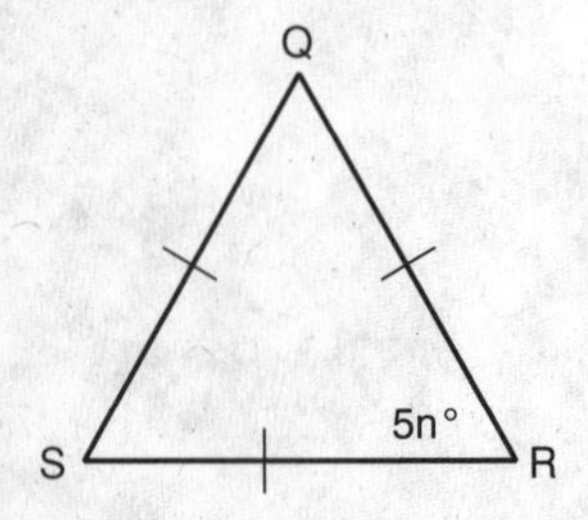

5. n = ____________

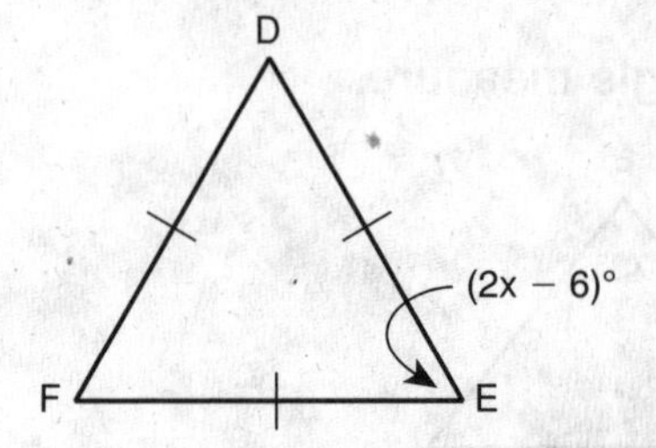

6. x = ____________

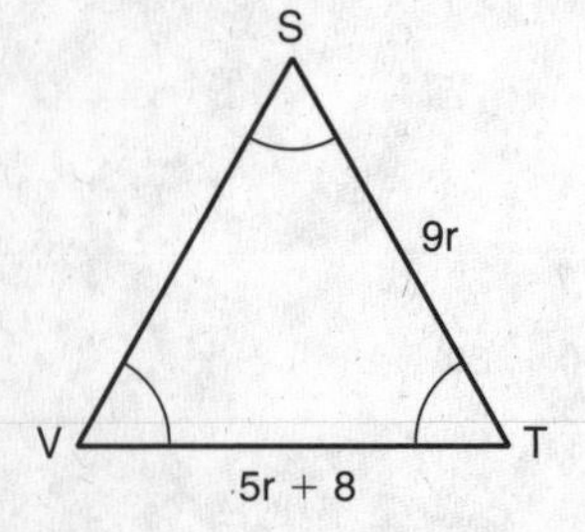

7. VT = ____________

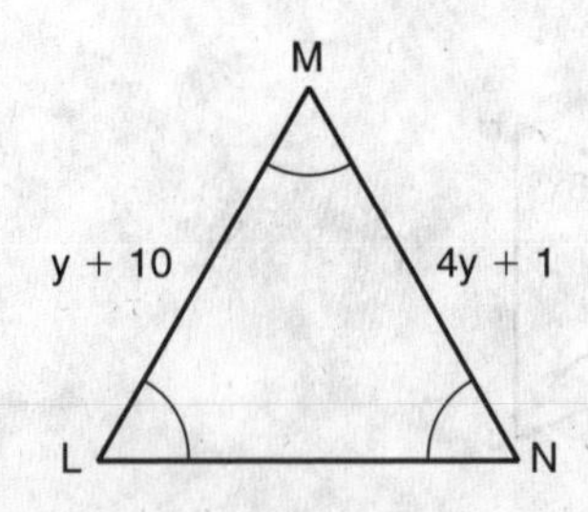

8. MN = ____________

Name ______________________ Date ____________ Class ____________

LESSON 5-1

Review for Mastery

Perpendicular and Angle Bisectors

Theorem	Example
Perpendicular Bisector Theorem If a point is on the perpendicular bisector of a segment, then it is **equidistant,** or the same distance, from the endpoints of the segment.	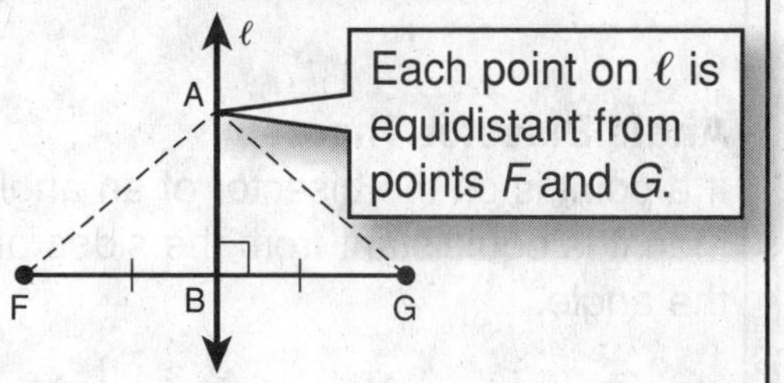 Each point on ℓ is equidistant from points F and G. **Given:** ℓ is the perpendicular bisector of $\overline{FG}$. **Conclusion:** $AF = AG$

The **Converse of the Perpendicular Bisector Theorem** is also true. If a point is equidistant from the endpoints of a segment, then it is on the perpendicular bisector of the segment.

You can write an equation for the perpendicular bisector of a segment. Consider the segment with endpoints $Q(-5, 6)$ and $R(1, 2)$.

Step 1 Find the midpoint of $\overline{QR}$.

$$\left(\frac{x_1 + x_2}{2}, \frac{y_1 + y_2}{2}\right) = \left(\frac{-5 + 1}{2}, \frac{6 + 2}{2}\right)$$
$$= (-2, 4)$$

Step 2 Find the slope of the $\perp$ bisector of $\overline{QR}$.

$$\frac{y_2 - y_1}{x_2 - x_1} = \frac{2 - 6}{1 - (-5)} \quad \text{Slope of } \overline{QR}$$
$$= -\frac{2}{3}$$

So the slope of the $\perp$ bisector of $\overline{QR}$ is $\frac{3}{2}$.

Step 3 Use the point-slope form to write an equation.

$y - y_1 = m(x - x_1)$ Point-slope form

$y - 4 = \frac{3}{2}(x + 2)$ Slope $= \frac{3}{2}$; line passes through $(-2, 4)$, the midpoint of $\overline{QR}$.

Find each measure.

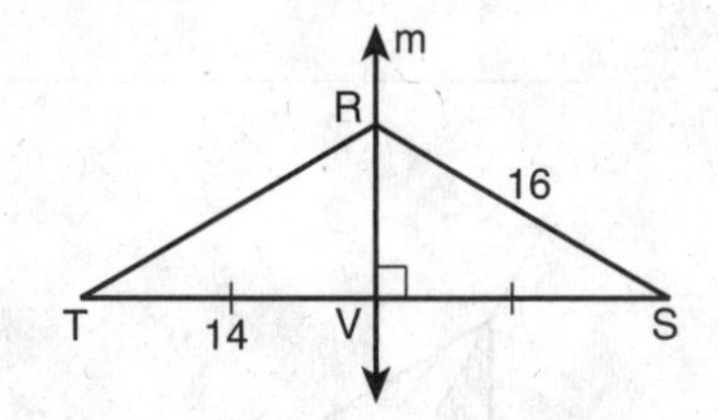

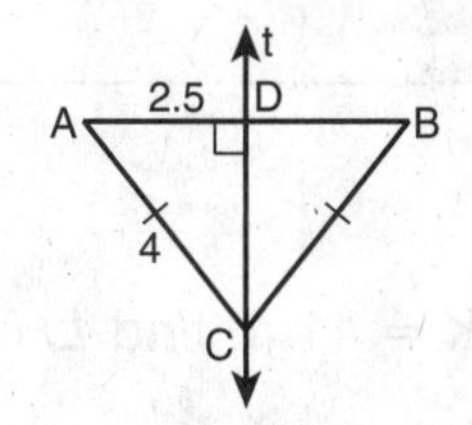

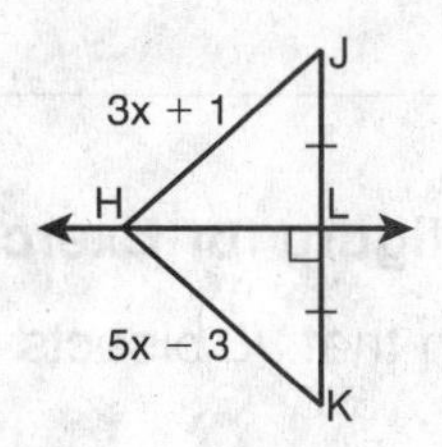

1. $RT =$ ____________ 2. $AB =$ ____________ 3. $HJ =$ ____________

Write an equation in point-slope form for the perpendicular bisector of the segment with the given endpoints.

4. $A(6, -3)$, $B(0, 5)$

5. $W(2, 7)$, $X(-4, 3)$

Name ______________________ Date ____________ Class ____________

LESSON 5-1

Review for Mastery

Perpendicular and Angle Bisectors continued

Theorem	Example
Angle Bisector Theorem If a point is on the bisector of an angle, then it is equidistant from the sides of the angle.	Point *P* is equidistant from sides $\overrightarrow{ML}$ and $\overrightarrow{MN}$. L M P N **Given:** $\overrightarrow{MP}$ is the angle bisector of $\angle LMN$. **Conclusion:** $LP = NP$
Converse of the Angle Bisector Theorem If a point in the interior of an angle is equidistant from the sides of the angle, then it is on the bisector of the angle.	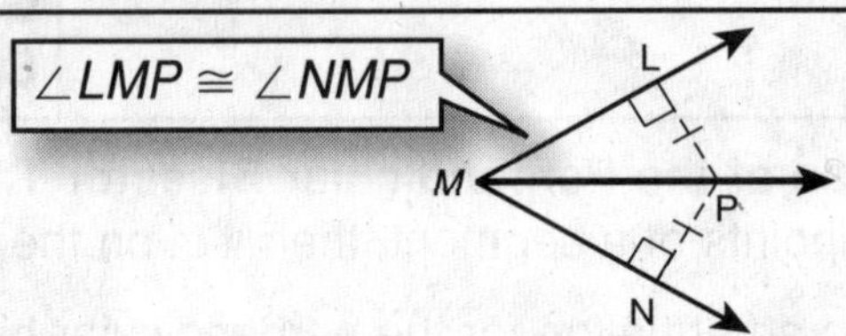 **Given:** $LP = NP$ **Conclusion:** $\overrightarrow{MP}$ is the angle bisector of $\angle LMN$.

Find each measure.

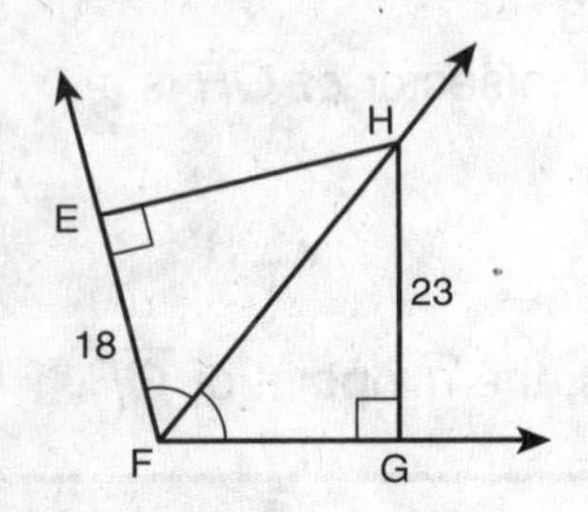

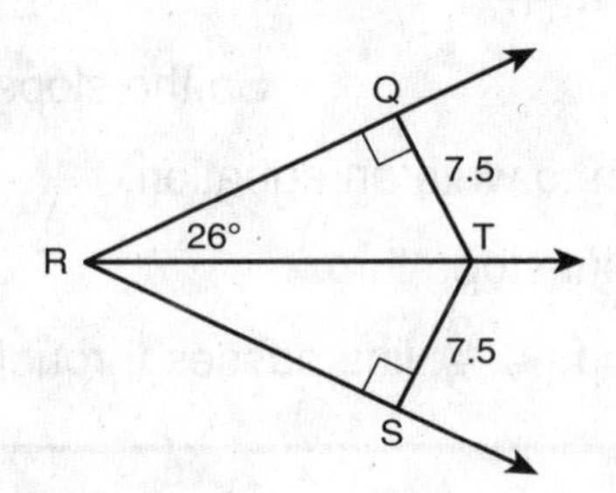

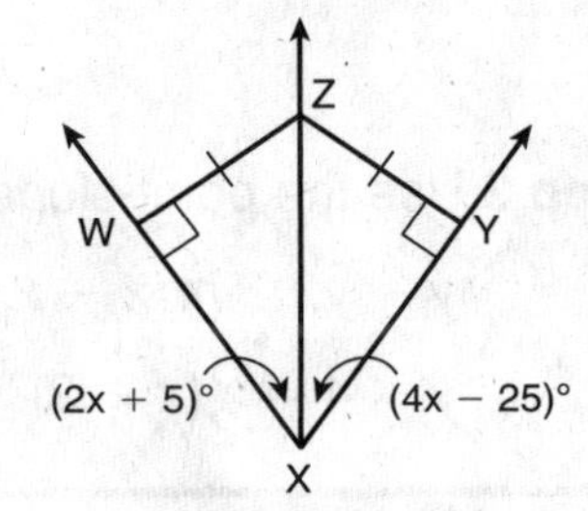

6. *EH* ____________

7. $m\angle QRS$ ____________

8. $m\angle WXZ$ ____________

Use the figure for Exercises 9–11.

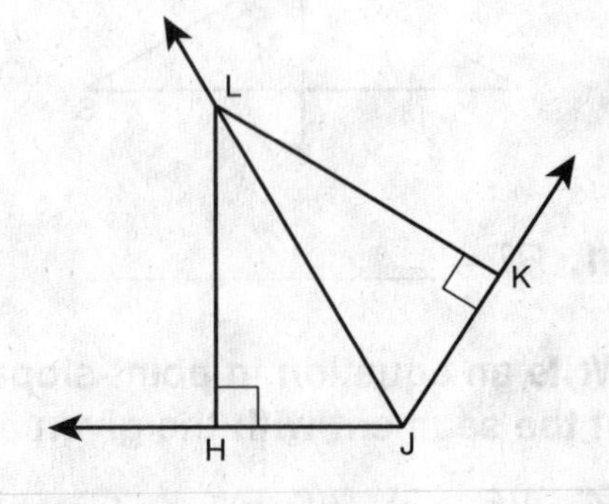

9. Given that $\overrightarrow{JL}$ bisects $\angle HJK$ and $LK = 11.4$, find *LH*.

____

10. Given that $LH = 26$, $LK = 26$, and $m\angle HJK = 122°$, find $m\angle LJK$.

____

11. Given that $LH = LK$, $m\angle HJL = (3y + 19)°$, and $m\angle LJK = (4y + 5)°$, find the value of *y*.

____

Name ______________________ Date __________ Class __________

LESSON 5-2

Review for Mastery

Bisectors of Triangles

Perpendicular bisectors $\overline{MR}$, $\overline{MS}$, and $\overline{MT}$ are **concurrent** because they intersect at one point.

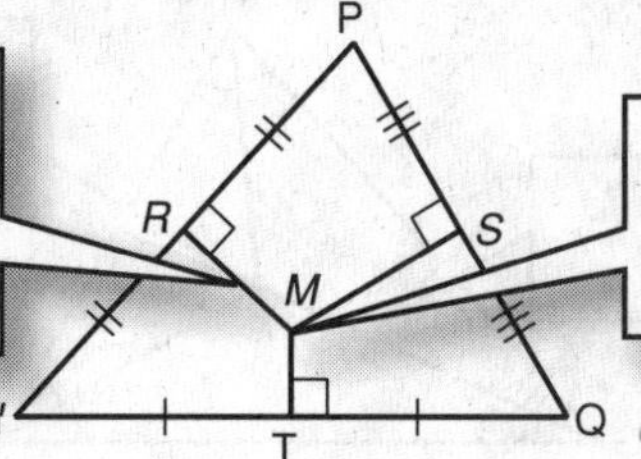

The point of intersection of $\overline{MR}$, $\overline{MS}$, *and* $\overline{MT}$ is called the **circumcenter** of $\triangle NPQ$.

Theorem	Example
Circumcenter Theorem The circumcenter of a triangle is equidistant from the vertices of the triangle.	**Given:** $\overline{MR}$, $\overline{MS}$, and $\overline{MT}$ are the perpendicular bisectors of $\triangle NPQ$. **Conclusion:** $MN = MP = MQ$ 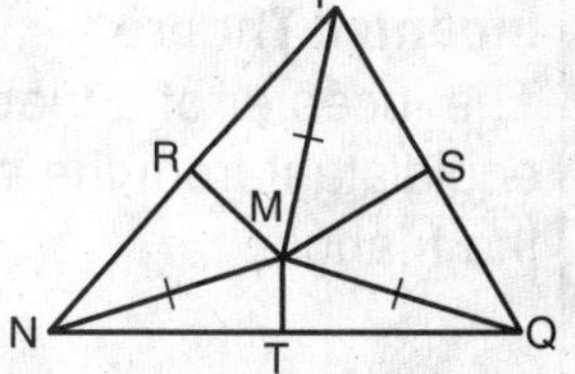

If a triangle on a coordinate plane has two sides that lie along the axes, you can easily find the circumcenter. Find the equations for the perpendicular bisectors of those two sides. The intersection of their graphs is the circumcenter.

$\overline{HD}$, $\overline{JD}$, and $\overline{KD}$ are the perpendicular bisectors of $\triangle EFG$. Find each length.

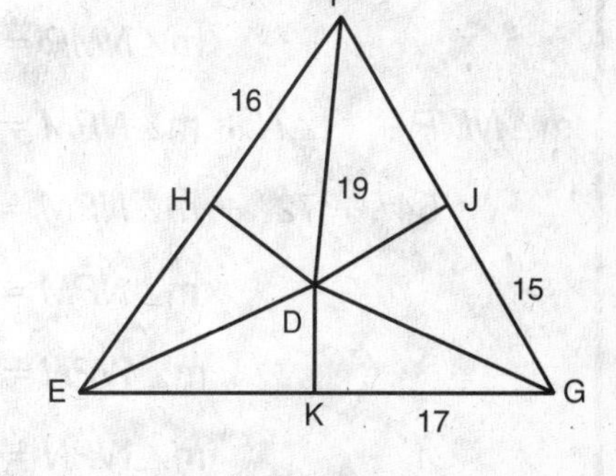

1. *DG* ________________

2. *EK* ________________

3. *FJ* ________________

4. *DE* ________________

Find the circumcenter of each triangle.

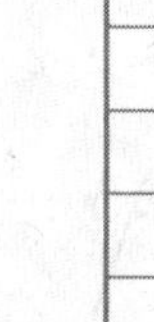

5.

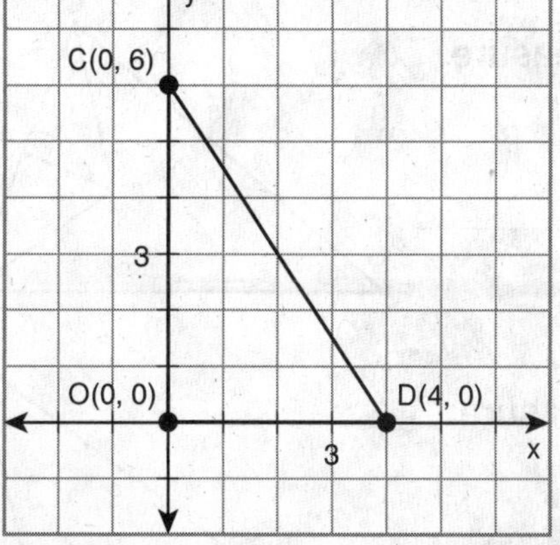

____

6.

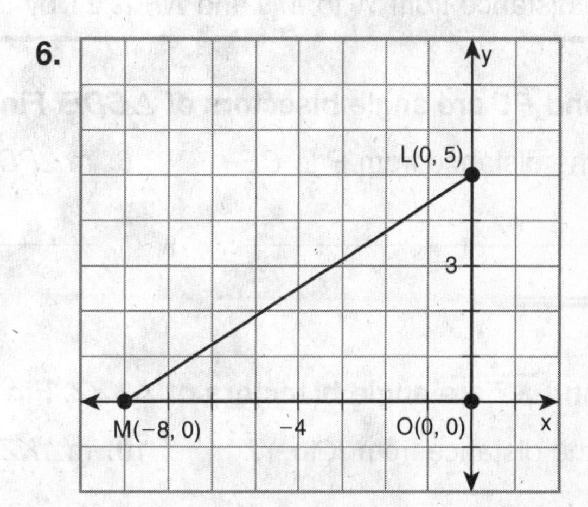

____

Name ______________________ Date ___________ Class ___________

LESSON 5-2

Review for Mastery

Bisectors of Triangles continued

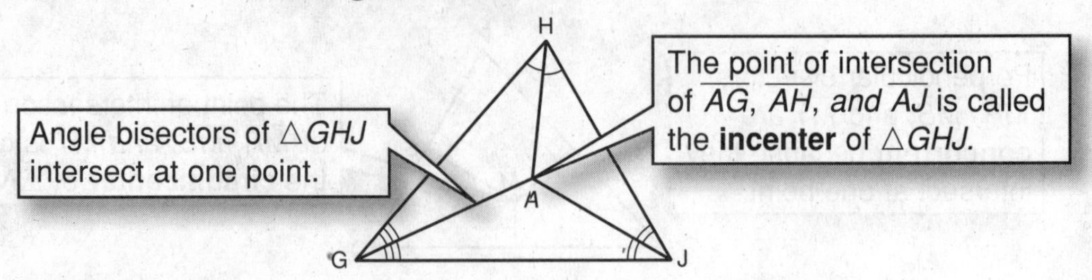

Theorem	Example
Incenter Theorem The incenter of a triangle is equidistant from the sides of the triangle.	**Given:** $\overline{AG}$, $\overline{AH}$, and $\overline{AJ}$ are the angle bisectors of $\triangle GHJ$. **Conclusion:** $AB = AC = AD$

$\overline{WM}$ and $\overline{WP}$ are angle bisectors of $\triangle MNP$, and $WK = 21$. Find $m\angle WPN$ and the distance from W to $\overline{MN}$ and $\overline{NP}$.

$m\angle NMP = 2m\angle NMW$	Def. of $\angle$ bisector
$m\angle NMP = 2(32°) = 64°$	Substitute.
$m\angle NMP + m\angle N + m\angle NPM = 180°$	$\triangle$ Sum Thm.
$64° + 72° + m\angle NPM = 180°$	Substitute.
$m\angle NPM = 44°$	Subtract 136° from each side.
$m\angle WPN = \frac{1}{2}m\angle NPM$	Def. of $\angle$ bisector
$m\angle WPN = \frac{1}{2}(44°) = 22°$	Substitute.

The distance from W to $\overline{MN}$ and $\overline{NP}$ is 21 by the Incenter Theorem.

$\overline{PC}$ and $\overline{PD}$ are angle bisectors of $\triangle CDE$. Find each measure.

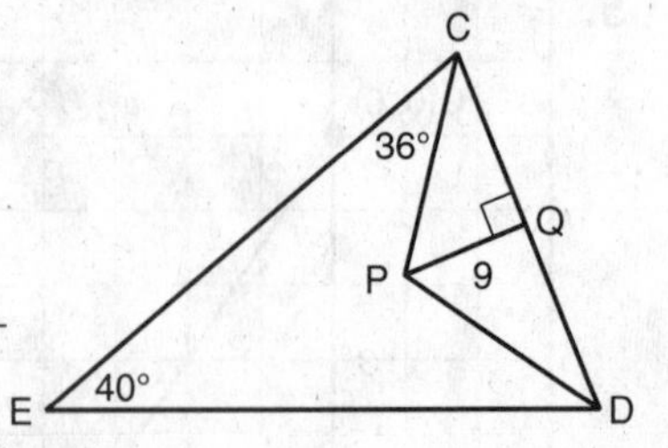

7. the distance from P to $\overline{CE}$ ____________

8. $m\angle PDE$ ____________

$\overline{KX}$ and $\overline{KZ}$ are angle bisectors of $\triangle XYZ$. Find each measure.

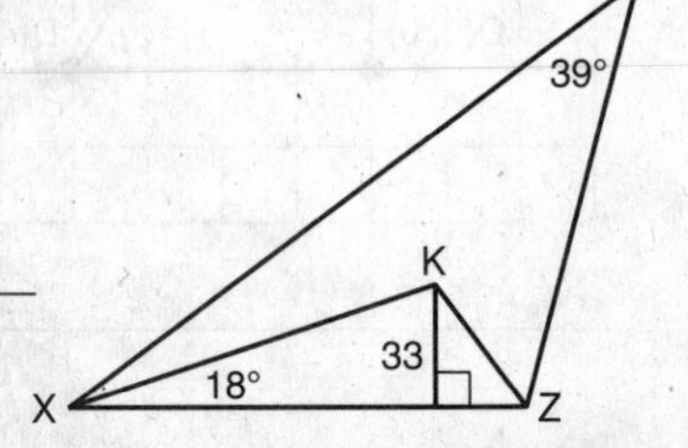

9. the distance from K to $\overline{YZ}$ ____________

10. $m\angle KZY$ ____________

Name ______________________ Date ____________ Class ____________

LESSON 5-3

Review for Mastery

Medians and Altitudes of Triangles

$\overline{AH}$, $\overline{BJ}$, and $\overline{CG}$ are **medians of a triangle.** They each join a vertex and the midpoint of the opposite side.

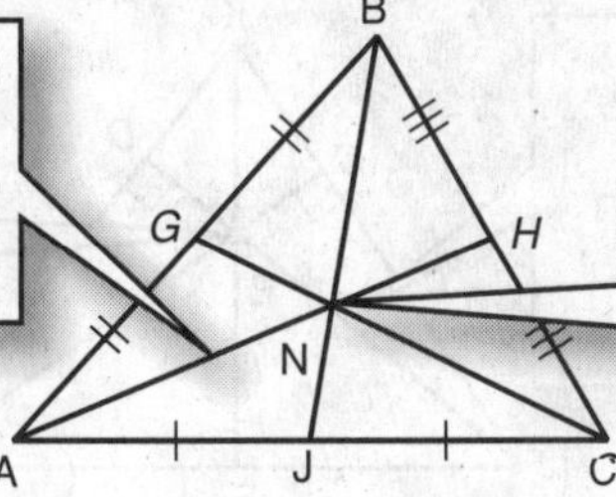

The point of intersection of the medians is called the **centroid** of $\triangle ABC$.

Theorem	Example
Centroid Theorem The centroid of a triangle is located $\frac{2}{3}$ of the distance from each vertex to the midpoint of the opposite side.	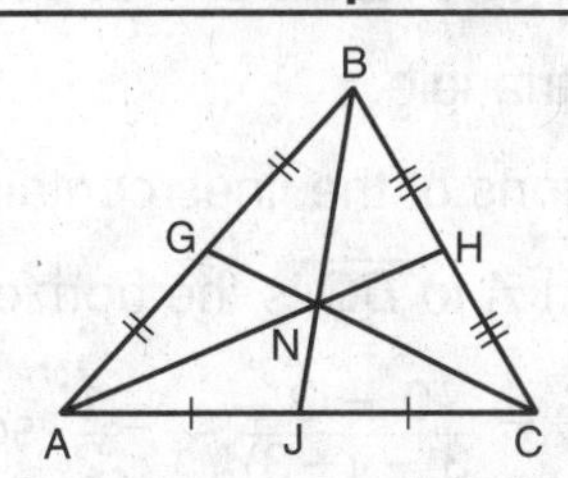 **Given:** $\overline{AH}$, $\overline{CG}$, and $\overline{BJ}$ are medians of $\triangle ABC$. **Conclusion:** $AN = \frac{2}{3}AH$, $CN = \frac{2}{3}CG$, $BN = \frac{2}{3}BJ$

In $\triangle ABC$ above, suppose $AH = 18$ and $BN = 10$. You can use the Centroid Theorem to find AN and BJ.

$AN = \frac{2}{3}AH$	Centroid Thm.	$BN = \frac{2}{3}BJ$	Centroid Thm.
$AN = \frac{2}{3}(18)$	Substitute 18 for AH.	$10 = \frac{2}{3}BJ$	Substitute 10 for BN.
$AN = 12$	Simplify.	$15 = BJ$	Simplify.

In $\triangle QRS$, $RX = 48$ and $QW = 30$. Find each length.

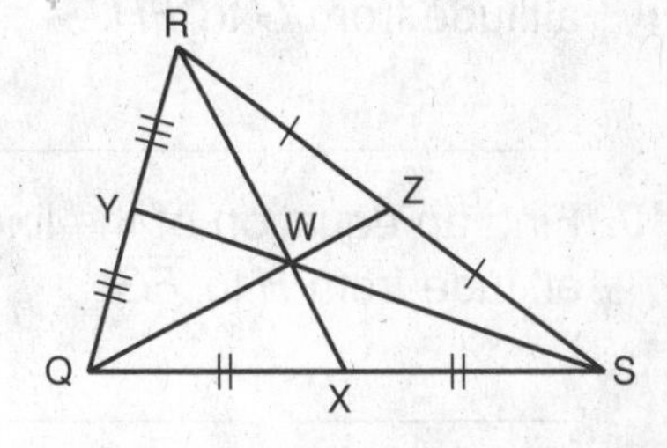

1. RW ____________

2. WX ____________

3. QZ ____________

4. WZ ____________

In $\triangle HJK$, $HD = 21$ and $BK = 18$. Find each length.

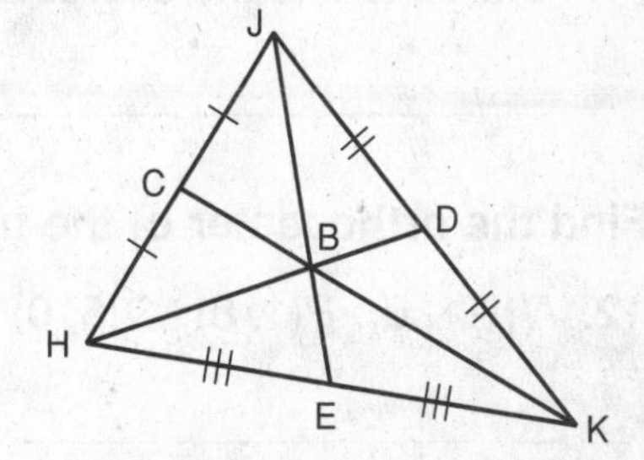

5. HB ____________

6. BD ____________

7. CK ____________

8. CB ____________

Name ______________________ Date __________ Class __________

LESSON 5-3

Review for Mastery

Medians and Altitudes of Triangles continued

$\overline{JD}$, $\overline{KE}$, and $\overline{LC}$ are **altitudes of a triangle.** They are perpendicular segments that join a vertex and the line containing the side opposite the vertex.

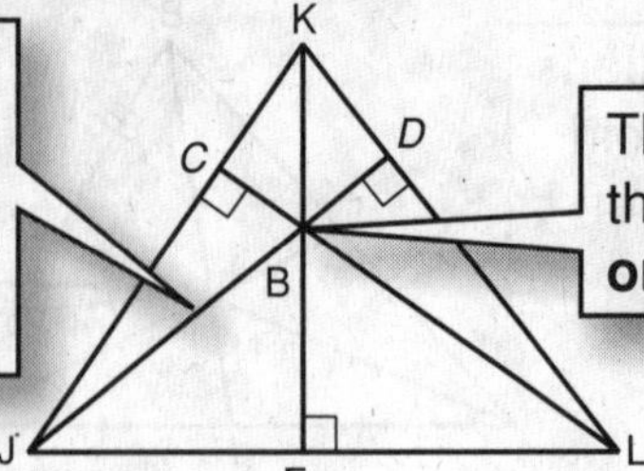

The point of intersection of the altitudes is called the **orthocenter** of $\triangle JKL$.

Find the orthocenter of $\triangle ABC$ with vertices $A(-3, 3)$, $B(3, 7)$, and $C(3, 0)$.

Step 1 Graph the triangle.

Step 2 Find equations of the lines containing two altitudes.

The altitude from A to $\overline{BC}$ is the horizontal line $y = 3$.

The slope of $\overleftrightarrow{AC} = \dfrac{0-3}{3-(-3)} = -\dfrac{1}{2}$, so the slope of the altitude from B to $\overline{AC}$ is 2. The altitude must pass through $B(3, 7)$.

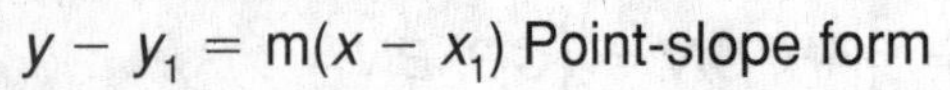

$y - y_1 = m(x - x_1)$ Point-slope form

$y - 7 = 2(x - 3)$ Substitute 2 for m and the coordinates of $B(3, 7)$ for (x_1, y_1).

$y = 2x + 1$ Simplify.

Step 3 Solving the system of equations $y = 3$ and $y = 2x + 1$, you find that the coordinates of the orthocenter are $(1, 3)$.

Triangle FGH has coordinates $F(-3, 1)$, $G(2, 6)$, and $H(4, 1)$.

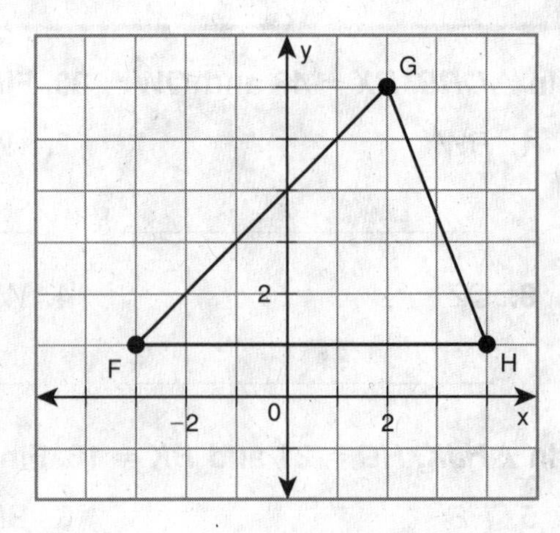

9. Find an equation of the line containing the altitude from G to $\overline{FH}$.

____

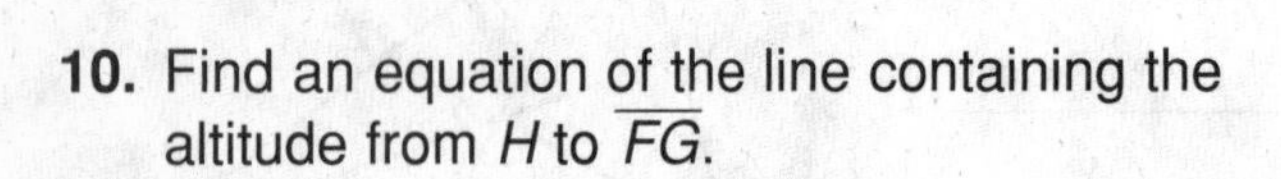

10. Find an equation of the line containing the altitude from H to $\overline{FG}$.

____

11. Solve the system of equations from Exercises 9 and 10 to find the coordinates of the orthocenter.

____

Find the orthocenter of the triangle with the given vertices.

12. $N(-1, 0)$, $P(1, 8)$, $Q(5, 0)$

____

13. $R(-1, 4)$, $S(5, -2)$, $T(-1, -6)$

____

Name ______________________ Date __________ Class __________

LESSON 5-4

Review for Mastery

The Triangle Midsegment Theorem

A **midsegment** of a triangle joins the midpoints of two sides of the triangle. Every triangle has three midsegments.

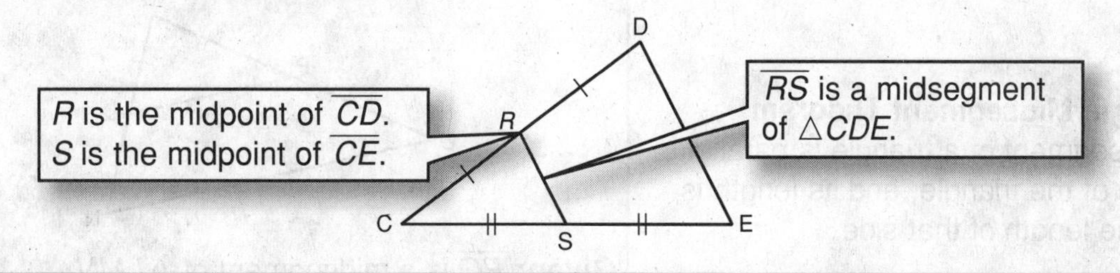

Use the figure for Exercises 1–4. $\overline{AB}$ is a midsegment of $\triangle RST$.

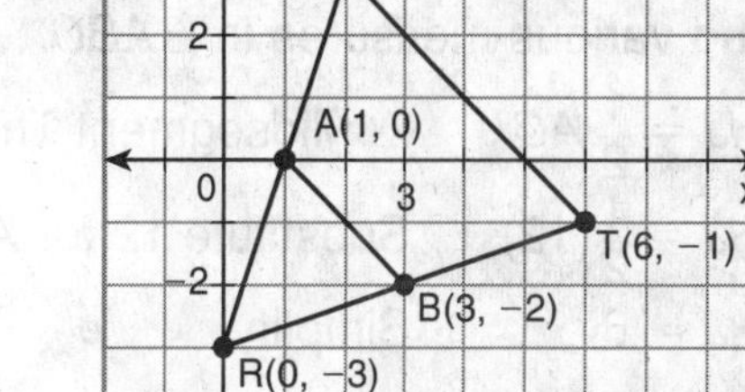

1. What is the slope of midsegment $\overline{AB}$ and the slope of side $\overline{ST}$?

   ____

2. What can you conclude about $\overline{AB}$ and $\overline{ST}$?

   ____

3. Find AB and ST.

   ____

4. Compare the lengths of $\overline{AB}$ and $\overline{ST}$.

Use $\triangle MNP$ for Exercises 5–7.

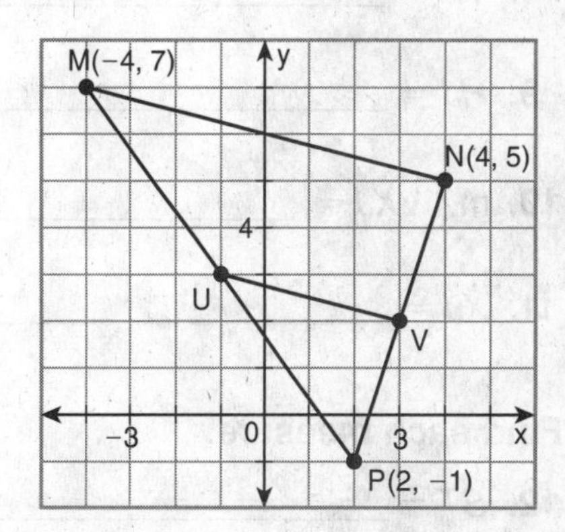

5. $\overline{UV}$ is a midsegment of $\triangle MNP$. Find the coordinates of U and V.

   ____

6. Show that $\overline{UV} \parallel \overline{MN}$.

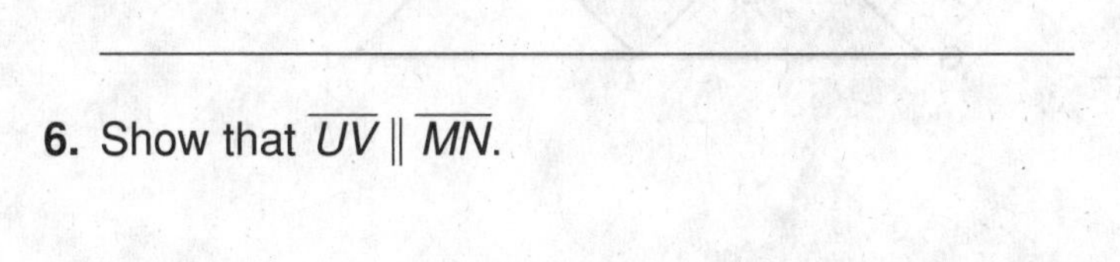

7. Show that $UV = \frac{1}{2}MN$.

   ____

Name ______________________ Date ____________ Class ____________

LESSON 5-4

Review for Mastery

The Triangle Midsegment Theorem continued

Theorem	Example
Triangle Midsegment Theorem A midsegment of a triangle is parallel to a side of the triangle, and its length is half the length of that side.	**Given:** $\overline{PQ}$ is a midsegment of $\triangle LMN$. **Conclusion:** $\overline{PQ} \parallel \overline{LN}$, $PQ = \frac{1}{2}LN$

You can use the Triangle Midsegment Theorem to find various measures in $\triangle ABC$.

$HJ = \frac{1}{2}AC$ $\triangle$ Midsegment Thm.

$HJ = \frac{1}{2}(12)$ Substitute 12 for AC.

$HJ = 6$ Simplify.

$JK = \frac{1}{2}AB$ $\triangle$ Midsegment Thm.

$4 = \frac{1}{2}AB$ Substitute 4 for JK.

$8 = AB$ Simplify.

$\overline{HJ} \parallel \overline{AC}$ Midsegment Thm.

$m\angle BCA = m\angle BJH$ Corr. ∠s Thm.

$m\angle BCA = 35°$ Substitute 35° for $m\angle BJH$.

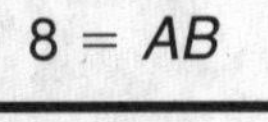

Find each measure.

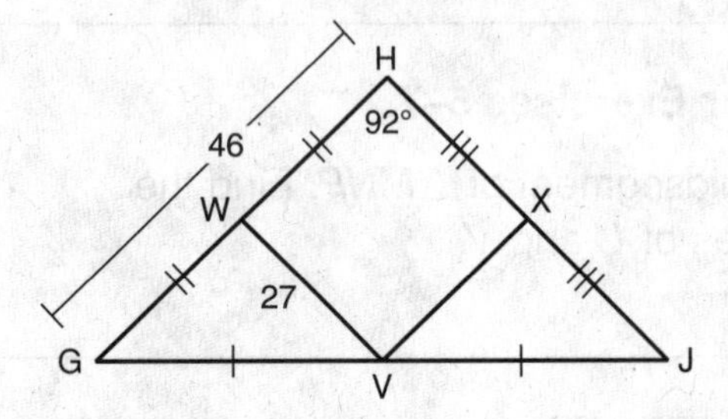

8. $VX =$ ______________

9. $HJ =$ ______________

10. $m\angle VXJ =$ ______________

11. $XJ =$ ______________

Find each measure.

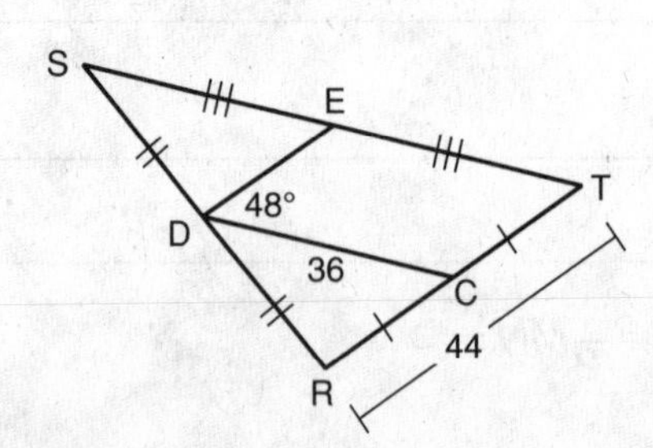

12. $ST =$ ______________

13. $DE =$ ______________

14. $m\angle DES =$ ______________

15. $m\angle RCD =$ ______________

Name ______________________ Date __________ Class __________

LESSON 5-5

Review for Mastery

Indirect Proof and Inequalities in One Triangle

In a direct proof, you begin with a true hypothesis and prove that a conclusion is true. In an **indirect proof,** you begin by assuming that the conclusion is false (that is, that the opposite of the conclusion is true). You then show that this assumption leads to a contradiction.

Consider the statement "Two acute angles do not form a linear pair."

Writing an Indirect Proof	
Steps	**Example**
1. Identify the conjecture to be proven.	**Given:** $\angle 1$ and $\angle 2$ are acute angles. **Prove:** $\angle 1$ and $\angle 2$ do not form a linear pair.
2. Assume the opposite of the conclusion is true.	Assume $\angle 1$ and $\angle 2$ form a linear pair.
3. Use direct reasoning to show that the assumption leads to a contradiction.	$m\angle 1 + m\angle 2 = 180°$ by def. of linear pair. Since $m\angle 1 < 90°$ and $m\angle 2 < 90°$, $m\angle 1 + m\angle 2 < 180°$. This is a contradiction.
4. Conclude that the assumption is false and hence that the original conjecture must be true.	The assumption that $\angle 1$ and $\angle 2$ form a linear pair is false. Therefore $\angle 1$ and $\angle 2$ do not form a linear pair.

Use the following statement for Exercises 1–4.

An obtuse triangle cannot have a right angle.

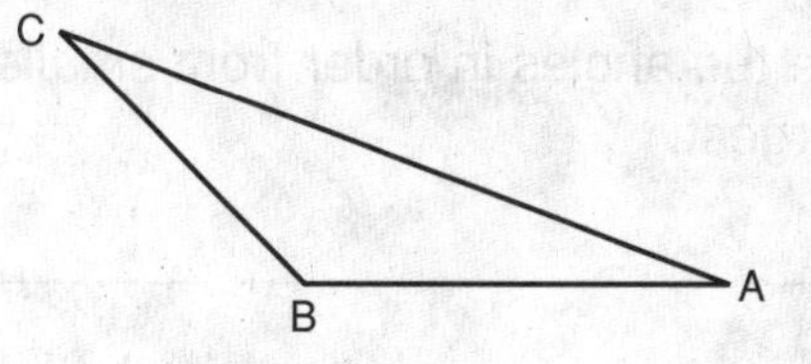

1. Identify the conjecture to be proven.

____

____

2. Assume the opposite of the conclusion. Write this assumption.

____

3. Use direct reasoning to arrive at a contradiction.

____

____

4. What can you conclude?

____

____

Name ________________________ Date ____________ Class ____________

LESSON 5-5

Review for Mastery

Indirect Proof and Inequalities in One Triangle continued

Theorem	Example
If two sides of a triangle are not congruent, then the larger angle is opposite the longer side.	$\angle X$ is the largest angle. $\overline{WY}$ is the longest side. If $WY > XY$, then $m\angle X > m\angle W$.

Another similar theorem says that if two angles of a triangle are not congruent, then the longer side is opposite the larger angle.

Write the correct answer.

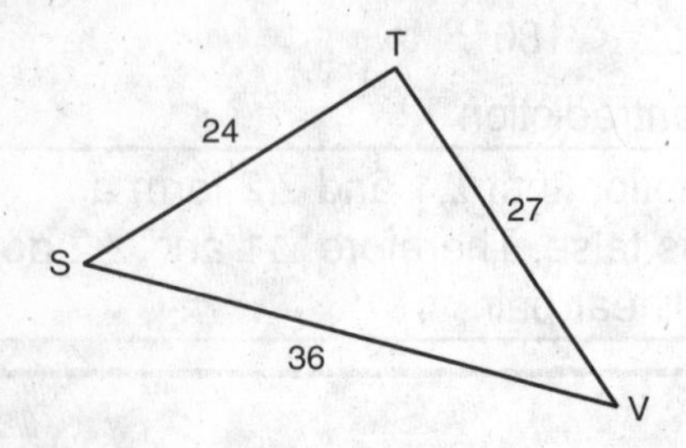

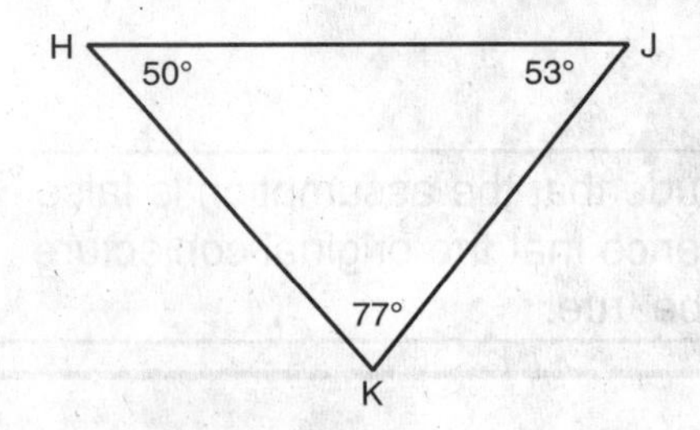

5. Write the angles in order from smallest to largest.

____

6. Write the sides in order from shortest to longest.

____

Theorem	Example
Triangle Inequality Theorem The sum of any two side lengths of a triangle is greater than the third side length.	$a + b > c$ $b + c > a$ $c + a > b$

Tell whether a triangle can have sides with the given lengths. Explain.

7. 3, 5, 8

____

8. 11, 15, 21

____

Name ________________________________ Date ______________ Class ______________

LESSON 5-6

Review for Mastery

Inequalities in Two Triangles

Theorem	Example
Hinge Theorem If two sides of one triangle are congruent to two sides of another triangle and the included angles are not congruent, then the included angle that is larger has the longer third side across from it.	L, K, M, H, G, J If $\angle K$ is larger than $\angle G$, then side $\overline{LM}$ is longer than side $\overline{HJ}$.

The **Converse of the Hinge Theorem** is also true. In the example above, if side $\overline{LM}$ is longer than side $\overline{HJ}$, then you can conclude that $\angle K$ is larger than $\angle G$. You can use both of these theorems to compare various measures of triangles.

Compare *NR* and *PQ* in the figure at right.

$PN = QR \quad PR = PR \qquad m\angle NPR < m\angle QRP$

Since two sides are congruent and $\angle NPR$ is smaller than $\angle QRP$, the side across from it is shorter than the side across from $\angle QRP$.

So $NR < PQ$ by the Hinge Theorem.

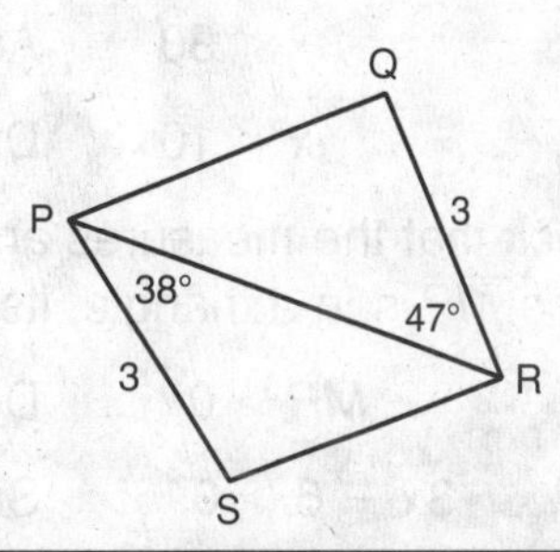

Compare the given measures.

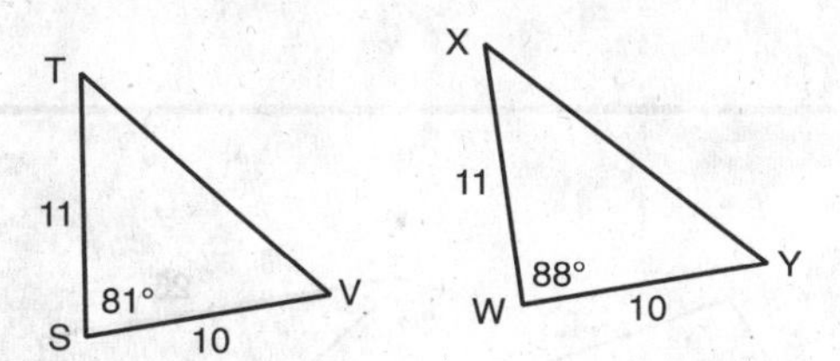

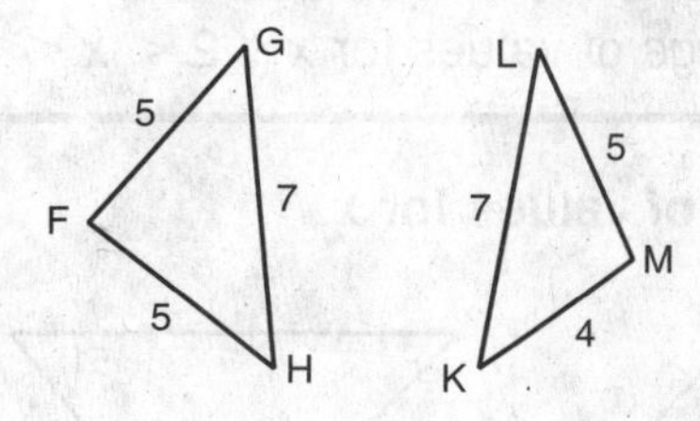

1. *TV* and *XY*

____

2. $m\angle G$ and $m\angle L$

____

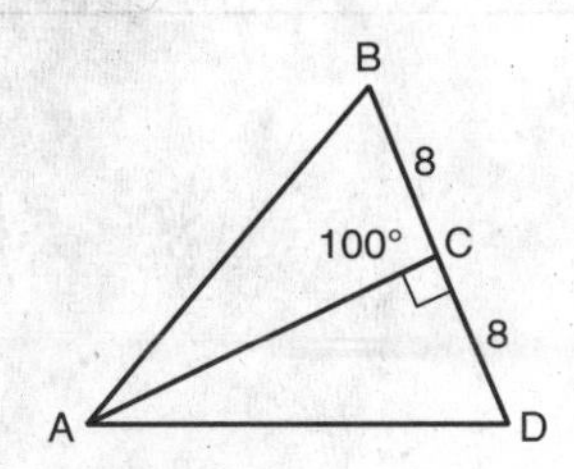

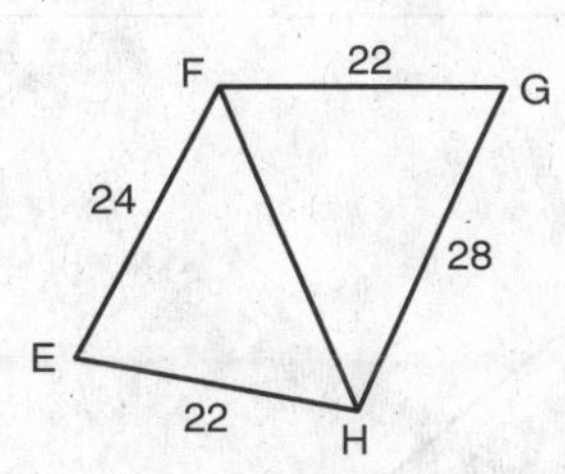

3. *AB* and *AD*

____

4. $m\angle FHE$ and $m\angle HFG$

____

Name ______________________________ Date ____________ Class ____________

LESSON 5-6

Review for Mastery

Inequalities in Two Triangles continued

You can use the Hinge Theorem and its converse to find a range of values in triangles.

Use $\triangle MNP$ and $\triangle QRS$ to find the range of values for *x*.

Step 1 Compare the side lengths in the triangles.

$NM = SR \quad NP = SQ \quad m\angle N < m\angle S$

Since two sides of $\triangle MNP$ are congruent to two sides of $\triangle QRS$ and $m\angle N < m\angle S$, then $MP < QR$ by the Hinge Theorem.

$MP < QR$	Hinge Thm.
$3x - 6 < 24$	Substitute the given values.
$3x < 30$	Add 6 to each side.
$x < 10$	Divide each side by 3.

Step 2 Check that the measures are possible for a triangle.
Since $\overline{MP}$ is in a triangle, its length must be greater than 0.

$MP > 0$	Def. of $\triangle$
$3x - 6 > 0$	Substitute $3x - 6$ for MP.
$x > 2$	Simplify.

Step 3 Combine the inequalities.
A range of values for x is $2 < x < 10$.

Find a range of values for *x*.

5.

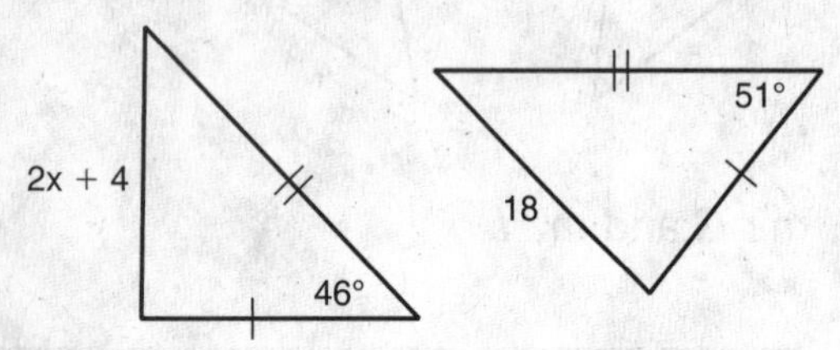

____

6.

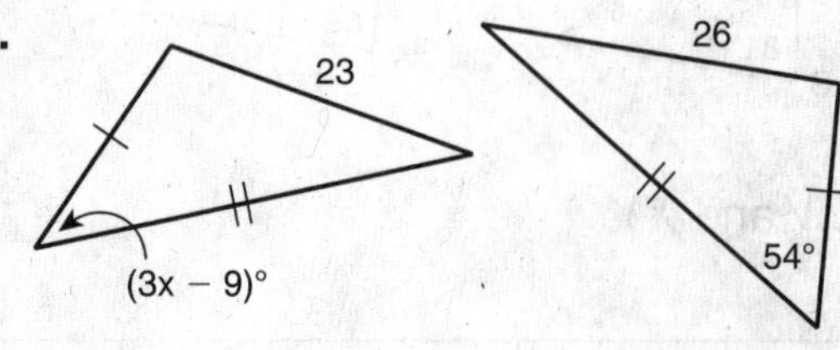

____

7.

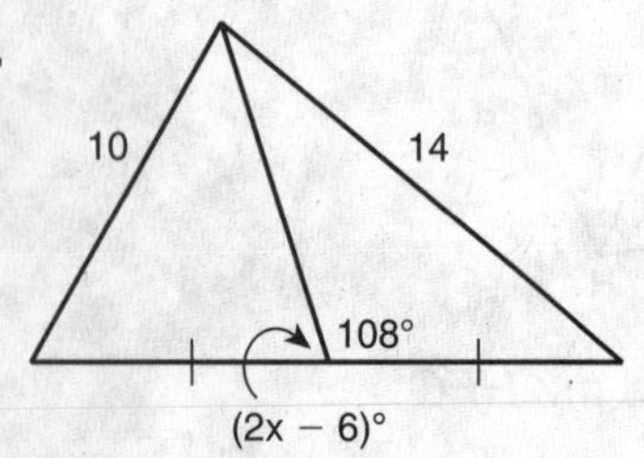

____

8.

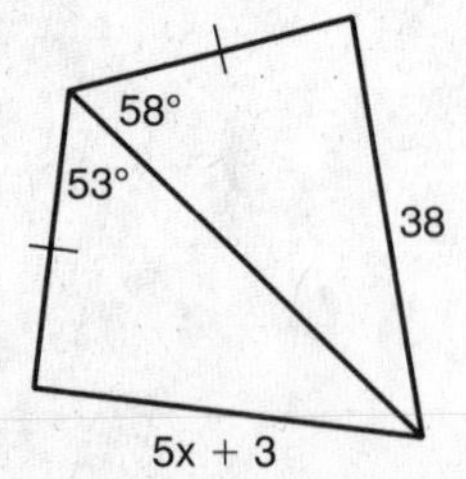

____

Name ______________________ Date __________ Class __________

LESSON 5-7

Review for Mastery

The Pythagorean Theorem

The **Pythagorean Theorem** states that the following relationship exists among the lengths of the legs, *a* and *b*, and the length of the hypotenuse, *c*, of any right triangle.

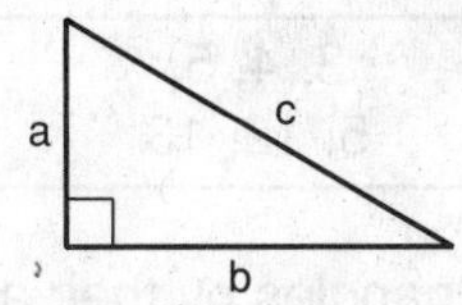

$a^2 + b^2 = c^2$

Use the Pythagorean Theorem to find the value of *x* in each triangle.

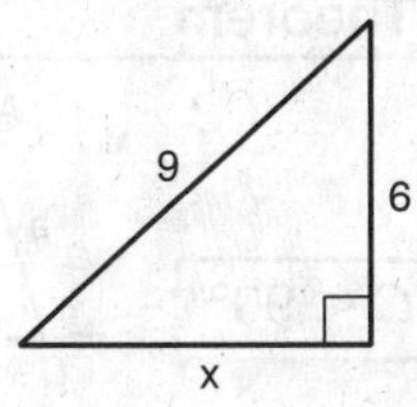

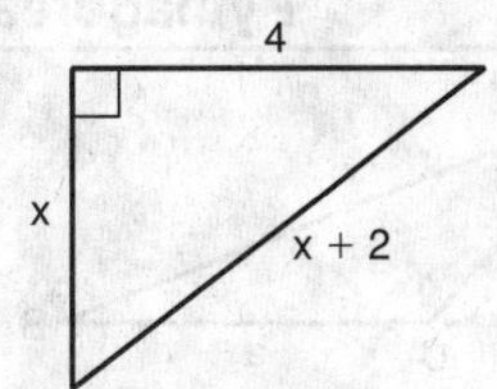

$a^2 + b^2 = c^2$	Pythagorean Theorem	$a^2 + b^2 = c^2$
$x^2 + 6^2 = 9^2$	Substitute.	$x^2 + 4^2 = (x + 2)^2$
$x^2 + 36 = 81$	Take the squares.	$x^2 + 16 = x^2 + 4x + 4$
$x^2 = 45$	Simplify.	$4x = 12$
$x = \sqrt{45}$	Take the positive square root and simplify.	$x = 3$
$x = 3\sqrt{5}$		

Find the value of *x*. Give your answer in simplest radical form.

1.

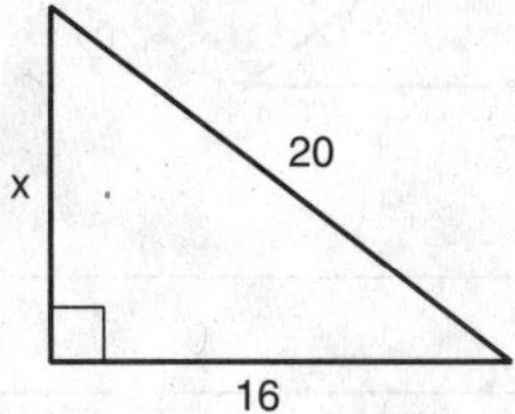

____

2.

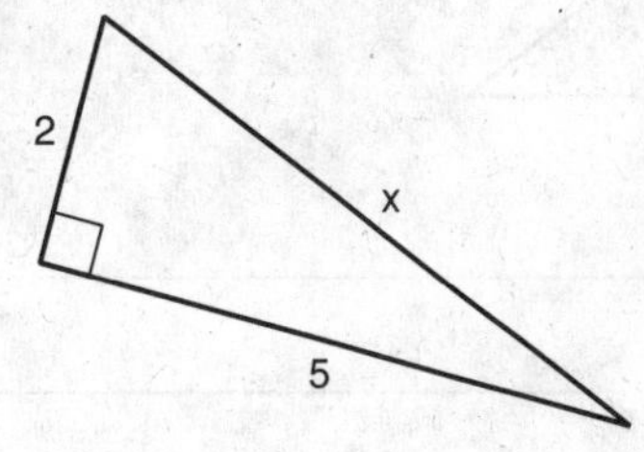

____

3.

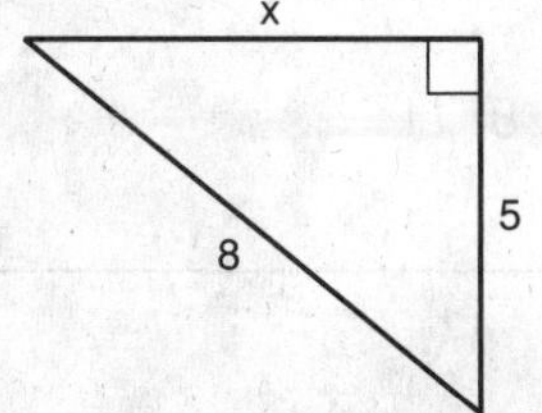

____

4.

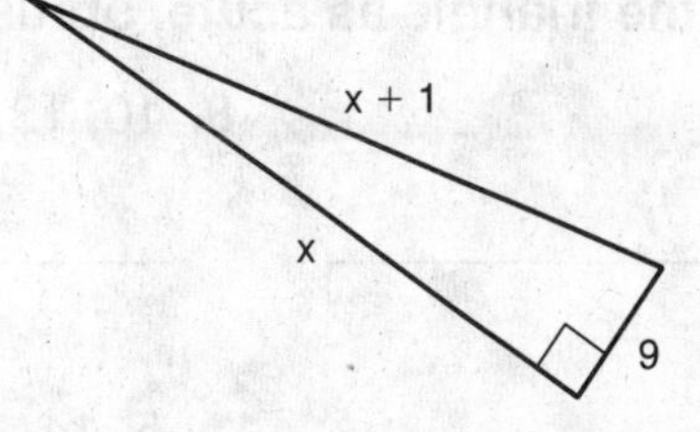

____

Name ______________________ Date __________ Class __________

LESSON 5-7

Review for Mastery

The Pythagorean Theorem continued

A **Pythagorean triple** is a set of three nonzero whole numbers *a*, *b*, and *c* that satisfy the equation $a^2 + b^2 = c^2$.

Pythagorean Triples	Not Pythagorean Triples
3, 4, 5, 5, 12, 13	2, 3, 4 6, 9, $\sqrt{117}$

You can use the following theorem to classify triangles by their angles if you know their side lengths. Always use the length of the longest side for *c*.

Pythagorean Inequalities Theorem

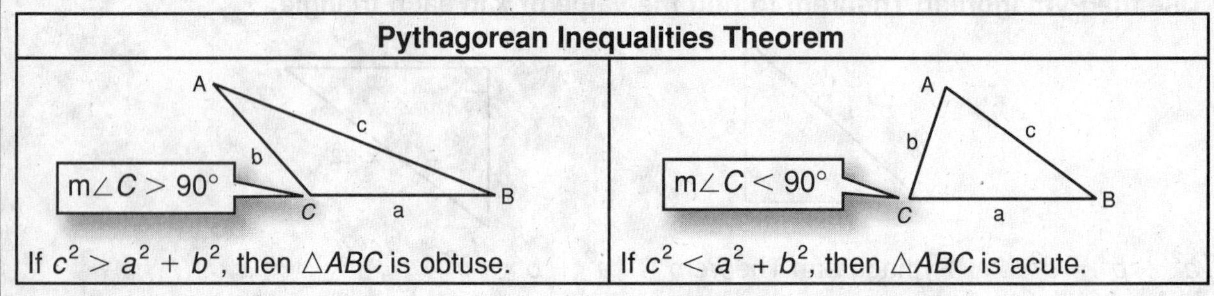

If $c^2 > a^2 + b^2$, then $\triangle ABC$ is obtuse. | If $c^2 < a^2 + b^2$, then $\triangle ABC$ is acute.

Consider the measures 2, 5, and 6. They can be the side lengths of a triangle since $2 + 5 > 6$, $2 + 6 > 5$, and $5 + 6 > 2$. If you substitute the values into $c^2 \stackrel{?}{=} a^2 + b^2$, you get $36 > 29$. Since $c^2 > a^2 + b^2$, a triangle with side lengths 2, 5, and 6 must be obtuse.

Find the missing side length. Tell whether the side lengths form a Pythagorean triple. Explain.

5.

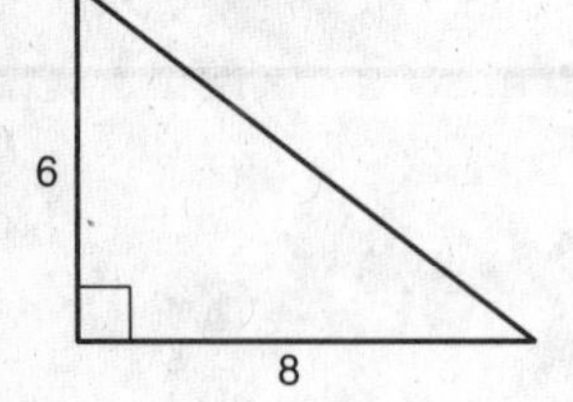

____

____

____

6.

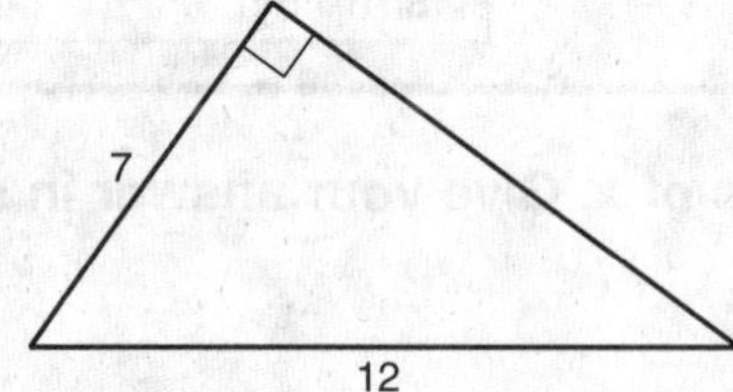

____

____

____

Tell whether the measures can be the side lengths of a triangle. If so, classify the triangle as acute, obtuse, or right.

7. 4, 7, 9 ______________

8. 10, 13, 16 ______________

9. 8, 8, 11 ______________

10. 9, 12, 15 ______________

11. 5, 14, 20 ______________

12. 4.5, 6, 10.2 ______________

Name ______________________________ Date ______________ Class ______________

LESSON 5-8

Review for Mastery

Applying Special Right Triangles

Theorem	Example
45°-45°-90° Triangle Theorem In a 45°-45°-90° triangle, both legs are congruent and the length of the hypotenuse is $\sqrt{2}$ times the length of a leg.	45°, 8, $8\sqrt{2}$, 45°, 8 23, 23, 45°, 45°, $23\sqrt{2}$

In a 45°-45°-90° triangle, if a leg length is x, then the hypotenuse length is $x\sqrt{2}$.

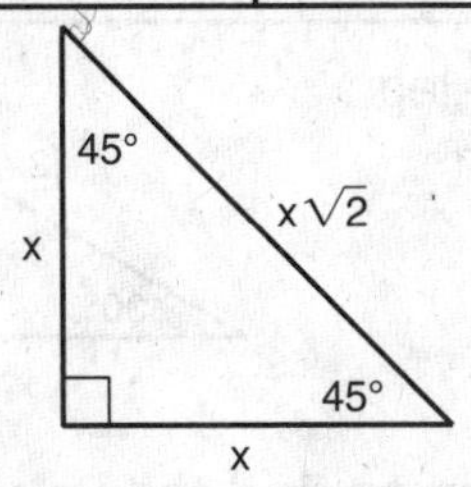

Use the 45°-45°-90° Triangle Theorem to find the value of x in $\triangle EFG$.

Every isosceles right triangle is a 45°-45°-90° triangle. Triangle *EFG* is a 45°-45°-90° triangle with a hypotenuse of length 10.

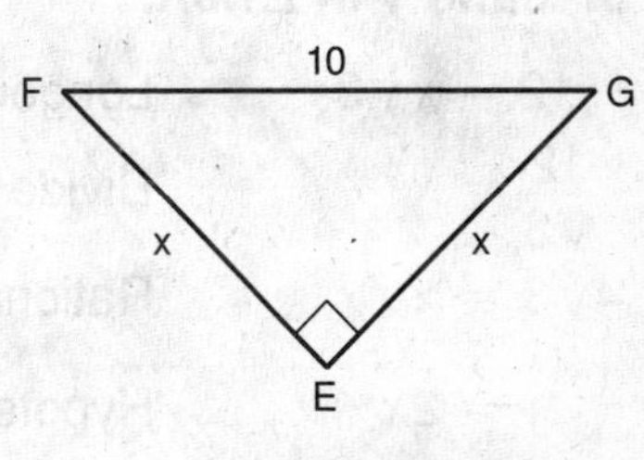

$10 = x\sqrt{2}$	Hypotenuse is $\sqrt{2}$ times the length of a leg.
$\dfrac{10}{\sqrt{2}} = \dfrac{x\sqrt{2}}{\sqrt{2}}$	Divide both sides by $\sqrt{2}$.
$5\sqrt{2} = x$	Rationalize the denominator.

Find the value of x. Give your answers in simplest radical form.

1.

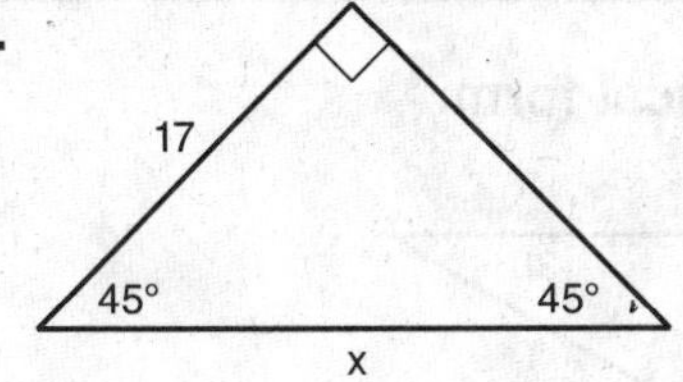

____

2.

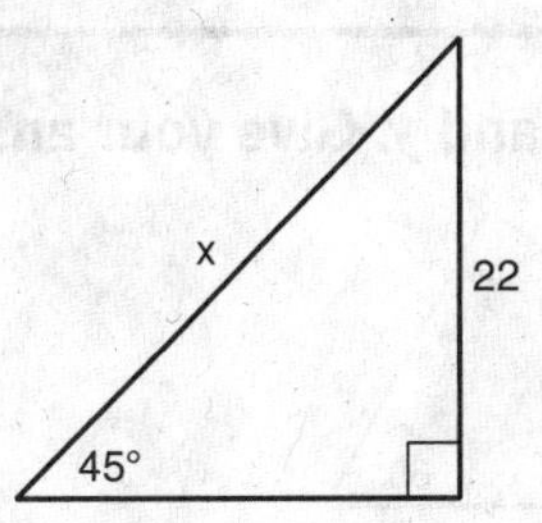

____

3.

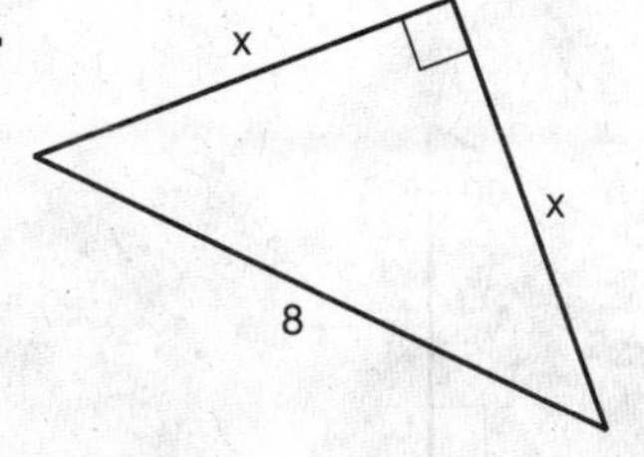

____

4.

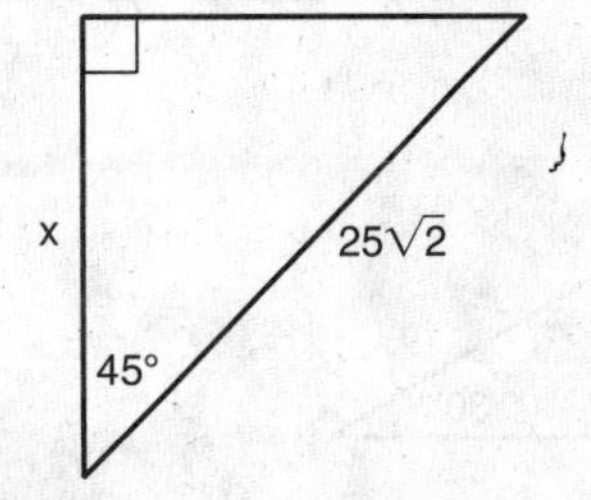

____

LESSON 5-8

Review for Mastery

Applying Special Right Triangles continued

Theorem	Examples
30°-60°-90° Triangle Theorem In a 30°-60°-90° triangle, the length of the hypotenuse is 2 multiplied by the length of the shorter leg, and the longer leg is $\sqrt{3}$ multiplied by the length of the shorter leg.	

In a 30°-60°-90° triangle, if the shorter leg length is x, then the hypotenuse length is $2x$ and the longer leg length is x.

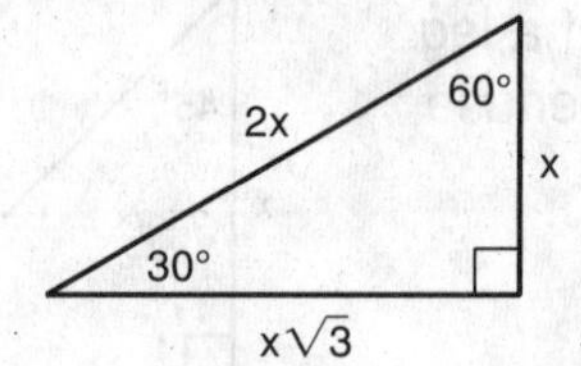

Use the 30°-60°-90° Triangle Theorem to find the values of x and y in $\triangle HJK$.

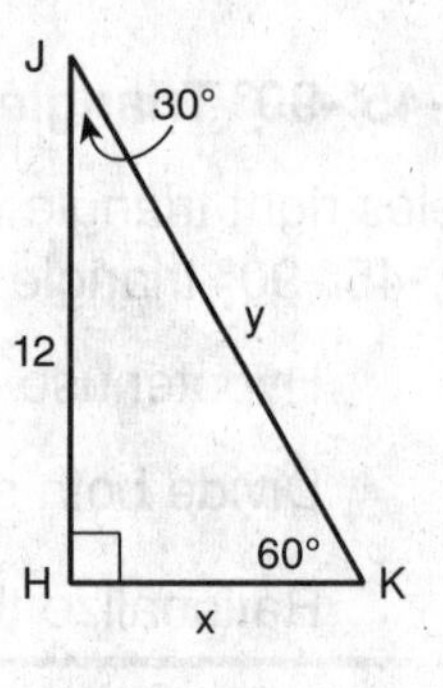

$12 = x\sqrt{3}$	Longer leg = shorter leg multiplied by $\sqrt{3}$.
$\frac{12}{\sqrt{3}} = x$	Divide both sides by $\sqrt{3}$.
$4\sqrt{3} = x$	Rationalize the denominator.
$y = 2x$	Hypotenuse = 2 multiplied by shorter leg.
$y = 2(4\sqrt{3})$	Substitute $4\sqrt{3}$ for x.
$y = 8\sqrt{3}$	Simplify.

Find the values of x and y. Give your answers in simplest radical form.

5.

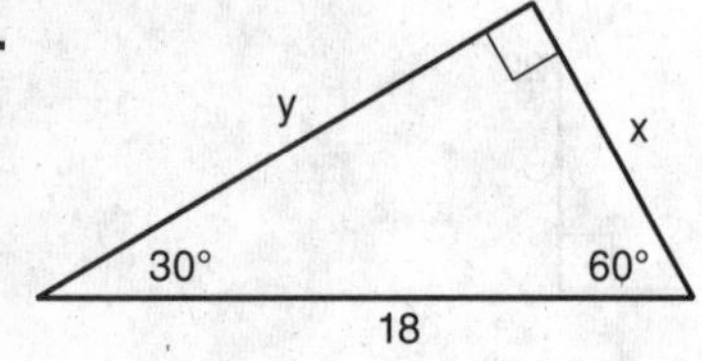

____

6.

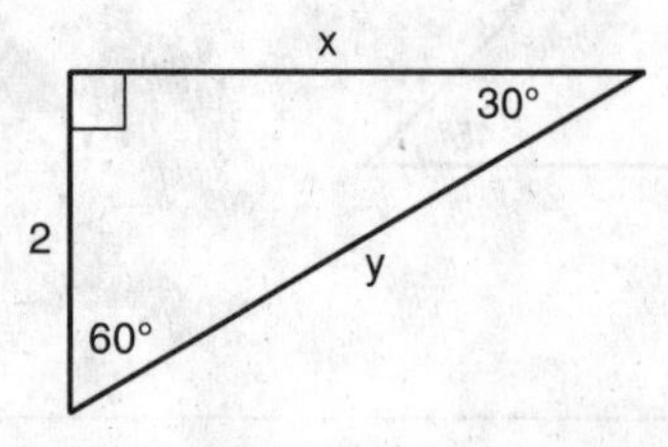

____

7.

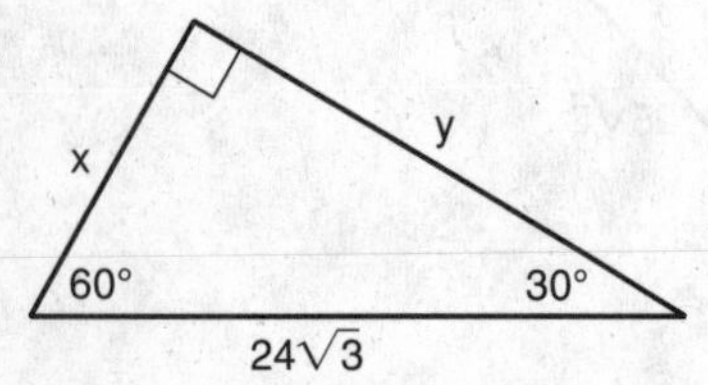

____

8.

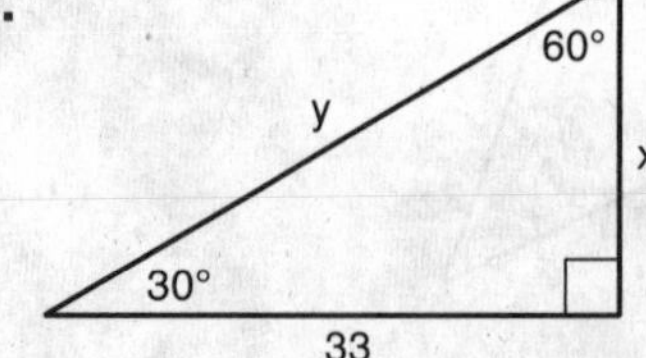

____

Name ______________________ Date ____________ Class ____________

LESSON 6-1

Review for Mastery

Properties and Attributes of Polygons

The parts of a polygon are named on the quadrilateral below.

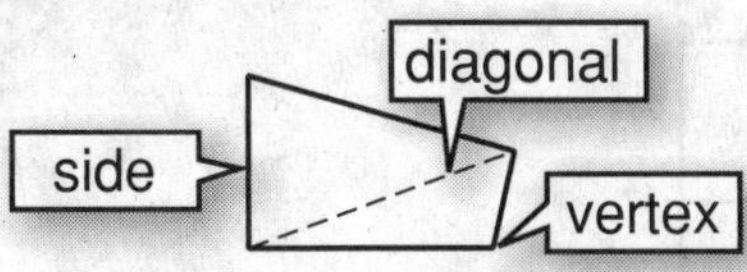

You can name a polygon by the number of its sides.

A **regular polygon** has all sides congruent and all angles congruent. A polygon is **convex** if all its diagonals lie in the interior of the polygon. A polygon is **concave** if all or part of at least one diagonal lies outside the polygon.

Number of Sides	Polygon
3	triangle
4	quadrilateral
5	pentagon
6	hexagon
7	heptagon
8	octagon
9	nonagon
10	decagon
n	*n*-gon

Types of Polygons

regular, convex	irregular, convex	irregular, concave

Tell whether each figure is a polygon. If it is a polygon, name it by the number of sides.

1. 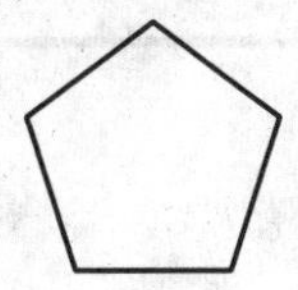
____

2. 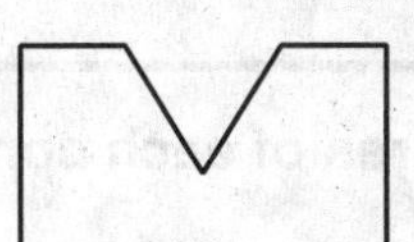
____

3. 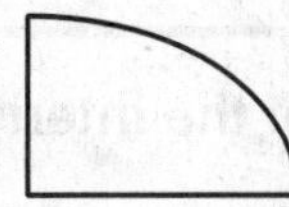
____

Tell whether each polygon is regular or irregular. Then tell whether it is concave or convex.

4. 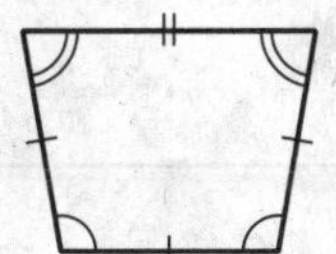
____

5. 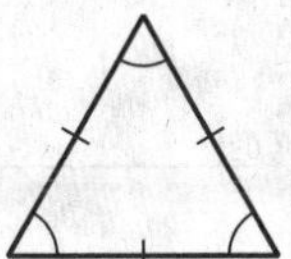
____

6. 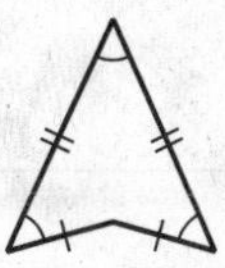
____

Name ______________________ Date __________ Class __________

LESSON 6-1

Review for Mastery

Properties and Attributes of Polygons continued

The **Polygon Angle Sum Theorem** states that the sum of the interior angle measures of a convex polygon with n sides is $(n - 2)180°$.

Convex Polygon	Number of Sides	Sum of Interior Angle Measures: $(n - 2)180°$
quadrilateral	4	$(4 - 2)180° = 360°$
hexagon	6	$(6 - 2)180° = 720°$
decagon	10	$(10 - 2)180° = 1440°$

If a polygon is a regular polygon, then you can divide the sum of the interior angle measures by the number of sides to find the measure of each interior angle.

Regular Polygon	Number of Sides	Sum of Interior Angle Measures	Measure of Each Interior Angle
quadrilateral	4	360°	$360° \div 4 = 90°$
hexagon	6	720°	$720° \div 6 = 120°$
decagon	10	1440°	$1440° \div 10 = 144°$

The **Polygon External Angle Sum Theorem** states that the sum of the exterior angle measures, one angle at each vertex, of a convex polygon is 360°.

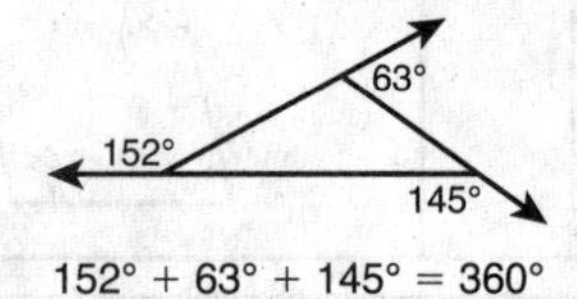

The measure of each exterior angle of a regular polygon with n exterior angles is $360° \div n$. So the measure of each exterior angle of a regular decagon is $360° \div 10 = 36°$.

Find the sum of the interior angle measures of each convex polygon.

7. pentagon ________

8. octagon ________

9. nonagon ________

Find the measure of each interior angle of each regular polygon. Round to the nearest tenth if necessary.

10. pentagon ________

11. heptagon ________

12. 15-gon ________

Find the measure of each exterior angle of each regular polygon.

13. quadrilateral ________

14. octagon ________

Name ______________________ Date ____________ Class ____________

LESSON 6-2

Review for Mastery

Properties of Parallelograms

A parallelogram is a quadrilateral with two pairs of parallel sides.
All parallelograms, such as ▱*FGHJ*, have the following properties.

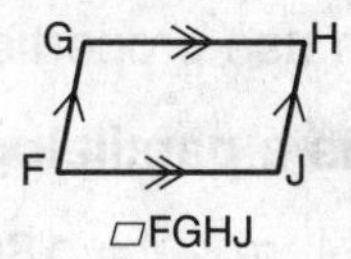

Properties of Parallelograms	
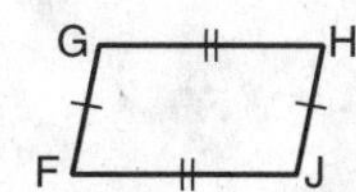 $\overline{FG} \cong \overline{HJ}$, $\overline{GH} \cong \overline{JF}$ Opposite sides are congruent.	$\angle F \cong \angle H$, $\angle G \cong \angle J$ Opposite angles are congruent.
$m\angle F + m\angle G = 180°$ $m\angle G + m\angle H = 180°$ $m\angle H + m\angle J = 180°$ $m\angle J + m\angle F = 180°$ Consecutive angles are supplementary.	$\overline{FP} \cong \overline{HP}$, $\overline{GP} \cong \overline{JP}$ The diagonals bisect each other.

Find each measure.

1. *AB*

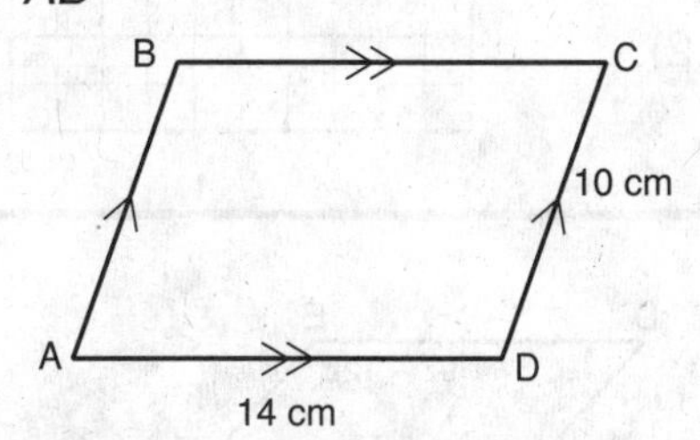

____

2. m∠*D*

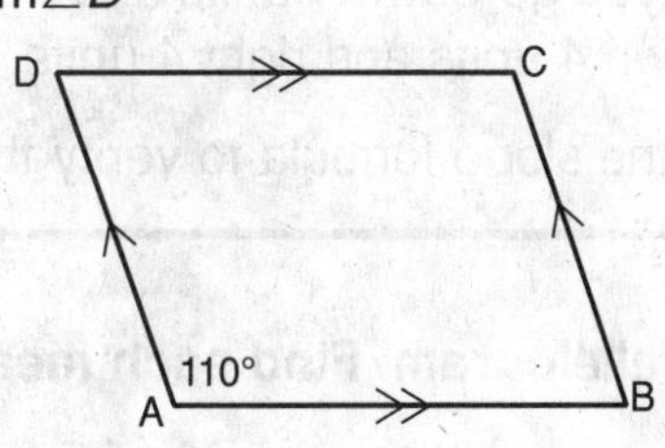

____

Find each measure in ▱*LMNP*.

3. *ML* ________________________

4. *LP* ________________________

5. m∠*LPM* ________________________

6. *LN* ________________________

7. m∠*MLN* ________________________

8. *QN* ________________________

Name ______________________ Date __________ Class __________

LESSON 6-2

Review for Mastery

Properties of Parallelograms continued

You can use properties of parallelograms to find measures.

***WXYZ* is a parallelogram. Find m∠*X*.**

$m\angle W + m\angle X = 180°$	If a quadrilateral is a ▱, then cons. ∠s are supp.
$(7x + 15) + 4x = 180°$	Substitute the given values.
$11x + 15 = 180$	Combine like terms.
$11x = 165$	Subtract 15° from both sides.
$x = 15$	Divide both sides by 11.

$m\angle X = (4x)° = [4(15)]° = 60°$

If you know the coordinates of three vertices of a parallelogram, you can use slope to find the coordinates of the fourth vertex.

Three vertices of ▱*RSTV* are *R*(3, 1), *S*(−1, 5), and *T*(3, 6). Find the coordinates of *V*.

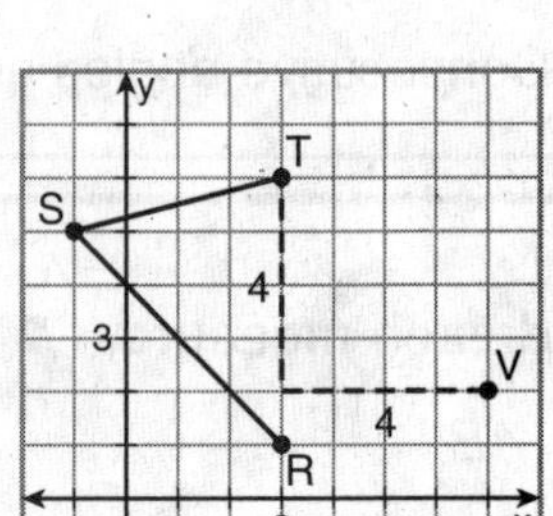

Since opposite sides must be parallel, the rise and the run from *S* to *R* must be the same as the rise and the run from *T* to *V*.

From *S* to *R*, you go down 4 units and right 4 units. So, from *T* to *V*, go down 4 units and right 4 units. Vertex *V* is at *V*(7, 2).

You can use the slope formula to verify that $\overline{ST} \parallel \overline{RV}$.

***CDEF* is a parallelogram. Find each measure.**

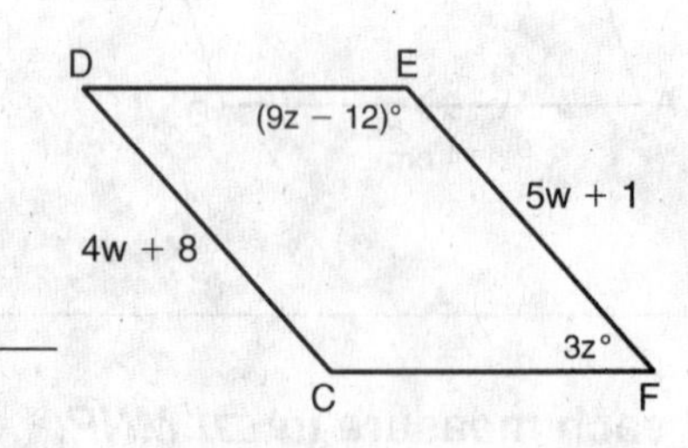

9. *CD* ____________

10. *EF* ____________

11. m∠*F* ____________

12. m∠*E* ____________

The coordinates of three vertices of a parallelogram are given. Find the coordinates of the fourth vertex.

13. ▱*ABCD* with *A*(0, 6), *B*(5, 8), *C*(5, 5)

____

14. ▱*KLMN* with *K*(−4, 7), *L*(3, 6), *M*(5, 3)

____

Name ______________________ Date ____________ Class ____________

LESSON 6-3

Review for Mastery

Conditions for Parallelograms

You can use the following conditions to determine whether a quadrilateral such as *PQRS* is a parallelogram.

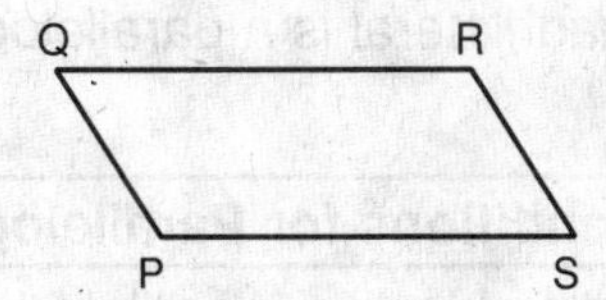

Conditions for Parallelograms	
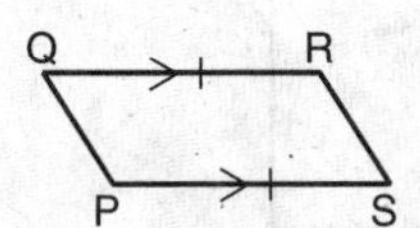 $\overline{QR} \parallel \overline{SP}$ $\overline{QR} \cong \overline{SP}$ If one pair of opposite sides is ∥ and ≅, then *PQRS* is a parallelogram.	$\overline{QR} \cong \overline{SP}$ $\overline{PQ} \cong \overline{RS}$ If both pairs of opposite sides are ≅, then *PQRS* is a parallelogram.
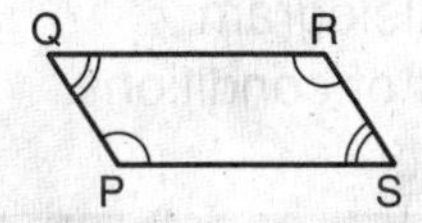 $\angle P \cong \angle R$ $\angle Q \cong \angle S$ If both pairs of opposite angles are ≅, then *PQRS* is a parallelogram.	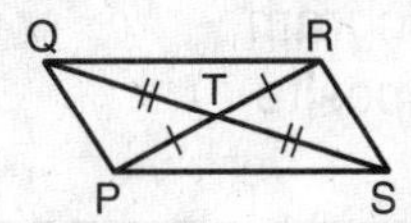 $\overline{PT} \cong \overline{RT}$ $\overline{QT} \cong \overline{ST}$ If the diagonals bisect each other, then *PQRS* is a parallelogram.

A quadrilateral is also a parallelogram if one of the angles is supplementary to both of its consecutive angles.

$65° + 115° = 180°$, so $\angle A$ is supplementary to $\angle B$ and $\angle D$.

Therefore, *ABCD* is a parallelogram.

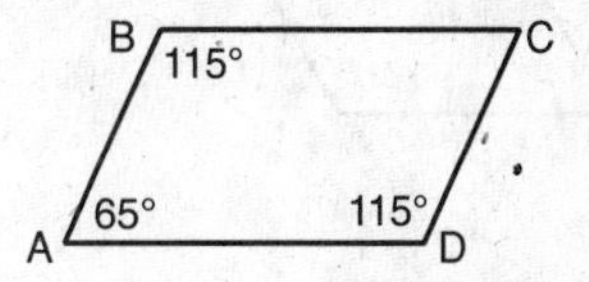

Show that each quadrilateral is a parallelogram for the given values. Explain.

1. Given: $x = 9$ and $y = 4$

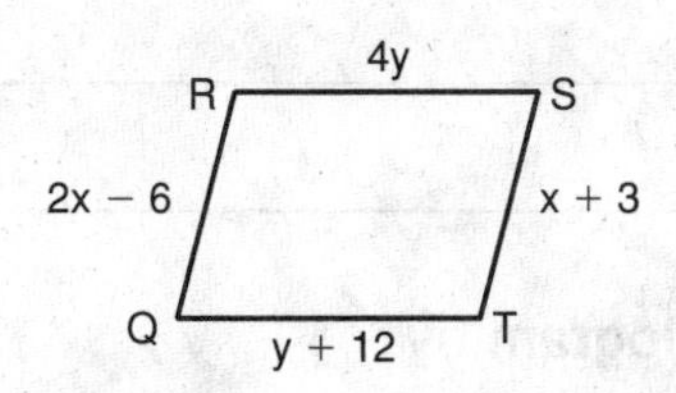

____

____

____

____

2. Given: $w = 3$ and $z = 31$

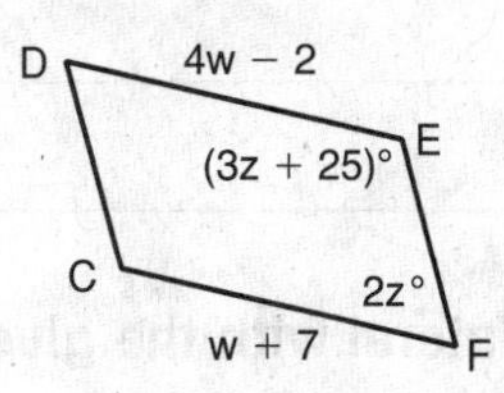

____

____

____

____

Name ______________________ Date __________ Class __________

LESSON 6-3

Review for Mastery

Conditions for Parallelograms continued

You can show that a quadrilateral is a parallelogram by using any of the conditions listed below.

Conditions for Parallelograms
• Both pairs of opposite sides are parallel (definition). • One pair of opposite sides is parallel and congruent. • Both pairs of opposite sides are congruent. • Both pairs of opposite angles are congruent. • The diagonals bisect each other. • One angle is supplementary to both its consecutive angles.

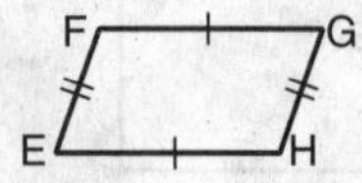

EFGH must be a parallelogram because both pairs of opposite sides are congruent.

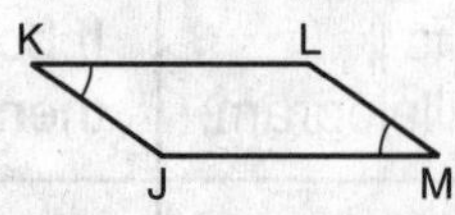

JKLM may not be a parallelogram because none of the sets of conditions for a parallelogram is met.

Determine whether each quadrilateral must be a parallelogram. Justify your answer.

3.

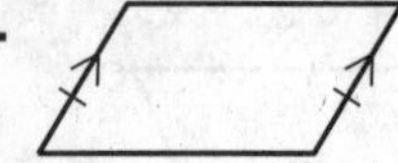

____

____

4.

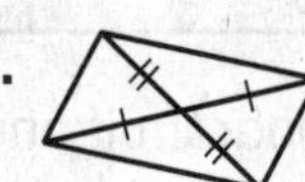

____

____

5.

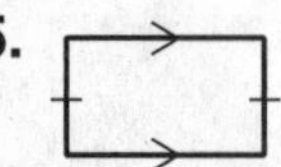

____

____

6.

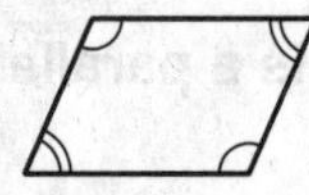

____

____

Show that the quadrilateral with the given vertices is a parallelogram by using the given definition or theorem.

7. $J(-2, -2)$, $K(-3, 3)$, $L(1, 5)$, $M(2, 0)$
Both pairs of opposite sides are parallel.

____

____

8. $N(5, 1)$, $P(2, 7)$, $Q(6, 9)$, $R(9, 3)$
Both pairs of opposite sides are congruent.

____

____

Name ______________________________ Date ______________ Class ______________

LESSON 6-4

Review for Mastery

Properties of Special Parallelograms

A **rectangle** is a quadrilateral with four right angles. A rectangle has the following properties.

Properties of Rectangles	
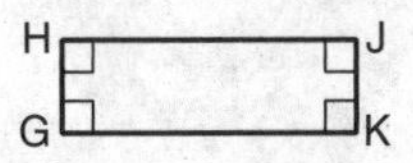 GHJK is a parallelogram.	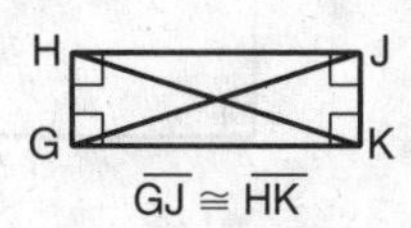 $\overline{GJ} \cong \overline{HK}$
If a quadrilateral is a rectangle, then it is a parallelogram.	If a parallelogram is a rectangle, then its diagonals are congruent.

Since a rectangle is a parallelogram, a rectangle also has all the properties of parallelograms.

A **rhombus** is a quadrilateral with four congruent sides. A rhombus has the following properties.

Properties of Rhombuses		
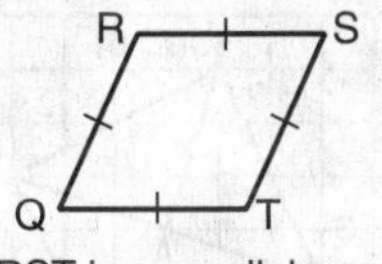 QRST is a parallelogram.	$\overline{QS} \perp \overline{RT}$	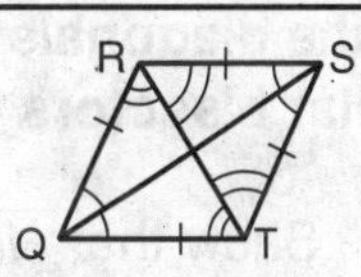 $\angle RQS \cong \angle SQT$
If a quadrilateral is a rhombus, then it is a parallelogram.	If a parallelogram is a rhombus, then its diagonals are perpendicular.	If a parallelogram is a rhombus, then each diagonal bisects a pair of opposite angles.

Since a rhombus is a parallelogram, a rhombus also has all the properties of parallelograms.

***ABCD* is a rectangle. Find each length.**

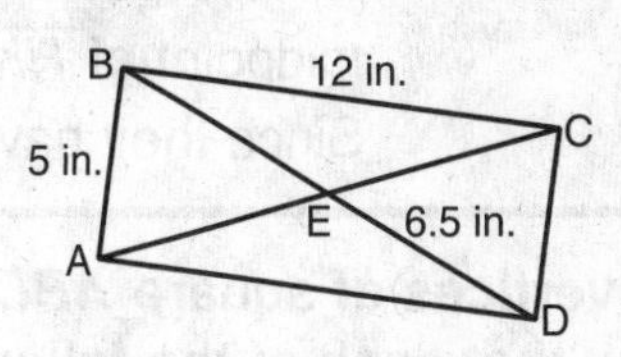

1. *BD* ______________________

2. *CD* ______________________

3. *AC* ______________________

4. *AE* ______________________

***KLMN* is a rhombus. Find each measure.**

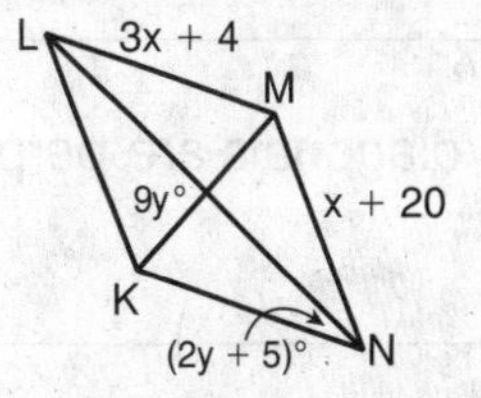

5. *KL* ______________________

6. m∠*MNK* ______________________

Name ______________________________ Date ______________ Class ______________

LESSON 6-4

Review for Mastery

Properties of Special Parallelograms continued

A **square** is a quadrilateral with four right angles and four congruent sides.
A square is a parallelogram, a rectangle, and a rhombus.

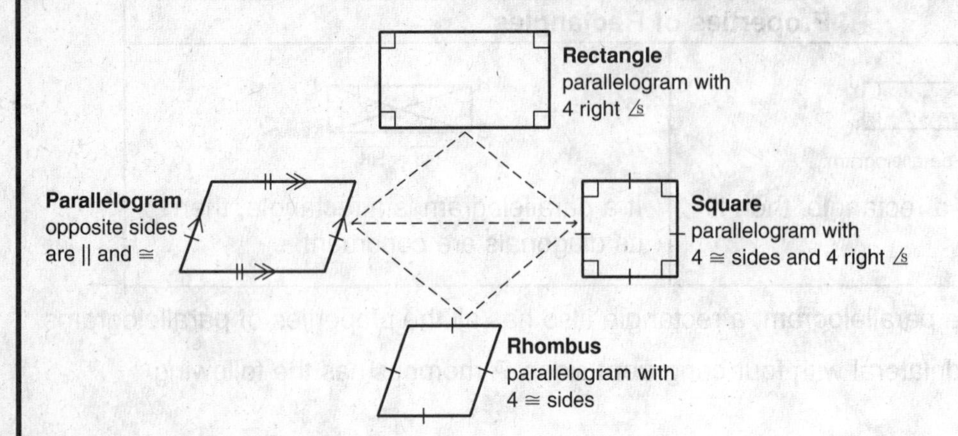

Show that the diagonals of square *HJKL* are congruent perpendicular bisectors of each other.

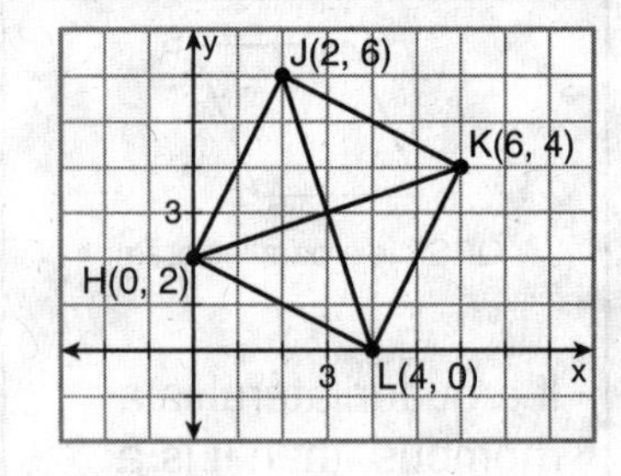

Step 1 Show that $\overline{HK} \cong \overline{JL}$.

$HK = \sqrt{(6-0)^2 + (4-2)^2} = 2\sqrt{10}$

$JL = \sqrt{(4-2)^2 + (0-6)^2} = 2\sqrt{10}$

$HK = JL = 2\sqrt{10}$, so $\overline{HK} \cong \overline{JL}$.

Step 2 Show that $\overline{HK} \perp \overline{JL}$.

slope of $\overline{HK} = \frac{4-2}{6-0} = \frac{1}{3}$ slope of $\overline{JL} = \frac{0-6}{4-2} = -3$

Since the product of the slopes is -1, $\overline{HK} \perp \overline{JL}$.

Step 3 Show that $\overline{HK}$ and $\overline{JL}$ bisect each other by comparing their midpoints.

midpoint of $\overline{HK} = (3, 3)$ midpoint of $\overline{JL} = (3, 3)$

Since they have the same midpoint, $\overline{HK}$ and $\overline{JL}$ bisect each other.

The vertices of square *ABCD* are *A*(−1, 0), *B*(−4, 5), *C*(1, 8), and *D*(4, 3). Show that each of the following is true.

7. The diagonals are congruent.

____

8. The diagonals are perpendicular bisectors of each other.

____

____

Name ______________________ Date __________ Class __________

LESSON 6-5

Review for Mastery

Conditions for Special Parallelograms

You can use the following conditions to determine whether a parallelogram is a rectangle.

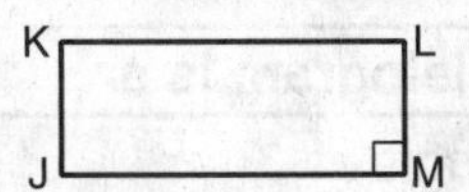 If one angle is a right angle, then ▱*JKLM* is a rectangle.	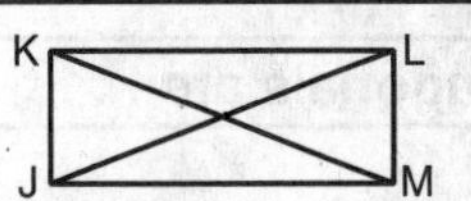 $\overline{JL} \cong \overline{KM}$ If the diagonals are congruent, then ▱*JKLM* is a rectangle.

You can use the following conditions to determine whether a parallelogram is a rhombus.

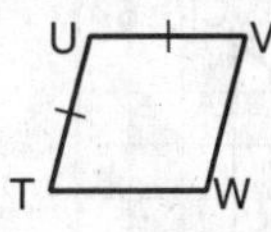 If one pair of consecutive sides are congruent, then ▱*TUVW* is a rhombus.	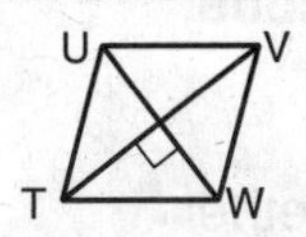 If the diagonals are perpendicular, then ▱*TUVW* is a rhombus.	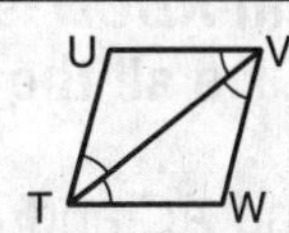 If one diagonal bisects a pair of opposite angles, then ▱*TUVW* is a rhombus.

Determine whether the conclusion is valid. If not, tell what additional information is needed to make it valid.

1. *EFGH* is a rectangle.

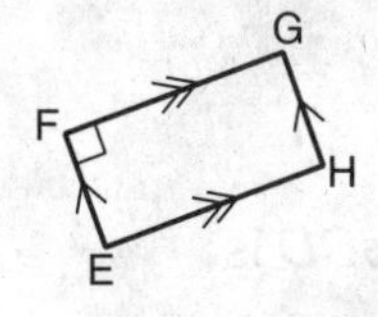

____

2. *MPQR* is a rhombus.

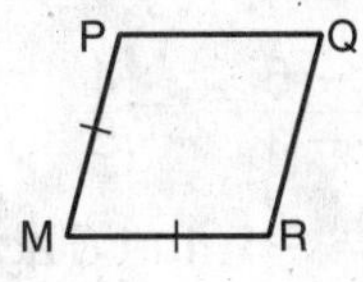

____

For Exercises 3 and 4, use the figure to determine whether the conclusion is valid. If not, tell what additional information is needed to make it valid.

3. **Given:** $\overline{EF} \parallel \overline{GH}$, $\overline{HE} \parallel \overline{FG}$, $\overline{EG} \cong \overline{FH}$
   Conclusion: *EFGH* is a rectangle.

____

____

4. **Given:** $m\angle EFG = 90°$
   Conclusion: *EFGH* is a rectangle.

____

____

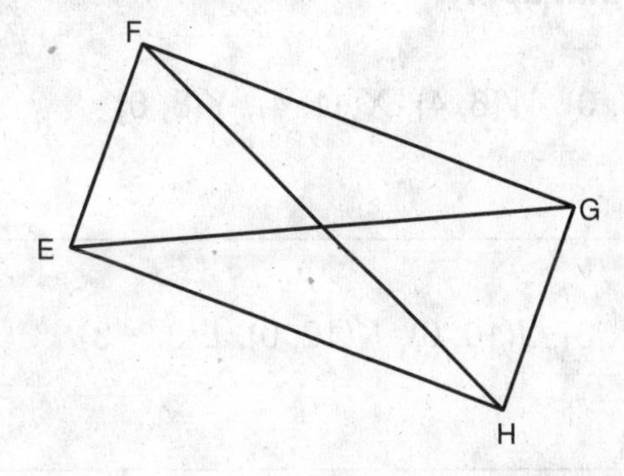

Name ______________________ Date __________ Class __________

LESSON 6-5

Review for Mastery

Conditions for Special Parallelograms continued

You can identify special parallelograms in the coordinate plane by examining their diagonals.

If the Diagonals are . . .	. . . the Parallelogram is a
congruent	rectangle
perpendicular	rhombus
congruent and perpendicular	square

Use the diagonals to determine whether parallelogram *ABCD* is a rectangle, rhombus, or square. Give all the names that apply.

B(1, 5) C(6, 5) A(1, 1) D(6, 1) 3 0 3 x y

Step 1 Find *AC* and *BD* to determine whether *ABCD* is a rectangle.

$AC = \sqrt{(6-1)^2 + (5-1)^2} = \sqrt{41}$

$BD = \sqrt{(6-1)^2 + (1-5)^2} = \sqrt{41}$

Since $\sqrt{41} = \sqrt{41}$, the diagonals are congruent. So *ABCD* is a rectangle.

Step 2 Find the slopes of $\overline{AC}$ and $\overline{BD}$ to determine whether *ABCD* is a rhombus.

slope of $\overline{AC} = \frac{5-1}{6-1} = \frac{4}{5}$

slope of $\overline{BD} = \frac{1-5}{6-1} = -\frac{4}{5}$

Since $\left(\frac{4}{5}\right)\left(-\frac{4}{5}\right) \neq -1$, the diagonals are not perpendicular. So *ABCD* is not a rhombus and cannot be a square.

Use the diagonals to determine whether a parallelogram with the given vertices is a rectangle, rhombus, or square. Give all the names that apply.

5. $V(3, 0)$, $W(6, 4)$, $X(11, 4)$, $Y(8, 0)$

____

6. $L(1, 2)$, $M(3, 5)$, $N(6, 3)$, $P(4, 0)$

____

7. $H(1, 3)$, $J(10, 6)$, $K(12, 0)$, $L(3, -3)$

____

8. $E(-4, 3)$, $F(-1, 2)$, $G(-2, -1)$, $H(-5, 0)$

____

Name ______________________ Date __________ Class __________

LESSON 6-6

Review for Mastery

Properties of Kites and Trapezoids

A **kite** is a quadrilateral with exactly two pairs of congruent consecutive sides. If a quadrilateral is a kite, such as *FGHJ*, then it has the following properties.

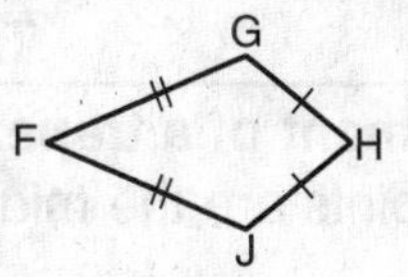

Properties of Kites	
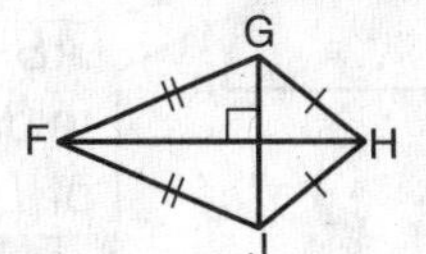 $\overline{FH} \perp \overline{GJ}$ The diagonals are perpendicular.	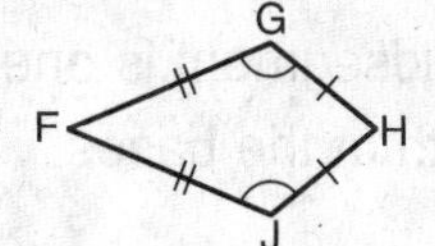 $\angle G \cong \angle J$ Exactly one pair of opposite angles is congruent.

A **trapezoid** is a quadrilateral with exactly one pair of parallel sides. If the legs of a trapezoid are congruent, the trapezoid is an **isosceles trapezoid.**

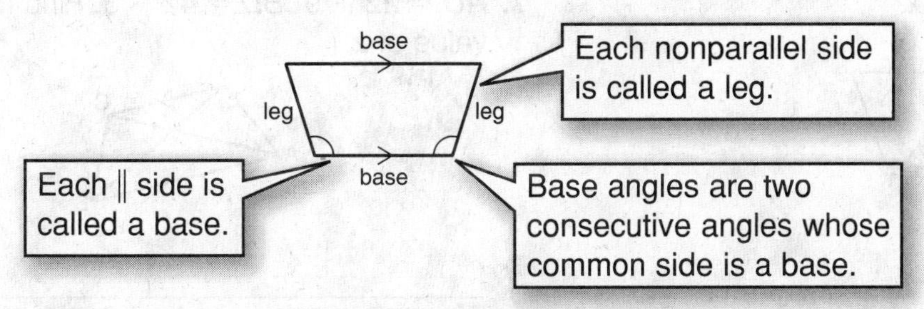

Isosceles Trapezoid Theorems
• In an isosceles trapezoid, each pair of base angles is congruent. • If a trapezoid has one pair of congruent base angles, then it is isosceles. • A trapezoid is isosceles if and only if its diagonals are congruent.

In kite *ABCD*, m∠*BCD* = 98°, and m∠*ADE* = 47°. Find each measure.

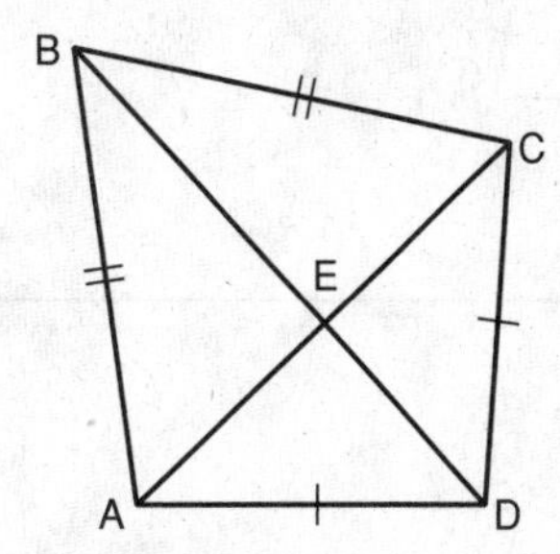

1. m∠*DAE*

____

2. m∠*BCE*

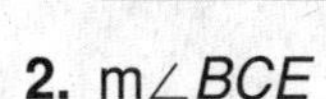

____

3. m∠*ABC*

____

4. Find m∠*J* in trapezoid *JKLM*.

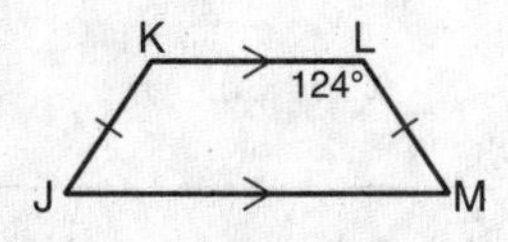

____

5. In trapezoid *EFGH*, *FH* = 9. Find *AG*.

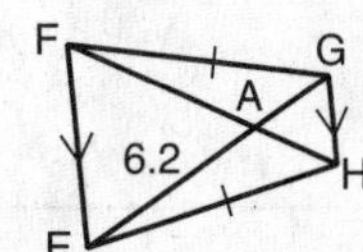

____

Name ______________________ Date ____________ Class ____________

LESSON 6-6

Review for Mastery

Properties of Kites and Trapezoids continued

Trapezoid Midsegment Theorem

The **midsegment of a trapezoid** is the segment whose endpoints are the midpoints of the legs.

- The midsegment of a trapezoid is parallel to each base. $\overline{AB} \parallel \overline{MN}$ and $\overline{AB} \parallel \overline{LP}$
- The length of the midsegment is one-half the sum of the length of the bases. $AB = \frac{1}{2}(MN + LP)$

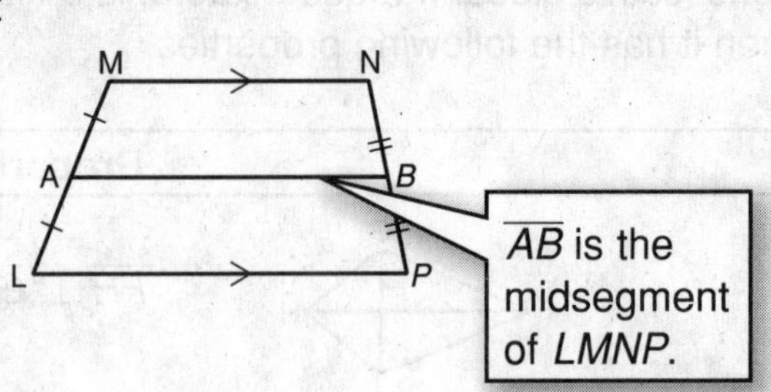

Find each value so that the trapezoid is isosceles.

6. Find the value of *x*.

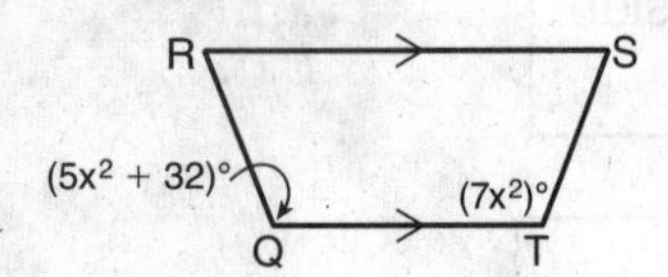

____

7. $AC = 2z + 9$, $BD = 4z - 3$. Find the value of *z*.

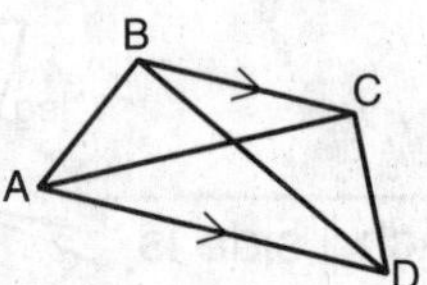

____

Find each length.

8. *KL*

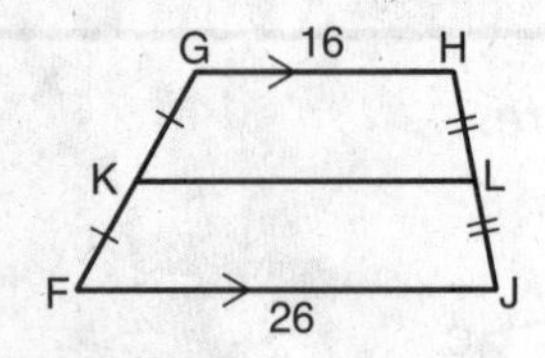

____

9. *PQ*

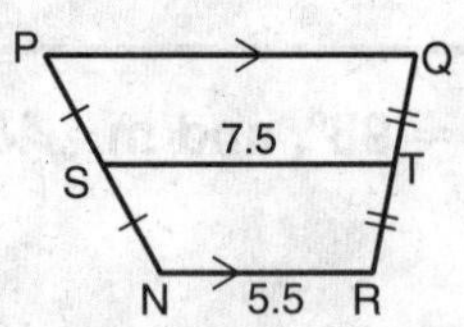

____

10. *EF*

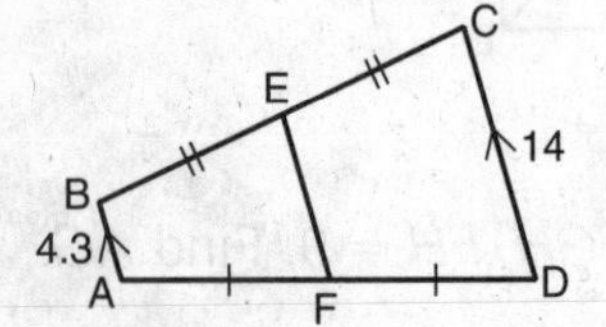

____

11. *WX*

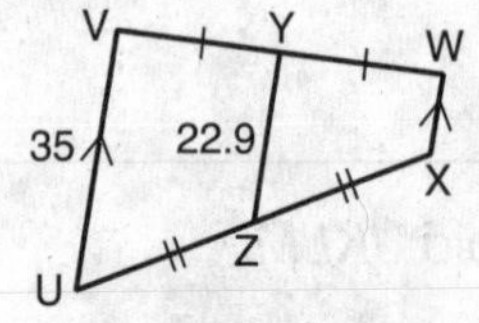

____

Name ________________________________ Date ____________ Class ____________

LESSON 7-1

Review for Mastery

Ratio and Proportion

A **ratio** is a comparison of two numbers by division. Ratios can be written in various forms.

Ratios comparing *x* and *y*			Ratios comparing 3 and 2		
x to y	$x : y$	$\frac{x}{y}$, where $y \neq 0$	3 to 2	3 : 2	$\frac{3}{2}$

Slope is a ratio that compares the rise, or change in *y*, to the run, or change in *x*.

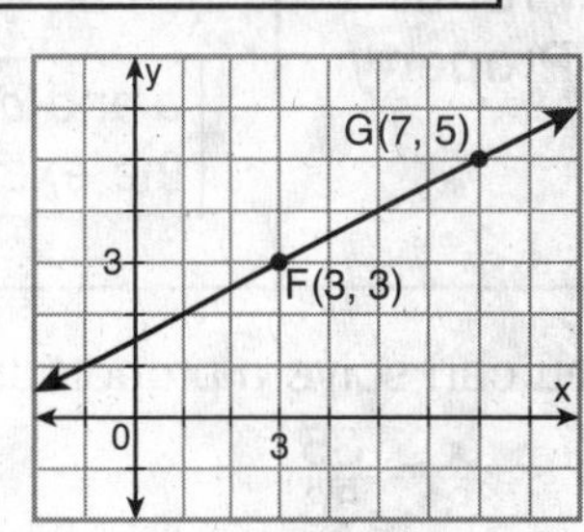

$$\text{Slope} = \frac{\text{rise}}{\text{run}} = \frac{y_2 - y_1}{x_2 - x_1} \quad \text{Definition of slope}$$

$$= \frac{5 - 3}{7 - 3} \quad \text{Substitution}$$

$$= \frac{2}{4} \text{ or } \frac{1}{2} \quad \text{Simplify.}$$

A ratio can involve more than two numbers.

The ratio of the angle measures in a triangle is 2 : 3 : 4. What is the measure of the smallest angle?

Let the angle measures be $2x°$, $3x°$, and $4x°$.

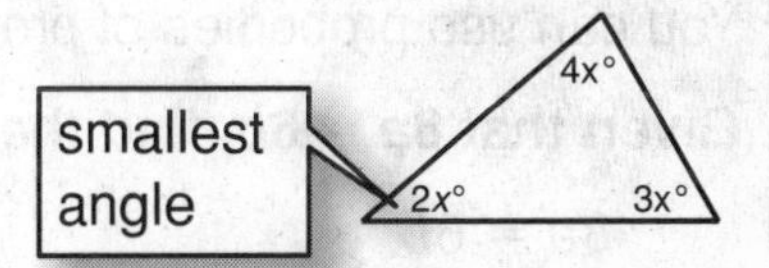

$2x + 3x + 4x = 180$	Triangle Sum Theorem
$9x = 180$	Simplify.
$x = 20$	Divide both sides by 9.

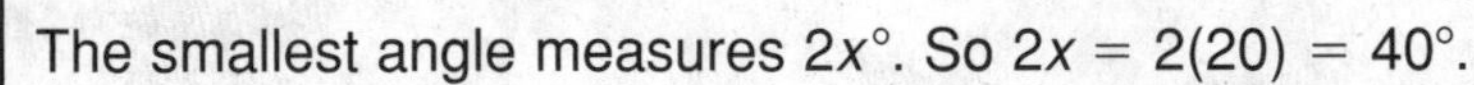

The smallest angle measures $2x°$. So $2x = 2(20) = 40°$.

Write a ratio expressing the slope of each line.

1. 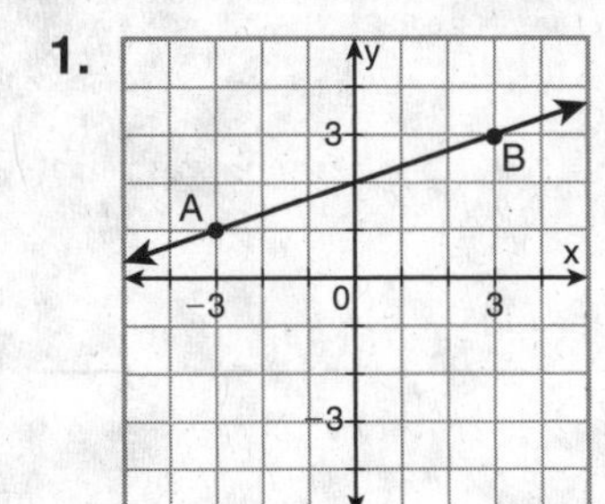

2. 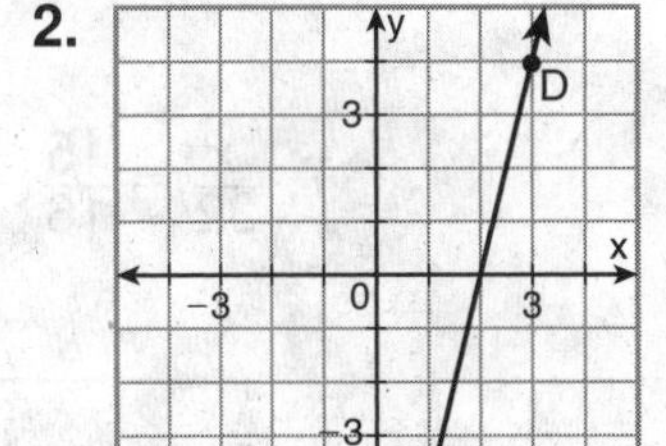

3. 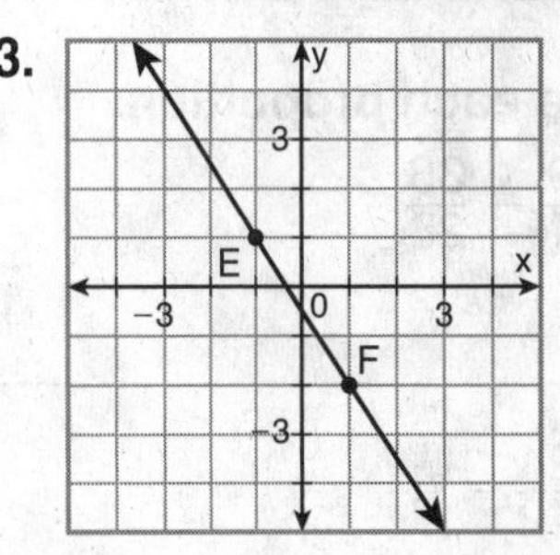

______________ ______________ ______________

4. The ratio of the side lengths of a triangle is 2 : 4 : 5, and the perimeter is 55 cm. What is the length of the shortest side?

____

5. The ratio of the angle measures in a triangle is 7 : 13 : 16. What is the measure of the largest angle?

____

Name ______________________ Date ___________ Class ___________

LESSON 7-1

Review for Mastery

Ratio and Proportion continued

A proportion is an equation stating that two ratios are equal. In every proportion, the product of the extremes equals the product of the means.

$a : b = c : d$ (means: b and c; extremes: a and d)

Cross Products Property	In a proportion, if $\frac{a}{b} = \frac{c}{d}$ and b and $d \neq 0$, then $ad = bc$. a and d are the *extremes*. $\frac{a}{b} = \frac{c}{d}$ b and c are the *means*.

You can solve a proportion like $\frac{x}{8} = \frac{35}{56}$ by finding the cross products.

$\frac{x}{8} = \frac{35}{56}$

$x(56) = 8(35)$ Cross Products Property

$56x = 280$ Simplify.

$x = 5$ Divide both sides by 56.

You can use properties of proportions to find ratios.

Given that $8a = 6b$, find the ratio of a to b in simplest form.

$8a = 6b$

$\frac{a}{b} = \frac{6}{8}$ Divide both sides by b.

$\frac{a}{b} = \frac{3}{4}$ Simplify $\frac{6}{8}$.

The ratio of a to b in simplest form is 3 to 4.

Solve each proportion.

6. $\frac{9}{t} = \frac{36}{28}$

____

7. $\frac{x}{32} = \frac{15}{16}$

____

8. $\frac{24}{42} = \frac{y}{7}$

____

9. $\frac{2a}{3} = \frac{8}{3a}$

____

10. Given that $5b = 20c$, find the ratio $\frac{b}{c}$ in simplest form.

____

11. Given that $24x = 9y$, find the ratio $x : y$ in simplest form.

____

Name ______________________ Date __________ Class __________

LESSON 7-2

Review for Mastery

Ratios in Similar Polygons

Similar polygons are polygons that have the same shape but not necessarily the same size.

Similar Polygons	
$\triangle ABC \sim \triangle DEF$	Corresponding angles are congruent. $\angle A \cong \angle D$ $\angle B \cong \angle E$ $\angle C \cong \angle F$ Corresponding sides are proportional. $\frac{AB}{DE} = \frac{6}{3} = 2$ $\frac{BC}{EF} = \frac{9}{4.5} = 2$ $\frac{CA}{FD} = \frac{10}{5} = 2$

A similarity ratio is the ratio of the lengths of the corresponding sides. So, for the similarity statement $\triangle ABC \sim \triangle DEF$, the similarity ratio is 2. For $\triangle DEF \sim \triangle ABC$, the similarity ratio is $\frac{1}{2}$.

Identify the pairs of congruent angles and corresponding sides.

1.

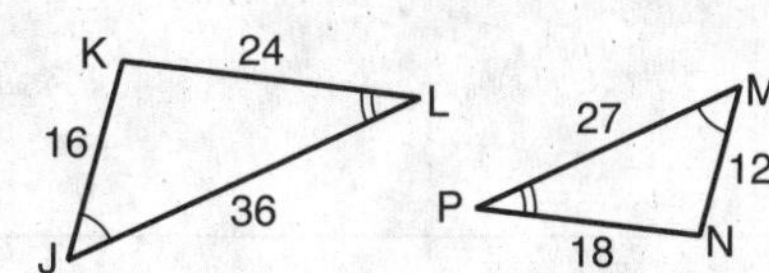

____

____

____

2.

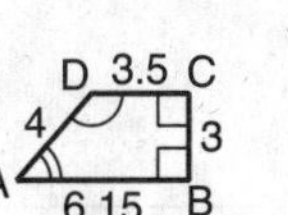

S 7 T, 6, 8, R 12.3 Q

____

____

____

Determine whether the polygons are similar. If so, write the similarity ratio and a similarity statement.

3. $\triangle EFG$ and $\triangle HJK$

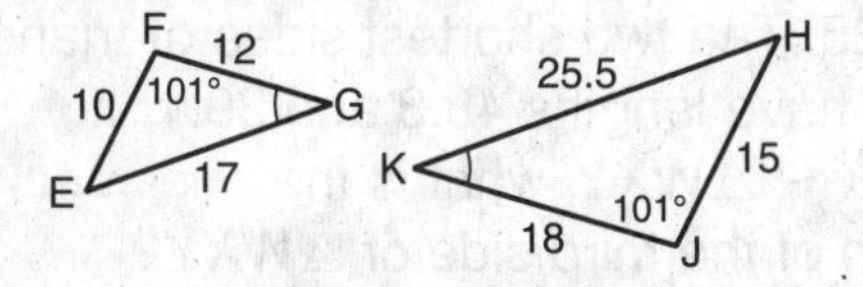

____

4. rectangles *QRST* and *UVWX*

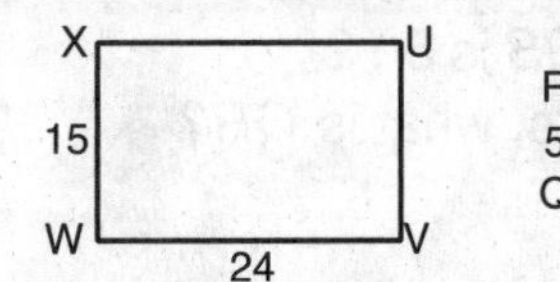

____

Name ________________________ Date ____________ Class ____________

LESSON 7-2

Review for Mastery

Ratios in Similar Polygons continued

You can use properties of similar polygons to solve problems.

Rectangle *DEFG* ~ rectangle *HJKL*. What is the length of *HJKL*?

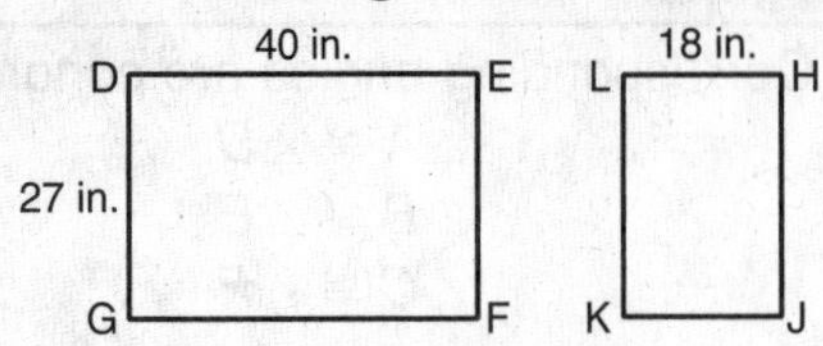

$\dfrac{\text{length of } DEFG}{\text{length of } HJKL} = \dfrac{\text{width of } DEFG}{\text{width of } HJKL}$	Write a proportion.
$\dfrac{40}{x} = \dfrac{27}{18}$	Substitute the known values.
$40(18) = 27(x)$	Cross Products Property
$720 = 27x$	Simplify.
$26\frac{2}{3} = x$	Divide both sides by 27.

The length of *HJKL* is $26\frac{2}{3}$ in.

5. A rectangle is 3.2 centimeters wide and 8 centimeters long. A similar rectangle is 5 centimeters long. What is the width of the second rectangle?

____

6. Rectangle *CDEF* ~ rectangle *GHJK*, and the similarity ratio of *CDEF* to *GHJK* is $\frac{1}{16}$. If $DE = 20$, what is *HK*?

____

7. $\triangle ABC$ is similar to $\triangle DEF$. What is *EF*?

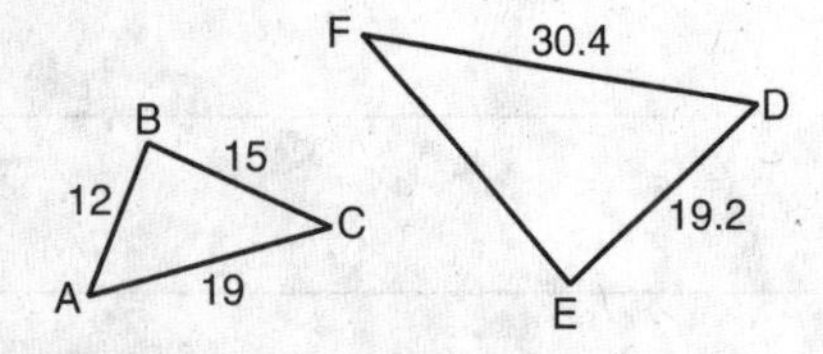

____

8. The two rectangles are similar. What is the value of *x* to the nearest tenth?

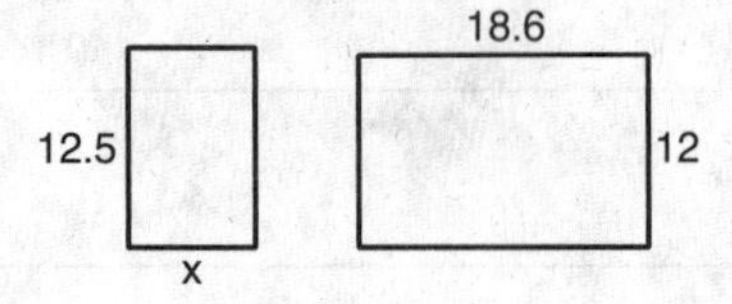

____

9. $\triangle MNP \sim \triangle QRS$, and the ratio of $\triangle MNP$ to $\triangle QRS$ is 5 : 2. If $MN = 42$ meters, what is *QR*?

____

10. Triangle *HJK* has side lengths 21, 17, and 25. The two shortest sides of triangle *WXY* have lengths 48.3 and 39.1. If $\triangle HJK \sim \triangle WXY$, what is the length of the third side of $\triangle WXY$?

____

Name ______________________________ Date ____________ Class ____________

LESSON 7-3

Review for Mastery

Triangle Similarity: AA, SSS, and SAS

Angle-Angle (AA) Similarity	If two angles of one triangle are congruent to two angles of another triangle, then the triangles are similar.	$\triangle ABC \sim \triangle DEF$
Side-Side-Side (SSS) Similarity	If the three sides of one triangle are proportional to the three corresponding sides of another triangle, then the triangles are similar.	$\triangle ABC \sim \triangle DEF$
Side-Angle-Side (SAS) Similarity	If two sides of one triangle are proportional to two sides of another triangle and their included angles are congruent, then the triangles are similar.	$\triangle ABC \sim \triangle DEF$

Explain how you know the triangles are similar, and write a similarity statement.

1.

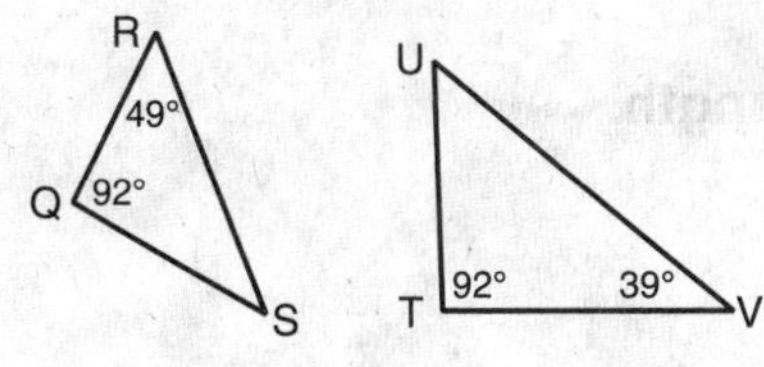

____

____

____

____

2.

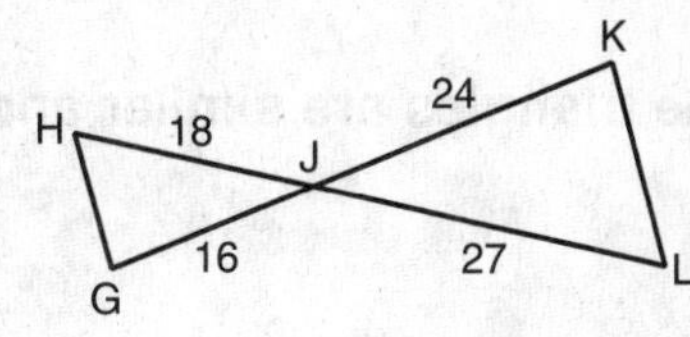

____

____

____

____

3. Verify that $\triangle ABC \sim \triangle MNP$.

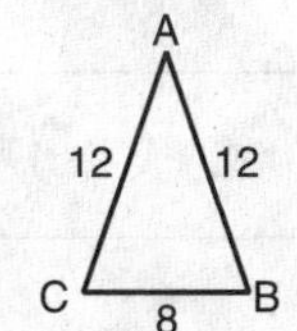

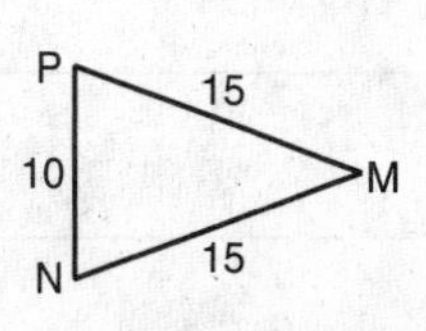

____

____

Name ______________________ Date __________ Class __________

LESSON 7-3

Review for Mastery

Triangle Similarity: AA, SSS, and SAS continued

You can use AA Similarity, SSS Similarity, and SAS Similarity to solve problems. First, prove that the triangles are similar. Then use the properties of similarity to find missing measures.

Explain why $\triangle ADE \sim \triangle ABC$ and then find BC.

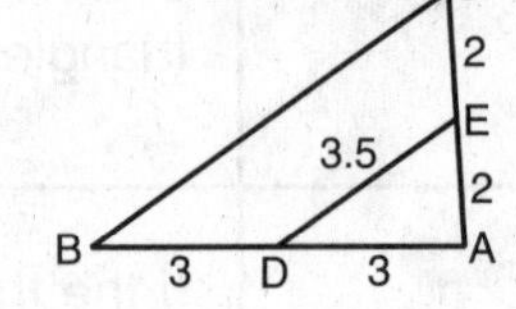

Step 1 Prove that the triangles are similar.

$\angle A \cong \angle A$ by the Reflexive Property of $\cong$.

$$\frac{AD}{AB} = \frac{3}{6} = \frac{1}{2}$$

$$\frac{AE}{AC} = \frac{2}{4} = \frac{1}{2}$$

Therefore, $\triangle ADE \sim \triangle ABC$ by SAS $\sim$.

Step 2 Find BC.

$\frac{AD}{AB} = \frac{DE}{BC}$	Corresponding sides are proportional.
$\frac{3}{6} = \frac{3.5}{BC}$	Substitute 3 for *AD*, 6 for *AB*, and 3.5 for *DE*.
$3(BC) = 6(3.5)$	Cross Products Property
$3(BC) = 21$	Simplify.
$BC = 7$	Divide both sides by 3.

Explain why the triangles are similar and then find each length.

4. *GK*

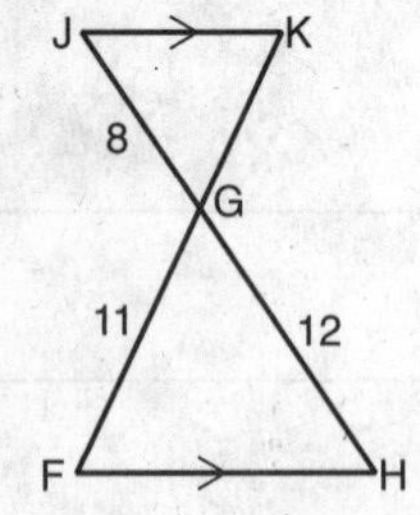

____

____

____

____

5. *US*

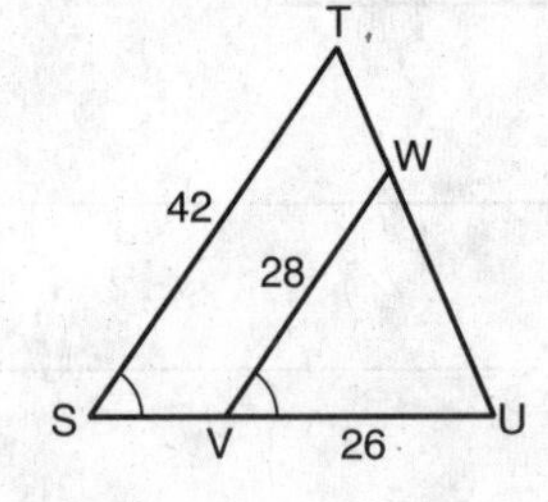

____

____

____

____

Name ______________________ Date __________ Class __________

LESSON 7-4

Review for Mastery

Applying Properties of Similar Triangles

Triangle Proportionality Theorem	Example
If a line parallel to a side of a triangle intersects the other two sides, then it divides those sides proportionally.	$\overleftrightarrow{XY} \parallel \overline{AC}$ So $\frac{BX}{XA} = \frac{BY}{YC}$.

You can use the Triangle Proportionality Theorem to find lengths of segments in triangles.

Find *EG*.

$\frac{EG}{GF} = \frac{DH}{HF}$	Triangle Proportionality Theorem
$\frac{EG}{6} = \frac{7.5}{5}$	Substitute the known values.
$EG(5) = 6(7.5)$	Cross Products Property
$5(EG) = 45$	Simplify.
$EG = 9$	Divide both sides by 5.

Converse of the Triangle Proportionality Theorem	Example
If a line divides two sides of a triangle proportionally, then it is parallel to the third side.	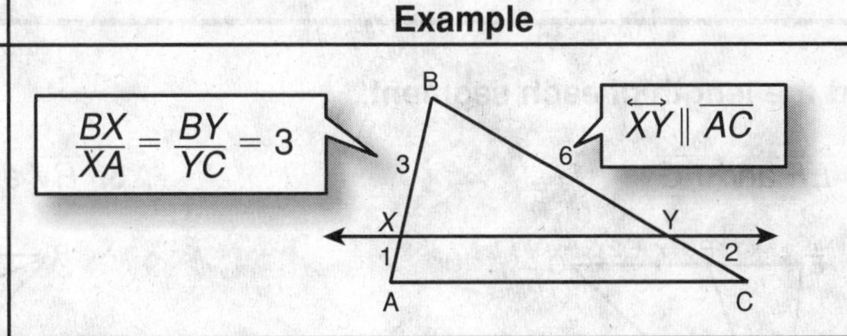

Find the length of each segment in Exercises 1 and 2.

1. $\overline{RQ}$

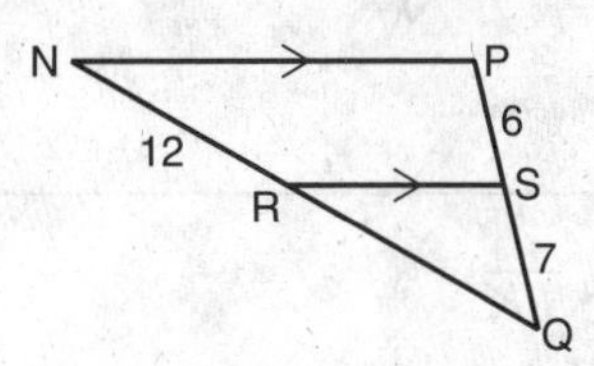

____

2. $\overline{JN}$

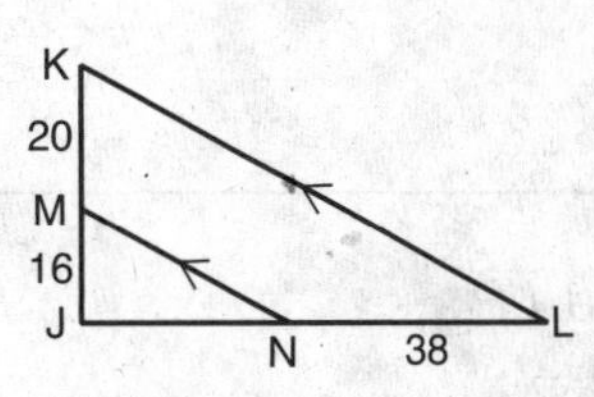

____

3. Show that $\overline{TU}$ and $\overline{WX}$ are parallel.

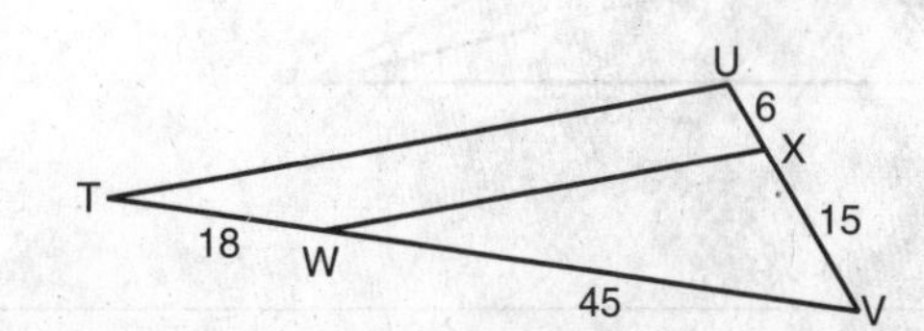

____

____

____

Name ______________________ Date __________ Class __________

LESSON 7-4

Review for Mastery

Applying Properties of Similar Triangles continued

Triangle Angle Bisector Theorem	Example
An angle bisector of a triangle divides the opposite side into two segments whose lengths are proportional to the lengths of the other two sides. (△ ∠ Bisector Thm.)	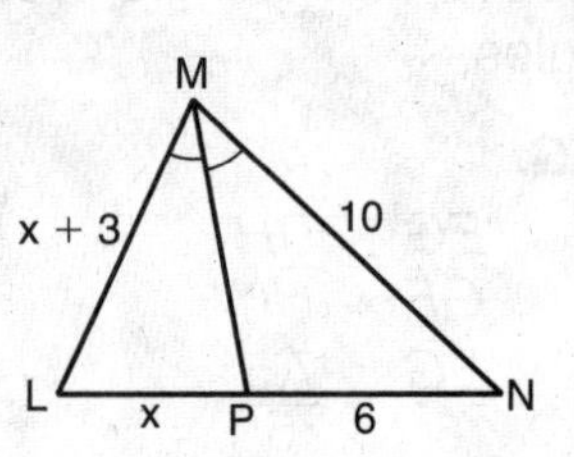 $\frac{BY}{YC} = \frac{15}{9} = \frac{5}{3}$ $\frac{AB}{AC} = \frac{40}{24} = \frac{5}{3}$

Find *LP* and *LM*.

$\frac{LP}{PN} = \frac{ML}{NM}$	△ ∠ Bisector Thm.
$\frac{x}{6} = \frac{x+3}{10}$	Substitute the given values.
$x(10) = 6(x + 3)$	Cross Products Property
$10x = 6x + 18$	Distributive Property
$4x = 18$	Simplify.
$x = 4.5$	Divide both sides by 4.

M, x + 3, 10, L, x, P, 6, N

$LP = x = 4.5$

$LM = x + 3 = 4.5 + 3 = 7.5$

Find the length of each segment.

4. $\overline{EF}$ and $\overline{FG}$

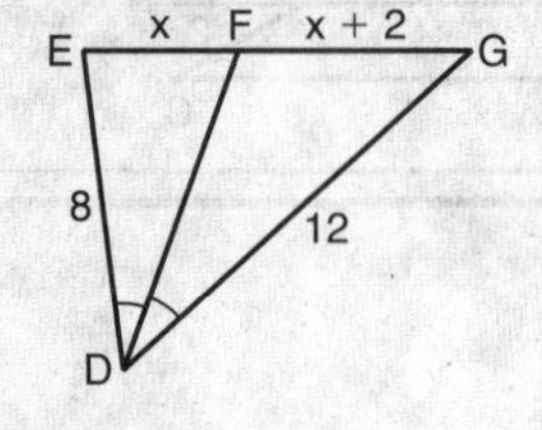

____

5. $\overline{RV}$ and $\overline{TV}$

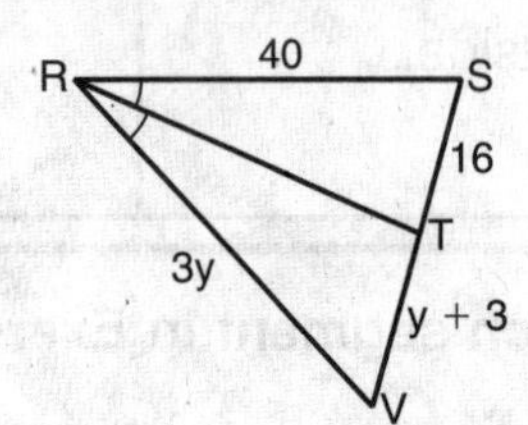

____

6. $\overline{NP}$ and $\overline{LP}$

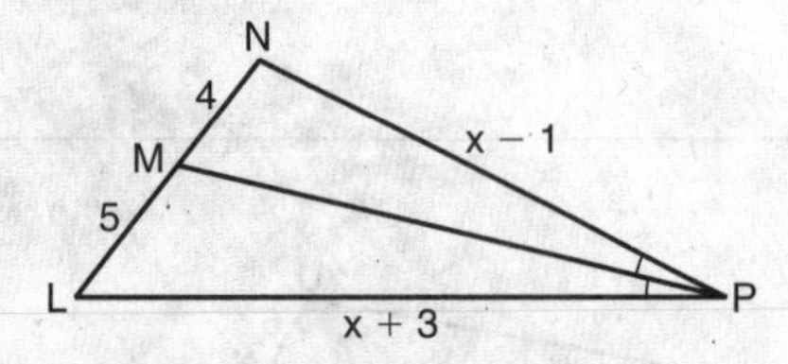

____

7. $\overline{JK}$ and $\overline{LK}$

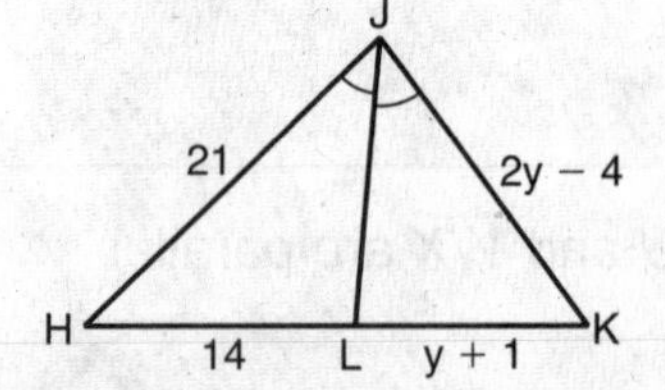

____

Name ______________________ Date ____________ Class ____________

LESSON 7-5

Review for Mastery

Using Proportional Relationships

A **scale drawing** is a drawing of an object that is smaller or larger than the object's actual size. The drawing's scale is the ratio of any length in the drawing to the actual length of the object.

The scale for the diagram of the doghouse is 1 in : 3 ft. Find the length of the actual doghouse.

First convert to equivalent units: 1 in : 36 in. (3 ft × 12 in./ft).

$$\text{diagram length} \rightarrow \quad \frac{1}{36} = \frac{0.75}{x} \quad \leftarrow \text{diagram length}$$
$$\text{actual length} \rightarrow \qquad\qquad \leftarrow \text{actual length}$$

$1x = 36(0.75)$ Cross Products Property

$x = 27$ in. Simplify.

The actual length of the doghouse is 27 in., or 2 ft 3 in.

The scale of the cabin shown in the blueprint is 1 cm : 2 m. Find the actual lengths of the following walls.

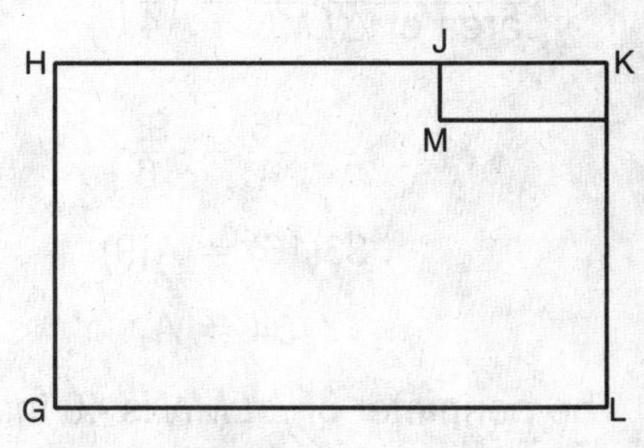

1. $\overline{HG}$ ________________

2. $\overline{GL}$ ________________

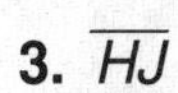

3. $\overline{HJ}$ ________________

4. $\overline{JM}$ ________________

A rectangular fitness room in a recreation center is 45 feet long and 28 feet wide. Find the length and width for a scale drawing of the room, using the following scales.

5. 1 in : 1 ft ________________

6. 1 in : 2 ft ________________

7. 1 in : 3 ft ________________

8. 1 in : 6 ft 8 in. ________________

Name ______________________________ Date ____________ Class ____________

LESSON 7-5

Review for Mastery

Using Proportional Relationships continued

Proportional Perimeters and Areas Theorem

If two figures are similar and their similarity ratio is $\frac{a}{b}$, then the ratio of their perimeters is $\frac{a}{b}$ and the ratio of their areas is $\left(\frac{a}{b}\right)^2$.

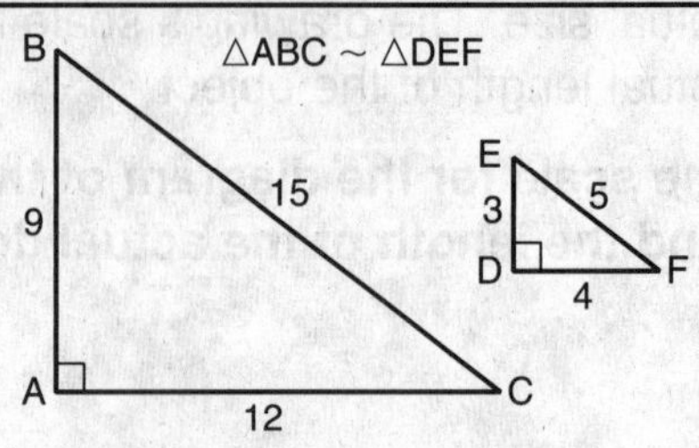

$$\frac{\text{perimeter of } \triangle ABC}{\text{perimeter of } \triangle DEF} = \frac{36}{12} = \frac{3}{1}$$

$$\frac{\text{area of } \triangle ABC}{\text{area of } \triangle DEF} = \frac{54}{6} = \frac{9}{1} = \left(\frac{3}{1}\right)^2$$

$$\frac{AB}{DE} = \frac{BC}{EF} = \frac{CA}{FD} = \frac{3}{1}$$

△*HJK* ~ △*LMN*. The perimeter of △*HJK* is 30 inches, and the area of △*HJK* is 36 square inches. Find the perimeter and area of △*LMN*.

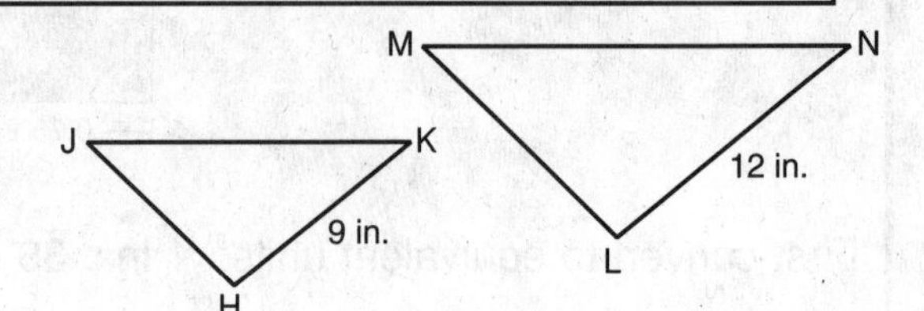

The similarity ratio of △*HJK* to △*LMN* = $\frac{9}{12} = \frac{3}{4}$.

$\frac{\text{perimeter of } \triangle HJK}{\text{perimeter of } \triangle LMN} = \frac{3}{4}$	The ratio of the perimeters equals the similarity ratio.
$\frac{30}{P} = \frac{3}{4}$	Substitute the known values.
$30(4) = P(3)$	Cross Products Property
$40 = P$	Simplify.
$\frac{\text{area of } \triangle HJK}{\text{area of } \triangle LMN} = \left(\frac{3}{4}\right)^2$	The ratio of the areas equals the square of the similarity ratio.
$\frac{36}{A} = \frac{9}{16}$	Substitute the known values.
$36(16) = A(9)$	Cross Products Property
$64 = A$	Simplify.

The perimeter of △*LMN* is 40 in., and the area is 64 in^2.

9. ▭*PQRS* ~ ▭*TUVW*. Find the perimeter and area of ▭*TUVW*.

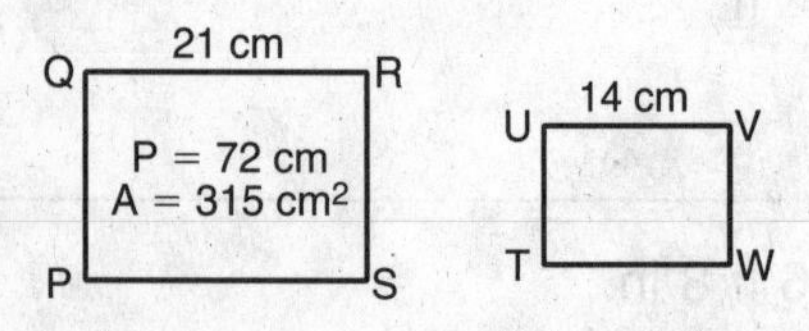

____

10. △*EFG* ~ △*HJK*. Find the perimeter and area of △*HJK*.

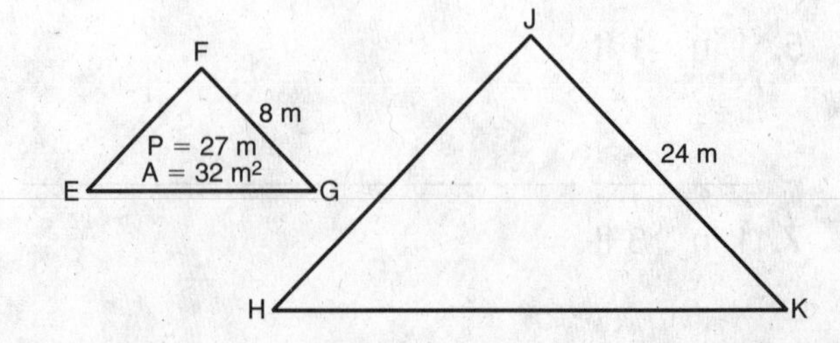

____

Name ________________________ Date ______________ Class ______________

LESSON 7-6

Review for Mastery

Dilations and Similarity in the Coordinate Plane

A **dilation** is a transformation that changes the size of a figure but not its shape. The preimage and image are always similar. A **scale factor** describes how much a figure is enlarged or reduced.

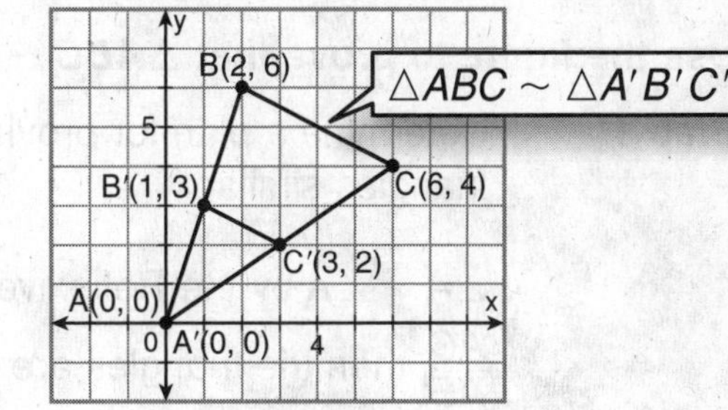

Triangle *ABC* has vertices *A*(0, 0), *B*(2, 6), and *C*(6, 4). Find the coordinates of the vertices of the image after a dilation with a scale factor $\frac{1}{2}$.

Preimage	**Image**
$\triangle ABC$	$\triangle A'B'C'$

$A(0, 0) \rightarrow \left(0 \cdot \frac{1}{2}, 0 \cdot \frac{1}{2}\right) \rightarrow A'(0, 0)$

$B(2, 6) \rightarrow \left(2 \cdot \frac{1}{2}, 6 \cdot \frac{1}{2}\right) \rightarrow B'(1, 3)$

$C(6, 4) \rightarrow \left(6 \cdot \frac{1}{2}, 4 \cdot \frac{1}{2}\right) \rightarrow C'(3, 2)$

$\triangle FEG \sim \triangle HEJ$. Find the coordinates of *F* and the scale factor.

y
H(0, 6)
F
x
E(0, 0)
G(4, 0)
J(8, 0)

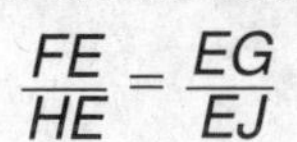

$\frac{FE}{HE} = \frac{EG}{EJ}$	Write a proportion.
$\frac{FE}{6} = \frac{4}{8}$	$HE = 6$, $EG = 4$, and $EJ = 8$.
$8(FE) = 24$	Cross Products Property
$FE = 3$	Divide both sides by 8.

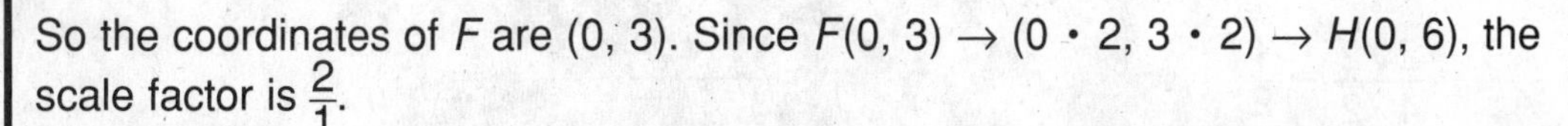

So the coordinates of *F* are (0, 3). Since $F(0, 3) \rightarrow (0 \cdot 2, 3 \cdot 2) \rightarrow H(0, 6)$, the scale factor is $\frac{2}{1}$.

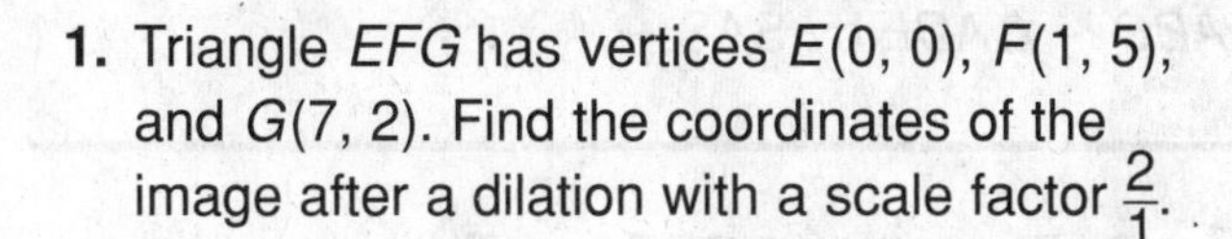

1. Triangle *EFG* has vertices *E*(0, 0), *F*(1, 5), and *G*(7, 2). Find the coordinates of the image after a dilation with a scale factor $\frac{2}{1}$.

____

____

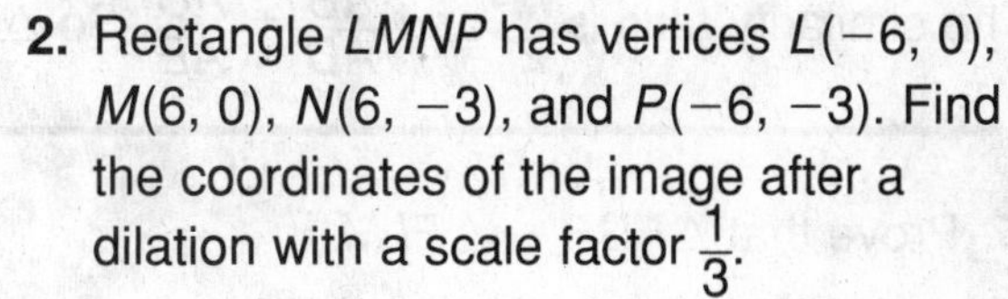

2. Rectangle *LMNP* has vertices *L*(−6, 0), *M*(6, 0), *N*(6, −3), and *P*(−6, −3). Find the coordinates of the image after a dilation with a scale factor $\frac{1}{3}$.

____

____

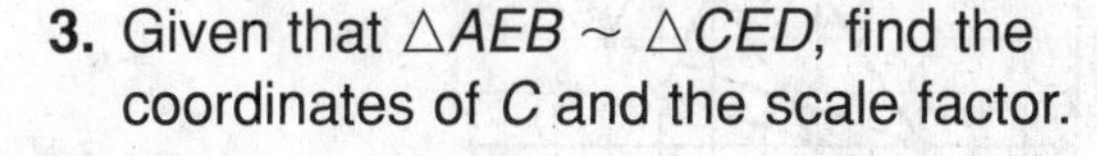

3. Given that $\triangle AEB \sim \triangle CED$, find the coordinates of *C* and the scale factor.

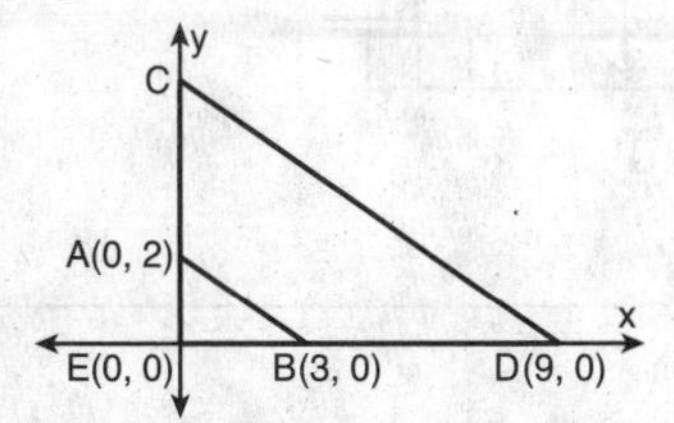

____

4. Given that $\triangle LKM \sim \triangle NKP$, find the coordinates of *P* and the scale factor.

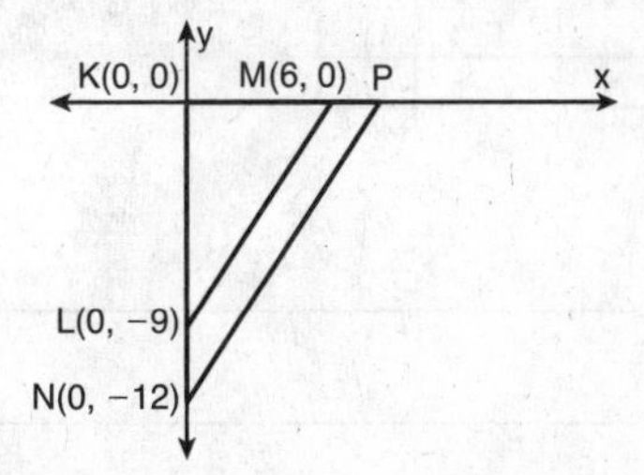

____

Name ______________________ Date __________ Class __________

LESSON 7-6

Review for Mastery

Dilations and Similarity in the Coordinate Plane continued

You can prove that triangles in the coordinate plane are similar by using the Distance Formula to find the side lengths. Then apply SSS Similarity or SAS Similarity.

Use the figure to prove that $\triangle ABC \sim \triangle ADE$.

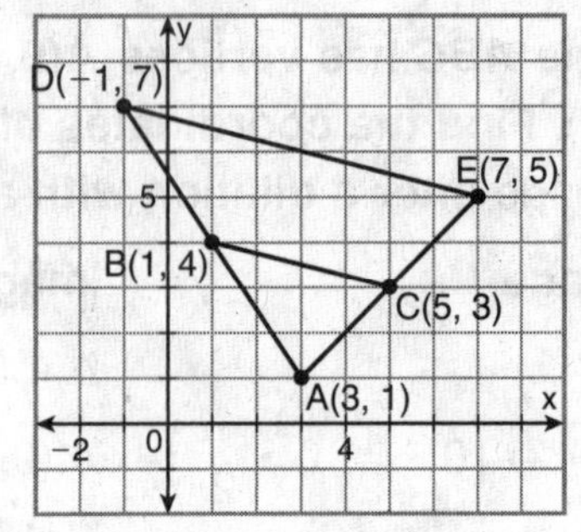

Step 1 Determine a plan for proving the triangles similar.

$\angle A \cong \angle A$ by the Reflexive Property. If $\frac{AB}{AD} = \frac{AC}{AE}$, then the triangles are similar by SAS $\sim$.

Step 2 Use the Distance Formula to find the side lengths.

$AB = \sqrt{(1-3)^2 + (4-1)^2}$
$= \sqrt{13}$

$AC = \sqrt{(5-3)^2 + (3-1)^2}$
$= \sqrt{8} = 2\sqrt{2}$

$AD = \sqrt{(-1-3)^2 + (7-1)^2}$
$= \sqrt{52} = 2\sqrt{13}$

$AE = \sqrt{(7-3)^2 + (5-1)^2}$
$= \sqrt{32} = 4\sqrt{2}$

Step 3 Compare the corresponding sides to determine whether they are proportional.

$\frac{AB}{AD} = \frac{\sqrt{13}}{2\sqrt{13}} = \frac{1}{2}$

$\frac{AC}{AE} = \frac{2\sqrt{2}}{4\sqrt{2}} = \frac{1}{2}$

The similarity ratio is $\frac{1}{2}$, and $\frac{AB}{AD} = \frac{AC}{AE}$. So $\triangle ABC \sim \triangle ADE$ by SAS $\sim$.

5. Prove that $\triangle FGH \sim \triangle FLM$.

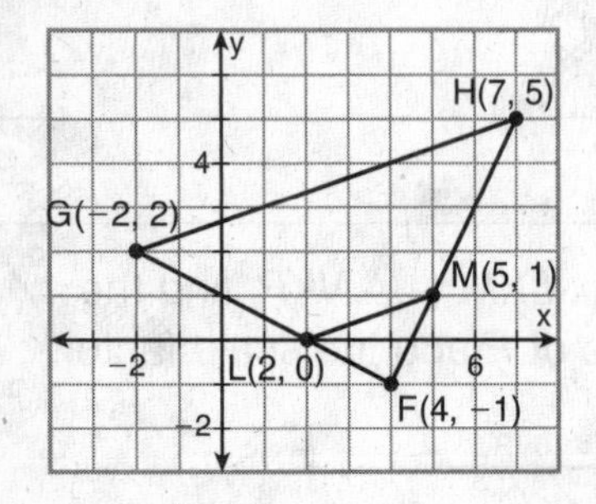

____

____

____

6. Prove that $\triangle QRS \sim \triangle TUV$.

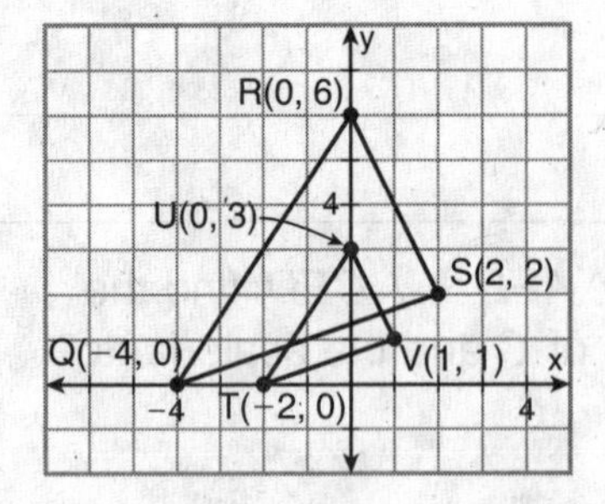

____

____

____

Name ______________________ Date __________ Class __________

LESSON 8-1

Review for Mastery

Similarity in Right Triangles

Altitudes and Similar Triangles

The altitude to the hypotenuse of a right triangle forms two triangles that are similar to each other and to the original triangle.

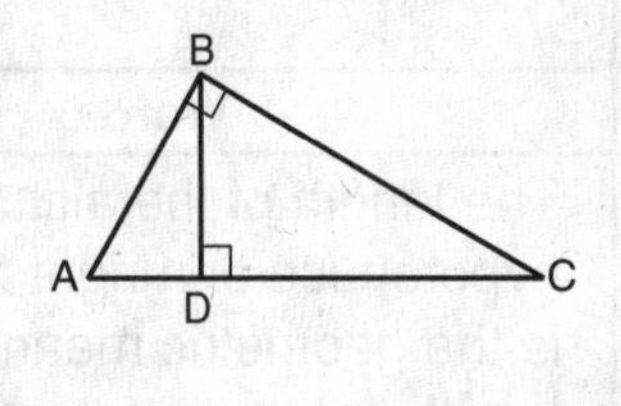

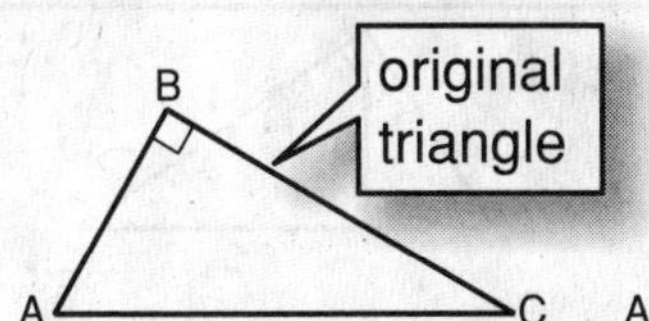

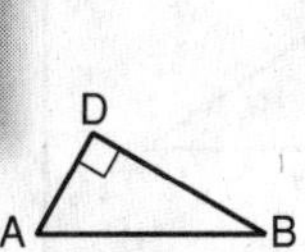

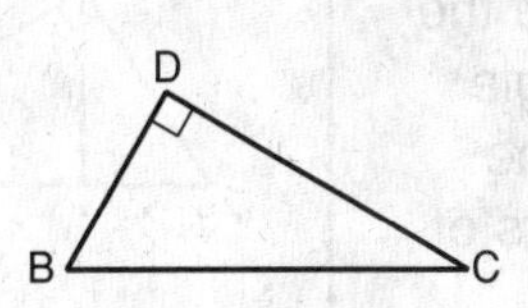

Similarity statement: $\triangle ABC \sim \triangle ADB \sim \triangle BDC$

The **geometric mean** of two positive numbers is the positive square root of their product.

Find the geometric mean of 5 and 20.

Let x be the geometric mean.

$x^2 = (5)(20)$	Definition of geometric mean	$\frac{a}{x} = \frac{x}{b}$
$x^2 = 100$	Simplify.	$x^2 = ab$
$x = 10$	Find the positive square root.	$x = \sqrt{ab}$

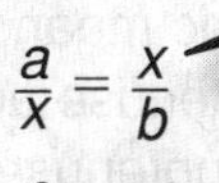

x is the geometric mean of *a* and *b*.

So 10 is the geometric mean of 5 and 20.

Write a similarity statement comparing the three triangles in each diagram.

1.

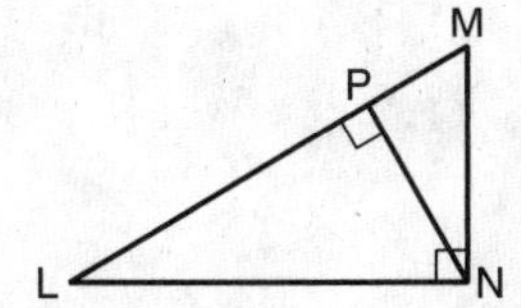

____

2.

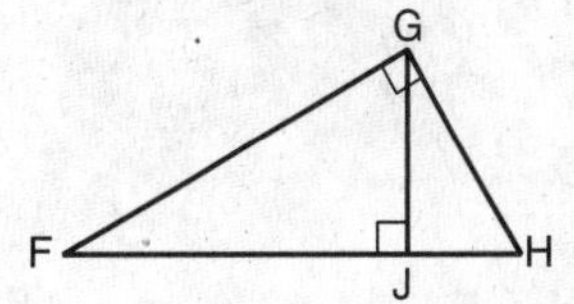

____

Find the geometric mean of each pair of numbers. If necessary, give the answer in simplest radical form.

3. 3 and 27

____

4. 9 and 16

____

5. 4 and 5

____

6. 8 and 12

____

Name ______________________ Date ____________ Class ____________

LESSON **8-1**

Review for Mastery

Similarity in Right Triangles continued

You can use geometric means to find side lengths in right triangles.

Geometric Means		
Words	**Symbols**	**Examples**
The length of the altitude to the hypotenuse of a right triangle is the geometric mean of the lengths of the two segments of the hypotenuse.	$h^2 = xy$	$h^2 = xy$ $6^2 = x(9)$ $36 = 9x$ $4 = x$
The length of a leg of a right triangle is the geometric mean of the length of the hypotenuse and the segment of the hypotenuse adjacent to that leg.	$a^2 = xc$ $\quad b^2 = yc$	$a^2 = xc$ $a^2 = 4(13)$ $a^2 = 52$ $a = \sqrt{52} = 2\sqrt{13}$

Find *x*, *y*, and *z*.

7.

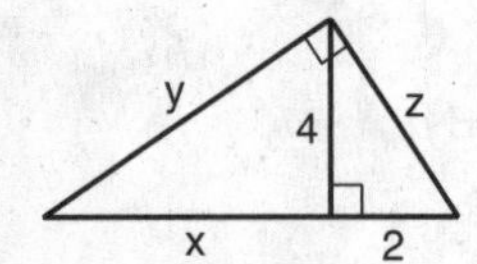

8.

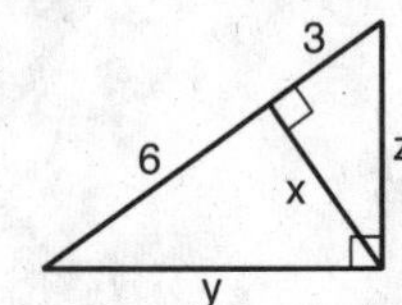

____

9.

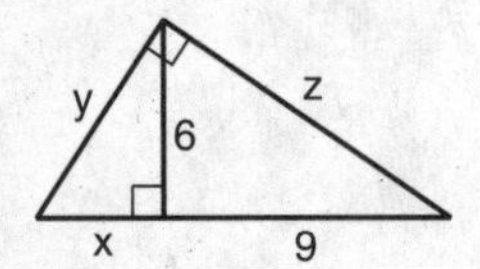

10.

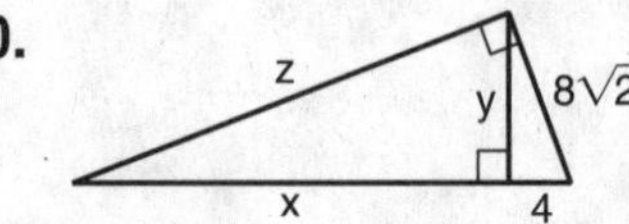

_______________________ _______________________

Name ______________________ Date __________ Class __________

LESSON 8-2

Review for Mastery

Trigonometric Ratios

Trigonometric Ratios

$\sin A = \dfrac{\text{leg opposite } \angle A}{\text{hypotenuse}} = \dfrac{4}{5} = 0.8$

$\cos A = \dfrac{\text{leg adjacent to } \angle A}{\text{hypotenuse}} = \dfrac{3}{5} = 0.6$

$\tan A = \dfrac{\text{leg opposite } \angle A}{\text{leg adjacent to } \angle A} = \dfrac{4}{3} \approx 1.33$

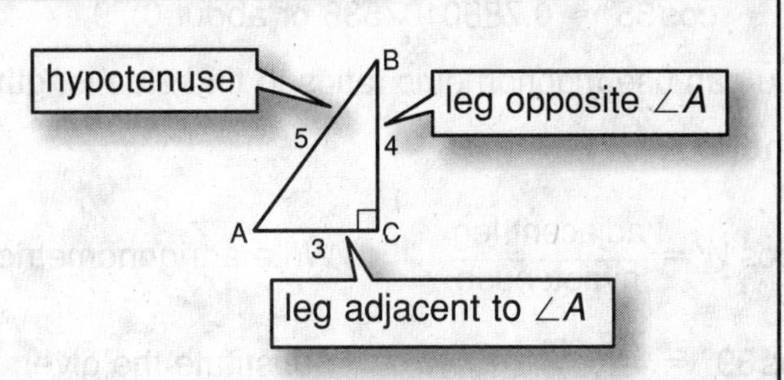

You can use special right triangles to write trigonometric ratios as fractions.

$\sin 45° = \sin Q = \dfrac{\text{leg opposite } \angle Q}{\text{hypotenuse}}$

$= \dfrac{x}{x\sqrt{2}} = \dfrac{1}{\sqrt{2}}$

$= \dfrac{\sqrt{2}}{2}$

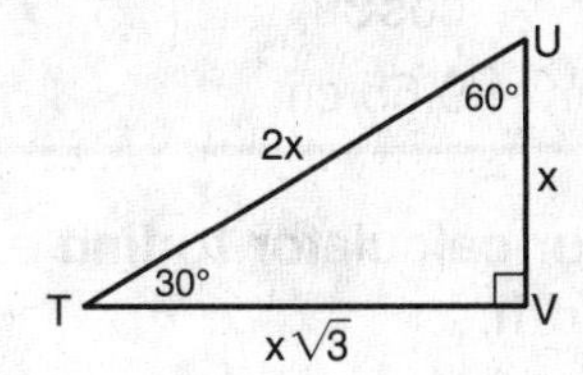

So $\sin 45° = \dfrac{\sqrt{2}}{2}$.

Write each trigonometric ratio as a fraction and as a decimal rounded to the nearest hundredth.

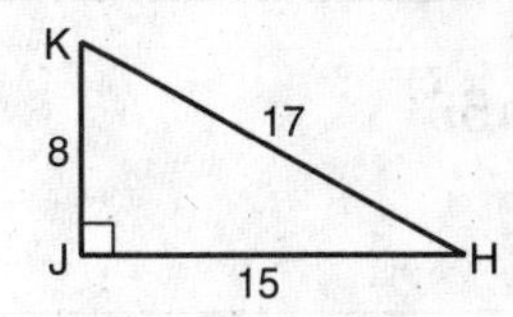

1. $\sin K$ ______________________

2. $\cos H$ ______________________

3. $\cos K$ ______________________

4. $\tan H$ ______________________

Use a special right triangle to write each trigonometric ratio as a fraction.

5. $\cos 45°$ ______________________

6. $\tan 45°$ ______________________

7. $\sin 60°$ ______________________

8. $\tan 30°$ ______________________

Name ____________________ Date __________ Class __________

LESSON 8-2

Review for Mastery

Trigonometric Ratios continued

You can use a calculator to find the value of trigonometric ratios.

$\cos 38° \approx 0.7880107536$ or about 0.79

You can use trigonometric ratios to find side lengths of triangles.

Find *WY*.

$\cos W = \dfrac{\text{adjacent leg}}{\text{hypotenuse}}$ Write a trigonometric ratio that involves *WY*.

$\cos 39° = \dfrac{7.5 \text{ cm}}{WY}$ Substitute the given values.

$WY = \dfrac{7.5}{\cos 39°}$ Solve for *WY*.

$WY \approx 9.65$ cm Simplify the expression.

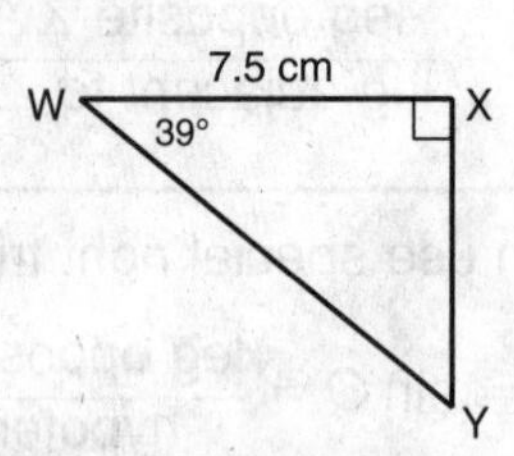

Use your calculator to find each trigonometric ratio. Round to the nearest hundredth.

9. $\sin 42°$ ____________

10. $\cos 89°$ ____________

11. $\tan 55°$ ____________

12. $\sin 6°$ ____________

Find each length. Round to the nearest hundredth.

13. *DE*

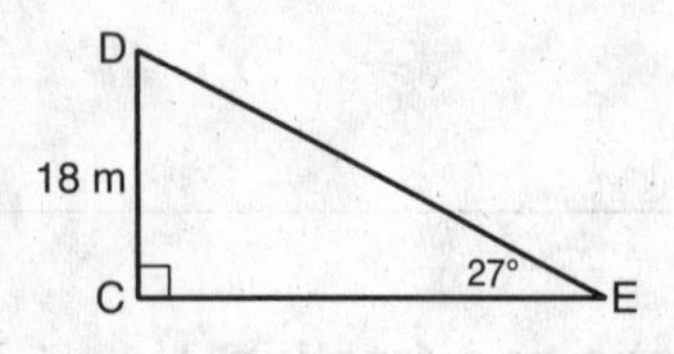

____

14. *FH*

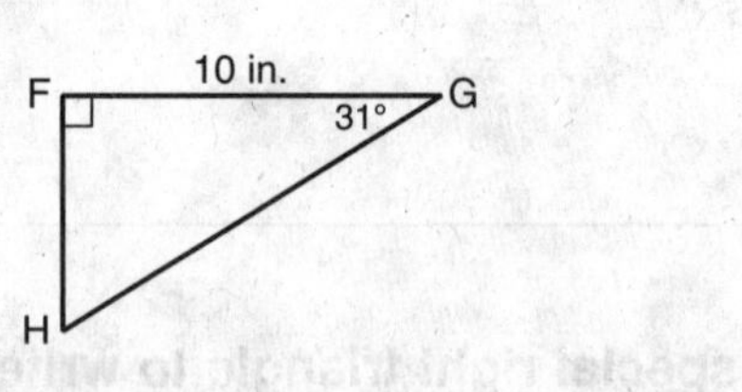

____

15. *JK*

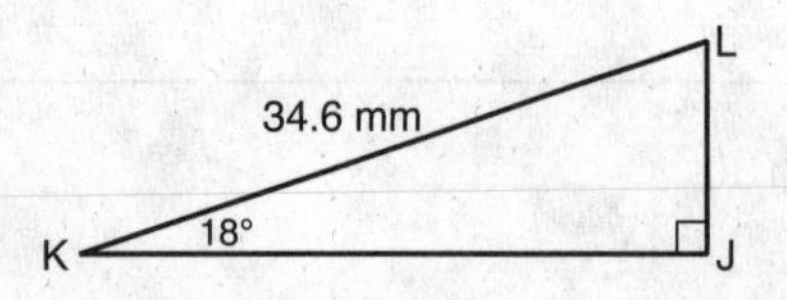

____

16. *US*

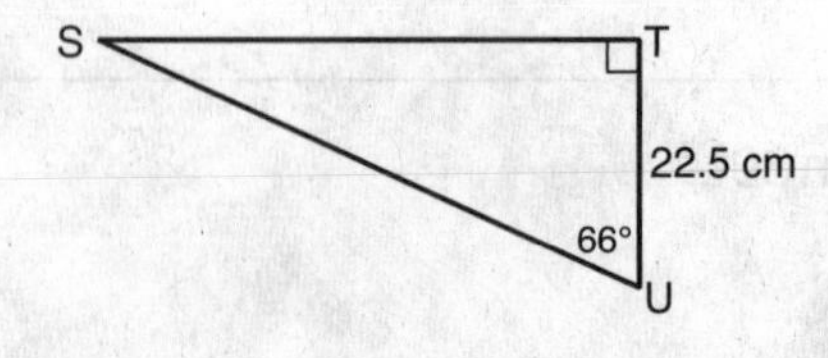

____

Name ______________________ Date ____________ Class ____________

LESSON 8-3

Review for Mastery

Solving Right Triangles

Use the trigonometric ratio $\sin A = 0.8$ to determine which angle of the triangle is $\angle A$.

$\sin\angle 1 = \dfrac{\text{leg opposite } \angle 1}{\text{hypotenuse}} = \dfrac{6}{10} = 0.6$

$\sin\angle 2 = \dfrac{\text{leg opposite } \angle 2}{\text{hypotenuse}} = \dfrac{8}{10} = 0.8$

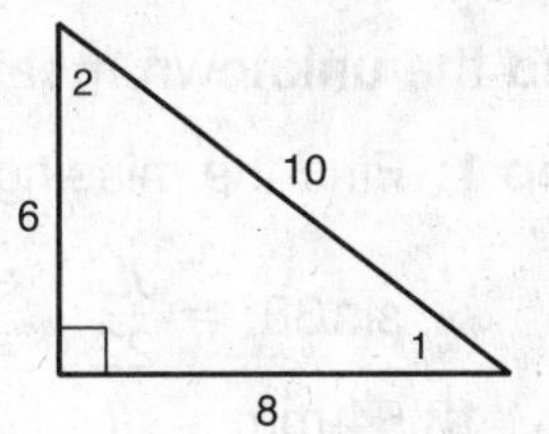

Since $\sin A = \sin\angle 2$, $\angle 2$ is $\angle A$.

If you know the sine, cosine, or tangent of an acute angle measure, then you can use your calculator to find the measure of the angle.

Inverse Trigonometric Functions	
Symbols	**Examples**
$\sin A = x \Rightarrow \sin^{-1} x = m\angle A$	$\sin 30° = \frac{1}{2} \Rightarrow \sin^{-1}\left(\frac{1}{2}\right) = 30°$
$\cos B = x \Rightarrow \cos^{-1} x = m\angle B$	$\cos 45° = \frac{\sqrt{2}}{2} \Rightarrow \cos^{-1}\left(\frac{\sqrt{2}}{2}\right) = 45°$
$\tan C = x \Rightarrow \tan^{-1} x = m\angle C$	$\tan 76° \approx 4.01 \Rightarrow \tan^{-1}(4.01) \approx 76°$

Use the given trigonometric ratio to determine which angle of the triangle is $\angle A$.

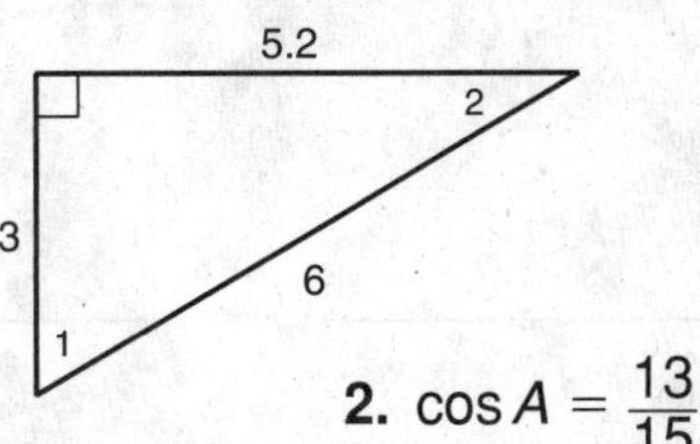

1. $\sin A = \frac{1}{2}$

____

2. $\cos A = \frac{13}{15}$

____

3. $\cos A = 0.5$

____

4. $\tan A = \frac{15}{26}$

____

Use your calculator to find each angle measure to the nearest degree.

5. $\sin^{-1}(0.8)$

____

6. $\cos^{-1}(0.19)$

____

7. $\tan^{-1}(3.4)$

____

8. $\sin^{-1}\left(\frac{1}{5}\right)$

____

Name ______________________________ Date ______________ Class ______________

LESSON 8-3

Review for Mastery

Solving Right Triangles continued

To *solve a triangle* means to find the measures of all the angles and all the sides of the triangle.

Find the unknown measures of $\triangle JKL$.

Step 1: Find the missing side lengths.

$\sin 38^\circ = \frac{JL}{22}$ ← leg opposite $\angle K$ / ← hypotenuse

$13.54 \text{ mm} \approx JL$

$JL^2 + LK^2 = JK^2$ — Pythagorean Theorem

$13.542 + LK^2 = 22^2$ — Substitute the known values.

$LK \approx 17.34 \text{ mm}$ — Simplify.

Step 2: Find the missing angle measures.

$m\angle J = 90^\circ - 38^\circ$ — Acute ∠s of a rt. △ are complementary.

$= 52^\circ$ — Simplify.

So $JL \approx 13.54$ mm, $LK \approx 17.34$ mm, and $m\angle J = 52^\circ$.

Find the unknown measures. Round lengths to the nearest hundredth and angle measures to the nearest degree.

9.

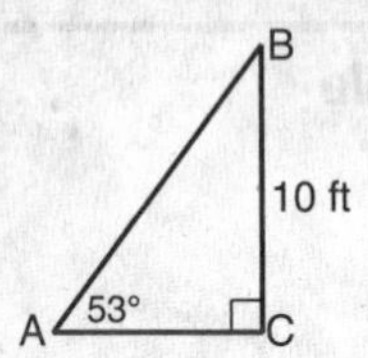

____

____

10.

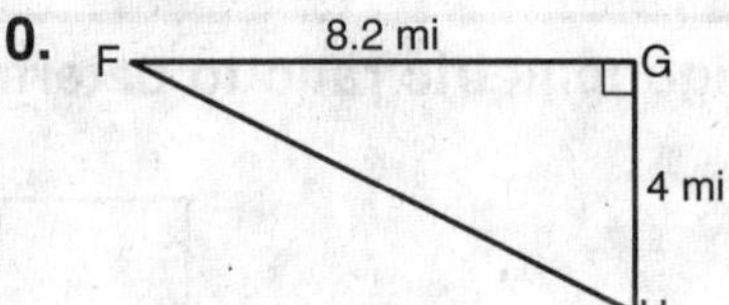

____

____

11.

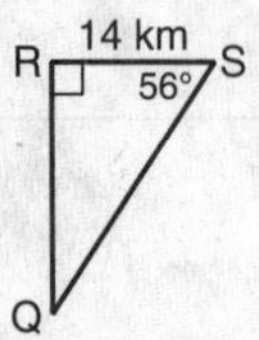

____

____

12.

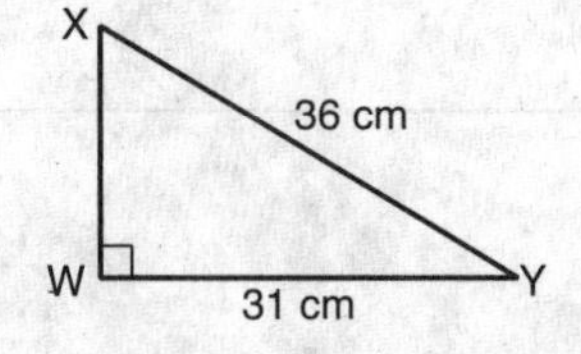

____

____

For each triangle, find the side lengths to the nearest hundredth and the angle measures to the nearest degree.

13. $M(-5, 1)$, $N(1, 1)$, $P(-5, 7)$

____

____

14. $J(2, 3)$, $K(-1, -4)$, $L(-1, 3)$

____

____

Name ______________________ Date __________ Class __________

LESSON 8-4

Review for Mastery

Angles of Elevation and Depression

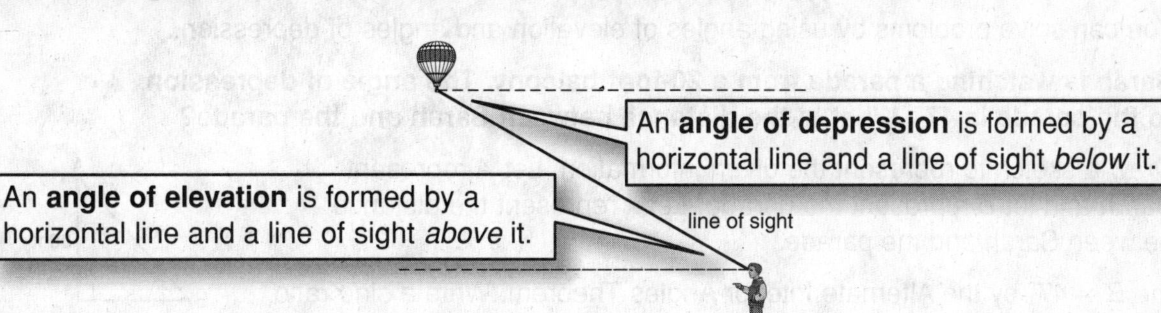

Classify each angle as an angle of elevation or an angle of depression.

1. $\angle 1$

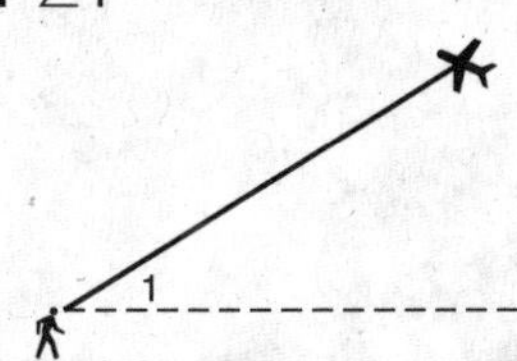

2. $\angle 2$

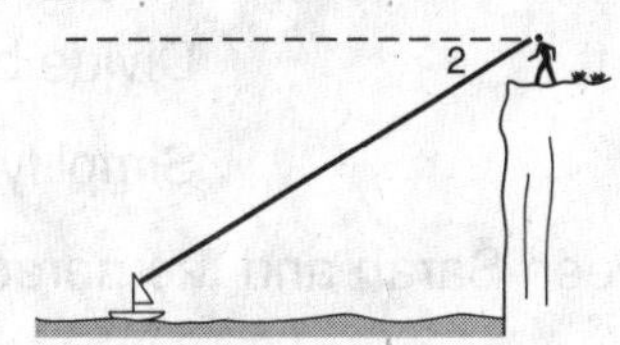

______________________ ______________________

Use the figure for Exercises 3 and 4. Classify each angle as an angle of elevation or an angle of depression.

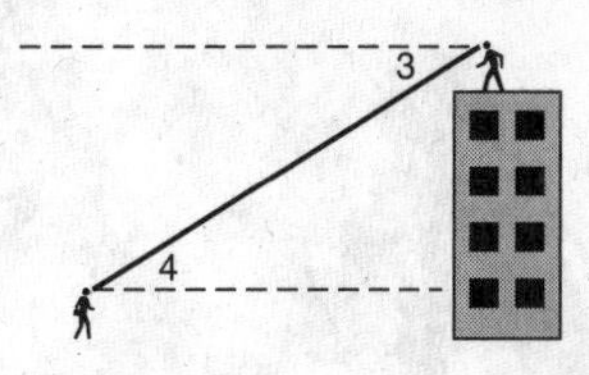

3. $\angle 3$

____

4. $\angle 4$

____

Use the figure for Exercises 5–8. Classify each angle as an angle of elevation or an angle of depression.

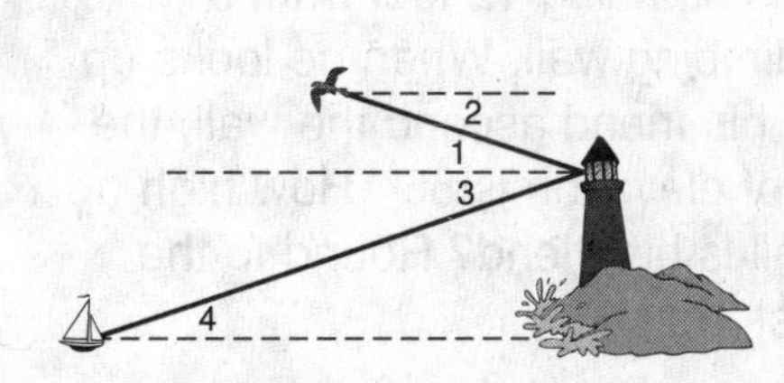

5. $\angle 1$

____

6. $\angle 2$

____

7. $\angle 3$

____

8. $\angle 4$

____

Name ______________________________ Date ______________ Class ____________

LESSON 8-4

Review for Mastery

Angles of Elevation and Depression continued

You can solve problems by using angles of elevation and angles of depression.

Sarah is watching a parade from a 20-foot balcony. The angle of depression to the parade is 47°. What is the distance between Sarah and the parade?

Draw a sketch to represent the given information. Let A represent Sarah and let B represent the parade. Let x represent the distance between Sarah and the parade.

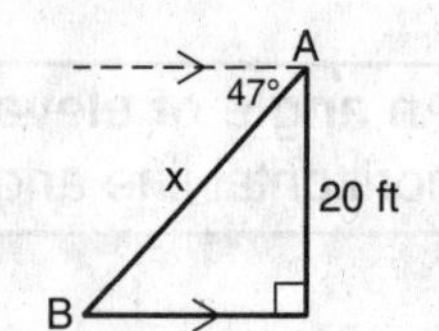

$m\angle B = 47°$ by the Alternate Interior Angles Theorem. Write a sine ratio using $\angle B$.

$\sin 47° = \frac{20}{x}$ ft ← leg opposite $\angle B$ ← hypotenuse

$x \sin 47° = 20$ ft — Multiply both sides by x.

$x = \frac{20}{\sin 47°}$ ft — Divide both sides by $\sin 47°$.

27 ft $\approx x$ — Simplify the expression.

The distance between Sarah and the parade is about 27 feet.

9. When the angle of elevation to the sun is 52°, a tree casts a shadow that is 9 meters long. What is the height of the tree? Round to the nearest tenth of a meter.

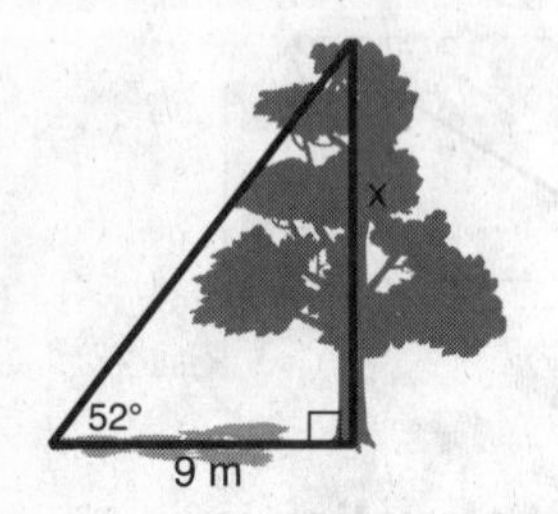

____

10. A person snorkeling sees a turtle on the ocean floor at an angle of depression of 38°. She is 14 feet above the ocean floor. How far from the turtle is she? Round to the nearest foot.

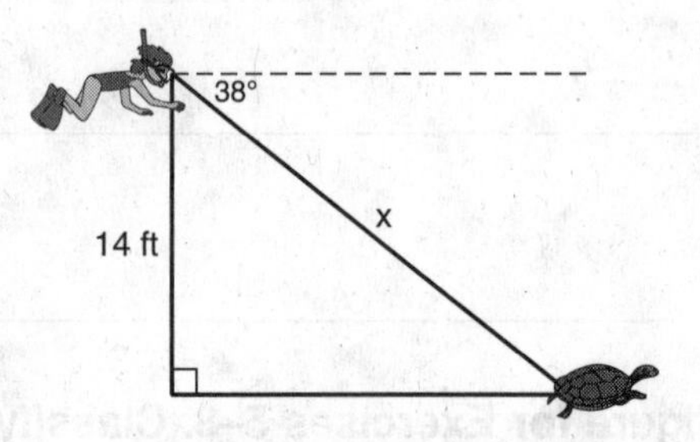

____

11. Jared is standing 12 feet from a rock-climbing wall. When he looks up to see his friend ascend the wall, the angle of elevation is 56°. How high up the wall is his friend? Round to the nearest foot.

____

12. Maria is looking out a 17-foot-high window and sees two deer. The angle of depression to the deer is 26°. What is the horizontal distance from Maria to the deer? Round to the nearest foot.

____

Name ______________________ Date __________ Class __________

LESSON 8-5

Review for Mastery

Law of Sines and Law of Cosines

You can use a calculator to find trigonometric ratios for obtuse angles.

$\sin 115° \approx 0.906307787$ $\cos 270° = 0$ $\tan 96° = -9.514364454$

The Law of Sines	
For any $\triangle ABC$ with side lengths a, b, and c that are opposite angles A, B, and C, respectively, $\frac{\sin A}{a} = \frac{\sin B}{b} = \frac{\sin C}{c}$.	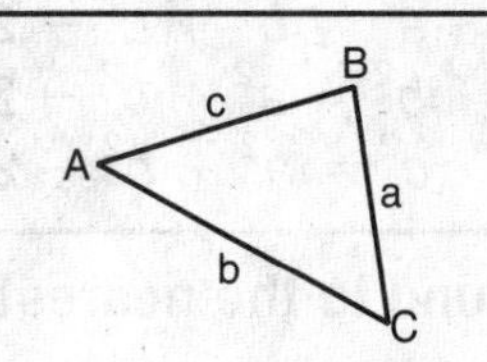

Find m∠P. Round to the nearest degree.

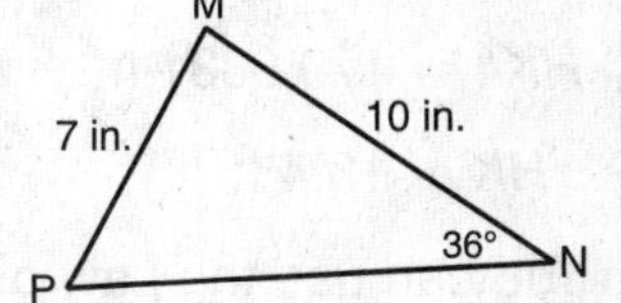

$\frac{\sin P}{MN} = \frac{\sin N}{PM}$	Law of Sines
$\frac{\sin P}{10 \text{ in.}} = \frac{\sin 36°}{7 \text{ in.}}$	$MN = 10$, $m\angle N = 36°$, $PM = 7$
$\sin P = 10 \text{ in.} \cdot \frac{\sin 36°}{7 \text{ in.}}$	Multiply both sides by 10 in.
$\sin P \approx 0.84$	Simplify.
$m\angle P \approx \sin^{-1}(0.84)$	Use the inverse sine function to find m∠P.
$m\angle P \approx 57°$	Simplify.

Use a calculator to find each trigonometric ratio. Round to the nearest hundredth.

1. $\cos 104°$ ______

2. $\tan 225°$ ______

3. $\sin 100°$ ______

Find each measure. Round the length to the nearest tenth and the angle measure to the nearest degree.

4. *TU*

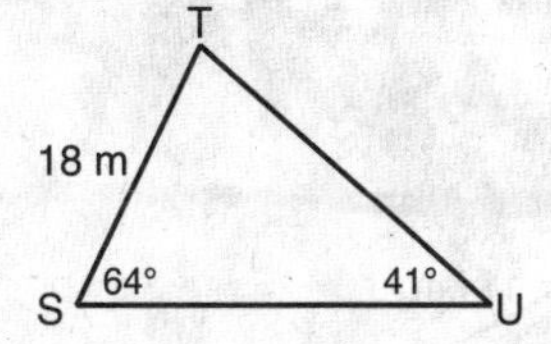

______

5. m∠*E*

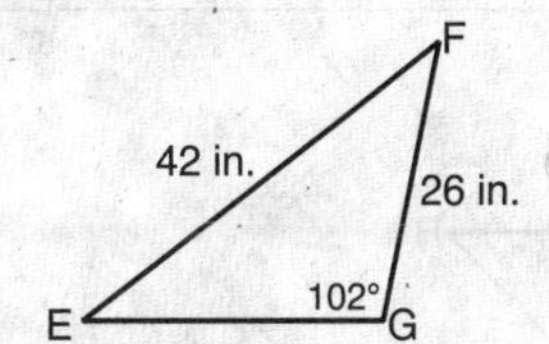

______

Name ________________________ Date ____________ Class ____________

LESSON 8-5

Review for Mastery

Law of Sines and Law of Cosines continued

The Law of Cosines	
For any $\triangle ABC$ with side lengths a, b, and c that are opposite angles A, B, and C, respectively, $a^2 = b^2 + c^2 - 2bc\cos A$, $b^2 = a^2 + c^2 - 2ac\cos B$, $c^2 = a^2 + b^2 - 2ab\cos C$.	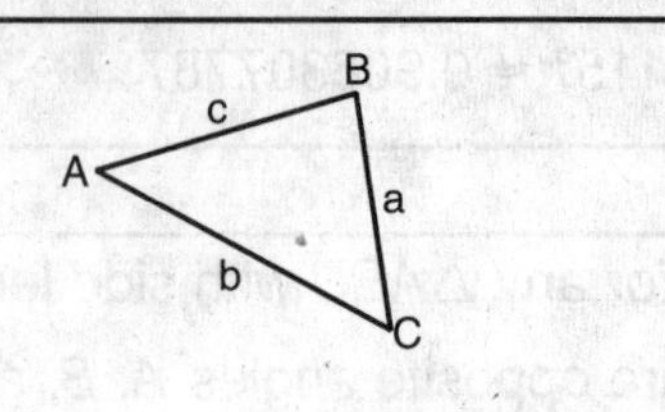

Find *HK*. Round to the nearest tenth.

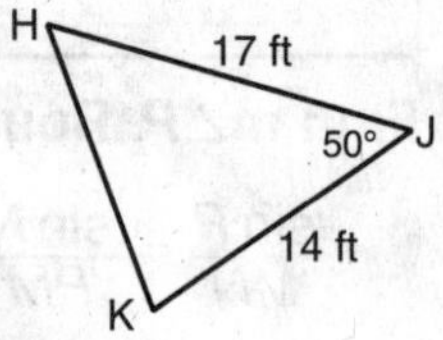

$HK^2 = HJ^2 + JK^2 - 2(HJ)(JK)\cos J$ — Law of Cosines

$= 289 + 196 - 2(17)(14)\cos 50°$ — Substitute the known values.

$HK^2 \approx 179.0331 \text{ ft}^2$ — Simplify.

$HK \approx 13.4$ ft — Find the square root of both sides.

You can use the Law of Sines and the Law of Cosines to solve triangles according to the information you have.

Use the Law of Sines if you know	Use the Law of Cosines if you know
• two angle measures and any side length, or • two side lengths and a nonincluded angle measure	• two side lengths and the included angle measure, or • three side lengths

Find each measure. Round lengths to the nearest tenth and angle measures to the nearest degree.

6. *EF*

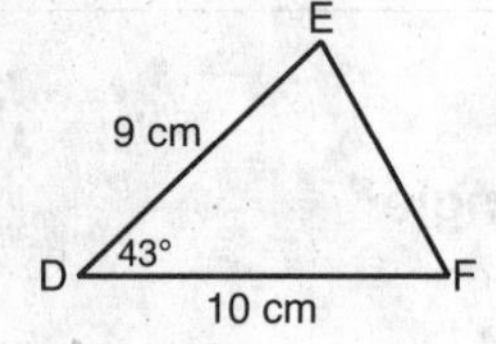

____

7. m∠*X*

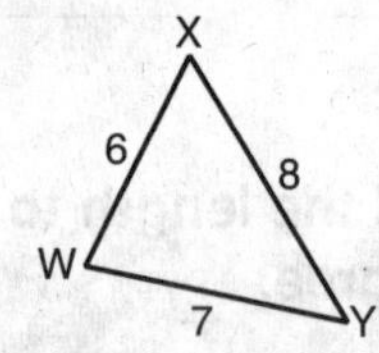

____

8. m∠*R*

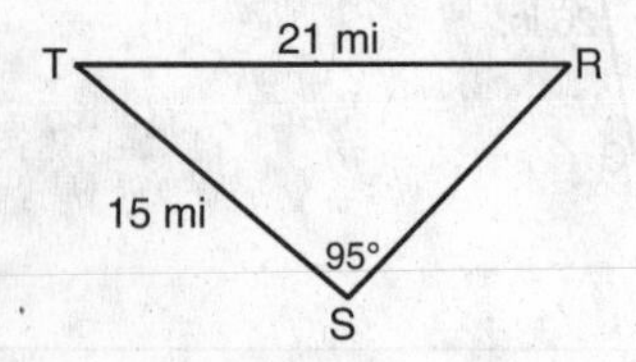

____

9. *AB*

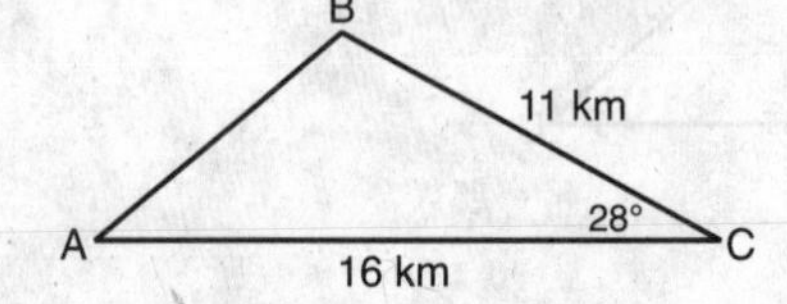

____

Name ______________________ Date __________ Class __________

LESSON 8-6

Review for Mastery

Vectors

A **vector** is a quantity that has both length and direction. The vector below may be named $\overrightarrow{HJ}$ or $\vec{v}$.

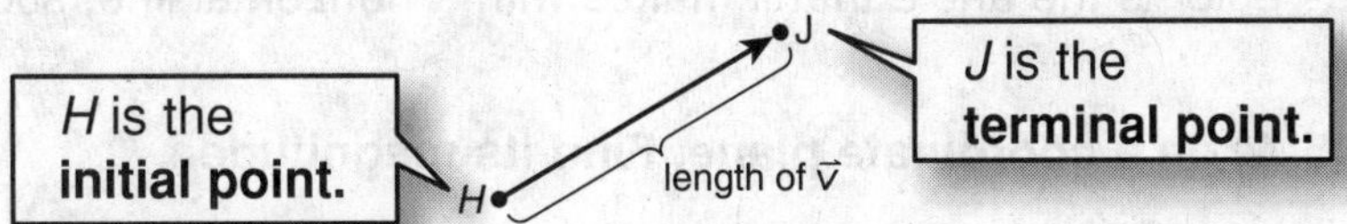

The **component form** of a vector lists the horizontal and vertical change from the initial point to the terminal point.

$\langle x, y \rangle$

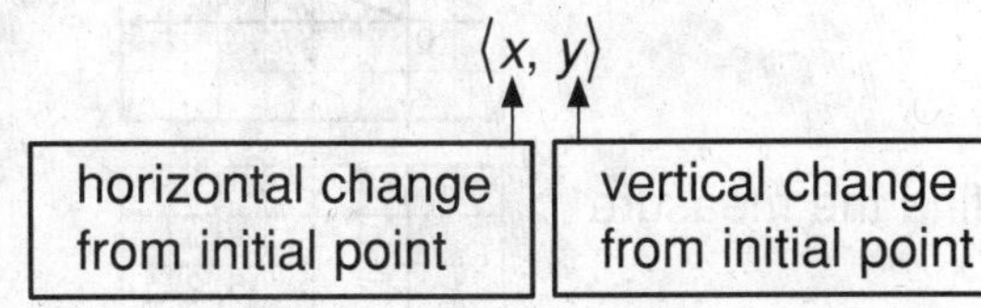

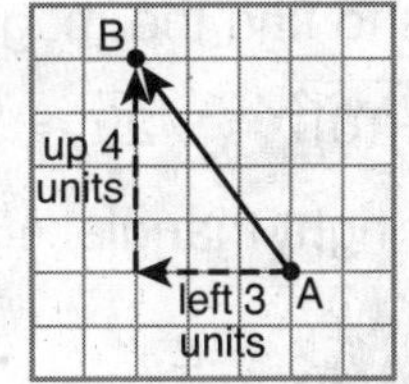

So the component form of $\overrightarrow{AB}$ is $\langle -3, 4 \rangle$.

You can also find the component form of a vector if you know the coordinates of the vector. Suppose $\overrightarrow{JK}$ has coordinates $J(6, 0)$ and $K(1, 3)$.

$\overrightarrow{JK} = \langle x_2 - x_1, y_2 - y_1 \rangle$	Subtract the coordinates of the initial point from the coordinates of the terminal point.
$\overrightarrow{JK} = \langle 1 - 6, 3 - 0 \rangle$	Substitute the coordinates of points *J* and *K*.
$\overrightarrow{JK} = \langle -5, 3 \rangle$	Simplify.

The component form of $\overrightarrow{JK}$ is $\langle -5, 3 \rangle$.

Write each vector in component form.

1. $\overrightarrow{FG}$

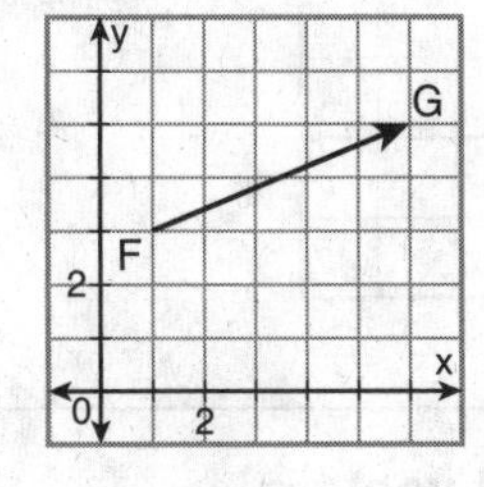

____

2. $\overrightarrow{QR}$

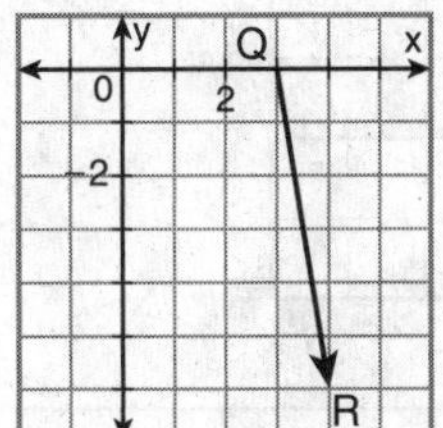

____

3. $\overrightarrow{LM}$ with initial point $L(6, 2)$ and terminal point $M(-1, 5)$

____

4. The vector with initial point $C(0, 5)$ and terminal point $D(2, -3)$

____

Name ______________________ Date __________ Class __________

LESSON 8-6

Review for Mastery

Vectors continued

The **magnitude** of a vector is its length. The magnitude of $\overrightarrow{AB}$ is written $|\overrightarrow{AB}|$. The **direction** of a vector is the angle that it makes with a horizontal line, such as the *x*-axis.

Draw the vector $\langle 5, 2 \rangle$ on a coordinate plane. Find its magnitude and direction.

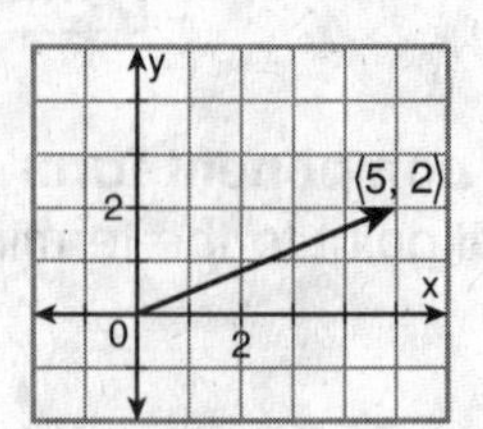

To draw the vector, use the origin as the initial point. Then (5, 2) is the terminal point.

Use the Distance Formula to find the magnitude.

$|\langle 5, 2 \rangle| = \sqrt{(5-0)^2 + (2-0)^2} = \sqrt{29} \approx 5.4$

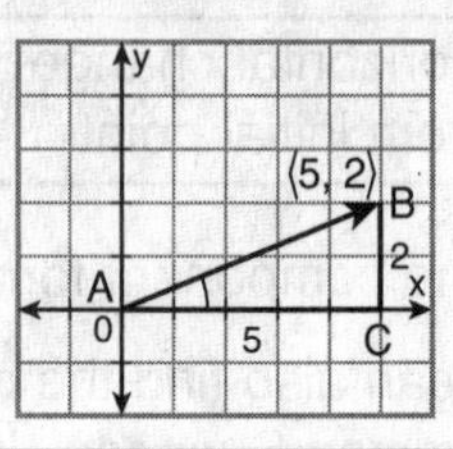

To find the direction, draw right triangle *ABC*. Then find the measure of $\angle A$.

$\tan A = \frac{2}{5}$

$m\angle A = \tan^{-1}\left(\frac{2}{5}\right) \approx 22°$

Find the magnitude of each vector to the nearest tenth.

5. $\langle 3, -1 \rangle$ ______________________

6. $\langle -4, 6 \rangle$ ______________________

Draw each vector on a coordinate plane. Find the direction of each vector to the nearest degree.

7. $\langle 4, 4 \rangle$

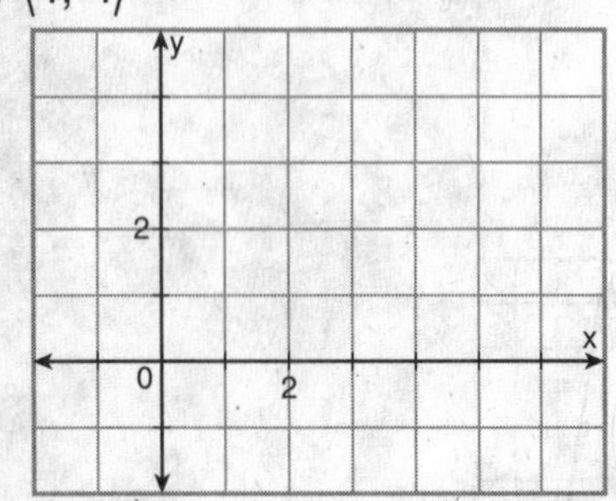

____

8. $\langle 6, 3 \rangle$

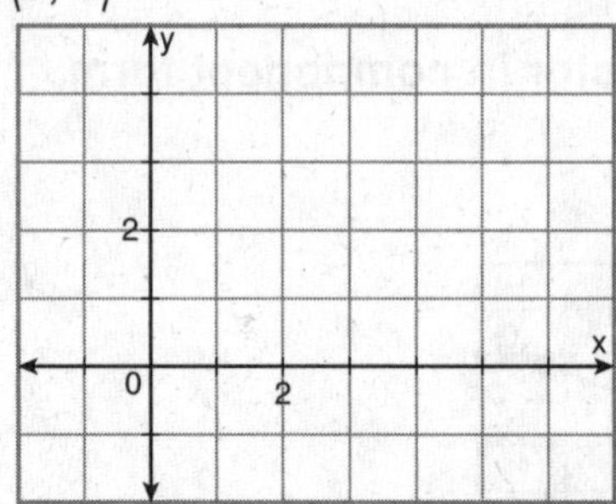

____

Equal vectors have the same magnitude and the same direction. **Parallel vectors** have the same direction or have opposite directions.

Identify each of the following.

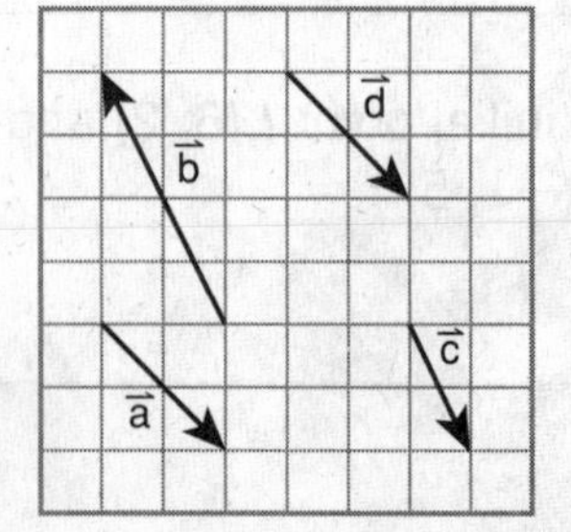

9. equal vectors

____

10. parallel vectors

____

Name ______________________ Date ____________ Class ____________

LESSON 9-1

Review for Mastery

Developing Formulas for Triangles and Quadrilaterals

Area of Triangles and Quadrilaterals		
Parallelogram	**Triangle**	**Trapezoid**
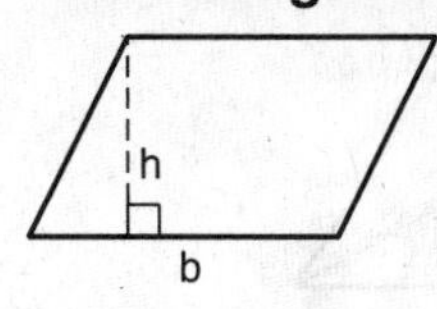	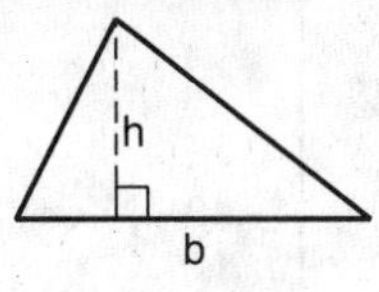	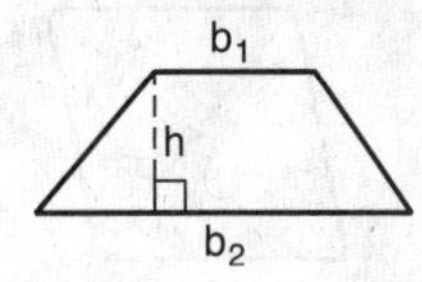
$A = bh$	$A = \frac{1}{2}bh$	$A = \frac{1}{2}(b_1 + b_2)h$

Find the perimeter of the rectangle in which $A = 27\ \text{mm}^2$.

3 mm

Step 1 Find the height.

$A = bh$	Area of a rectangle
$27 = 3h$	Substitute 27 for A and 3 for b.
$9\ \text{mm} = h$	Divide both sides by 3.

Step 2 Use the base and the height to find the perimeter.

$P = 2b + 2h$	Perimeter of a rectangle
$P = 2(3) + 2(9) = 24\ \text{mm}$	Substitute 3 for b and 9 for h.

Find each measurement.

1. the area of the parallelogram

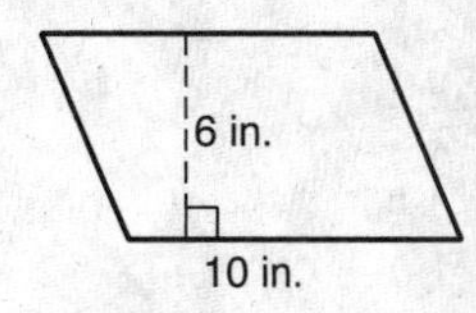

____

2. the base of the rectangle in which $A = 136\ \text{mm}^2$

8 mm

____

3. the area of the trapezoid

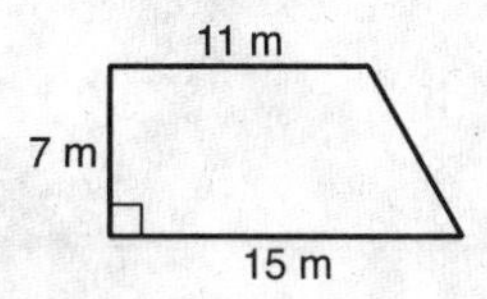

____

4. the height of the triangle in which $A = 192\ \text{cm}^2$

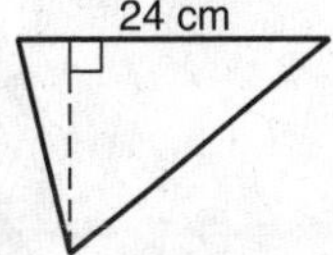

____

5. the perimeter of a rectangle in which $A = 154\ \text{in}^2$ and $h = 11$ in.

____

6. b_2 of a trapezoid in which $A = 5\ \text{ft}^2$, $h = 2$ ft, and $b_1 = 1$ ft

____

Name ______________________ Date ____________ Class ____________

LESSON 9-1

Review for Mastery

Developing Formulas for Triangles and Quadrilaterals continued

Area of Rhombuses and Kites	
Rhombus	**Kite**
$A = \frac{1}{2}d_1d_2$	$A = \frac{1}{2}d_1d_2$

Find d_2 of the kite in which $A = 156$ in².

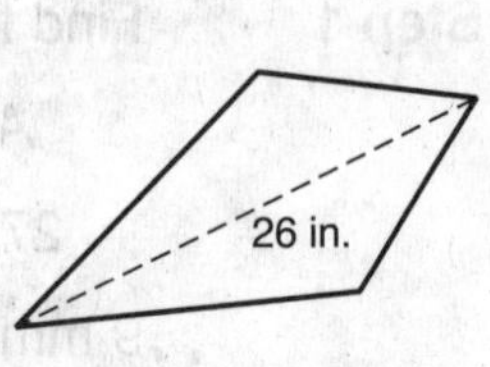

$A = \frac{1}{2}d_1d_2$	Area of a kite
$156 = \frac{1}{2}(26)d_2$	Substitute 156 in² for A and 26 in. for d_1.
$156 = 13d_2$	Simplify.
$12 \text{ in.} = d_2$	Divide both sides by 13.

Find each measurement.

7. the area of the rhombus

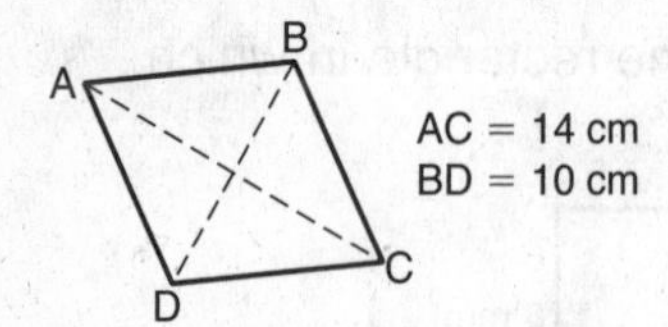

____

8. d_1 of the kite in which $A = 414$ ft²

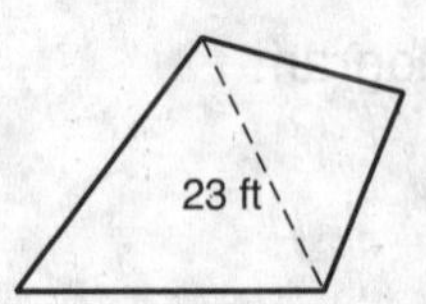

____

9. d_2 of the rhombus in which $A = 90$ m²

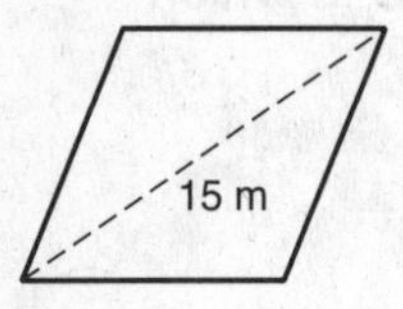

____

10. d_1 of the kite in which $A = 39$ mm²

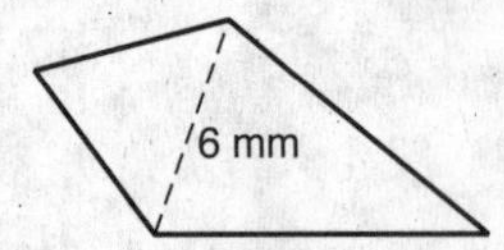

____

11. d_1 of a kite in which $A = 16x$ m² and $d_2 = 8$ m

____

12. the area of a rhombus in which $d_1 = 4ab$ in. and $d_2 = 7a$ in.

____

Name ______________________________ Date ____________ Class ____________

LESSON 9-2

Review for Mastery

Developing Formulas for Circles and Regular Polygons

Circumference and Area of Circles

A circle with diameter d and radius r has circumference $C = \pi d$ or $C = 2\pi r$.

A circle with radius r has area $A = \pi r^2$.

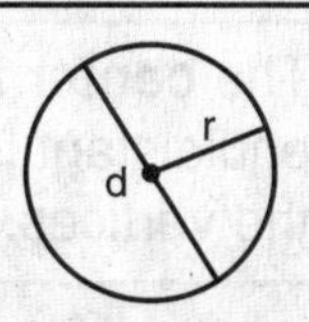

Find the circumference of circle S in which $A = 81\pi$ cm^2.

Step 1 Use the given area to solve for r.

$A = \pi r^2$	Area of a circle
$81\pi \text{ cm}^2 = \pi r^2$	Substitute 81π for A.
$81 \text{ cm}^2 = r^2$	Divide both sides by π.
$9 \text{ cm} = r$	Take the square root of both sides.

S (r cm)

Step 2 Use the value of r to find the circumference.

$C = 2\pi r$	Circumference of a circle
$C = 2\pi(9 \text{ cm}) = 18\pi \text{ cm}$	Substitute 9 cm for r and simplify.

Find each measurement.

1. the circumference of circle B

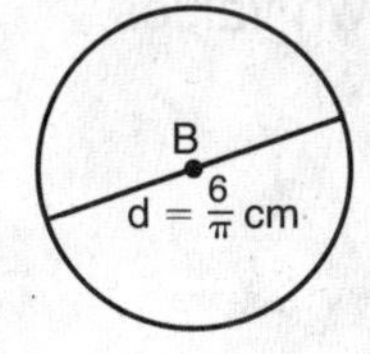

____

2. the area of circle R in terms of π

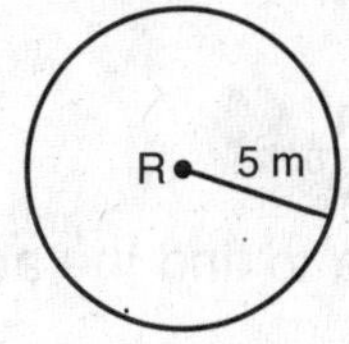

____

3. the area of circle Z in terms of π

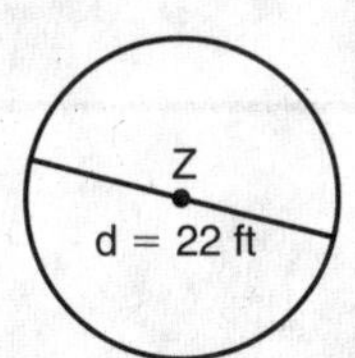

____

4. the circumference of circle T in terms of π

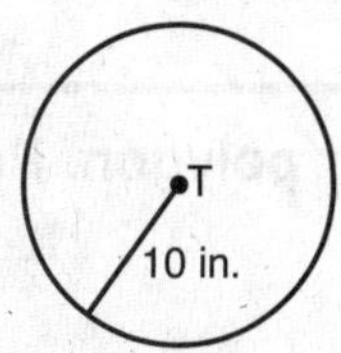

____

5. the circumference of circle X in which $A = 49\pi$ in^2

____

6. the radius of circle Y in which $C = 18\pi$ cm

____

Name ______________________ Date __________ Class __________

LESSON 9-2

Review for Mastery

Developing Formulas for Circles and Regular Polygons continued

Area of Regular Polygons

The area of a regular polygon with apothem a and perimeter P is $A = \frac{1}{2}aP$.

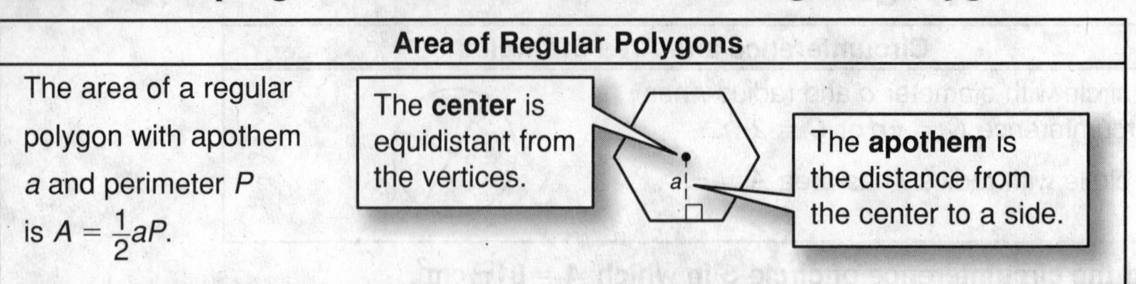

Find the area of a regular hexagon with side length 10 cm.

Step 1 Draw a figure and find the measure of a central angle. Each central angle measure of a regular n-gon is $\frac{360°}{n}$.

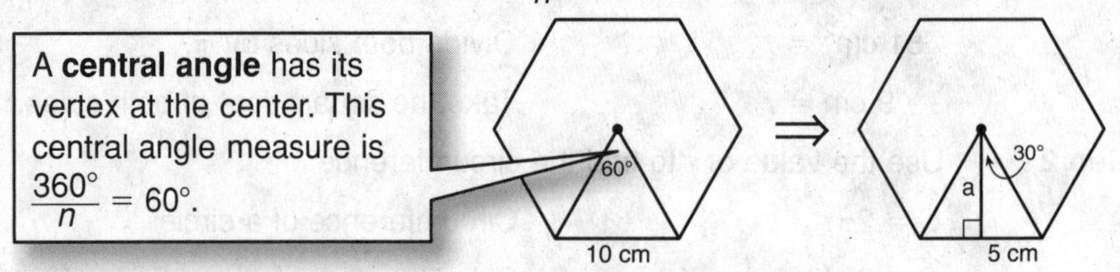

Step 2 Use the tangent ratio to find the apothem. You could also use the 30°-60°-90° △ Thm. in this case.

$\tan 30° = \frac{\text{leg opposite } 30° \text{ angle}}{\text{leg adjacent to } 30° \text{ angle}}$ Write a tangent ratio.

$\tan 30° = \frac{5 \text{ cm}}{a}$ Substitute the known values.

$a = \frac{5 \text{ cm}}{\tan 30°}$ Solve for a.

Step 3 Use the formula to find the area.

$A = \frac{1}{2}aP$

$A = \frac{1}{2}\left(\frac{5}{\tan 30°}\right)60$ $a = \frac{5}{\tan 30°}$, $P = 6 \times 10$ or 60 cm

$A \approx 259.8 \text{ cm}^2$ Simplify.

Find the area of each regular polygon. Round to the nearest tenth.

7.

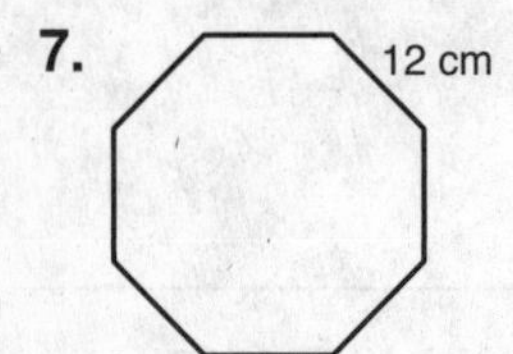

8.

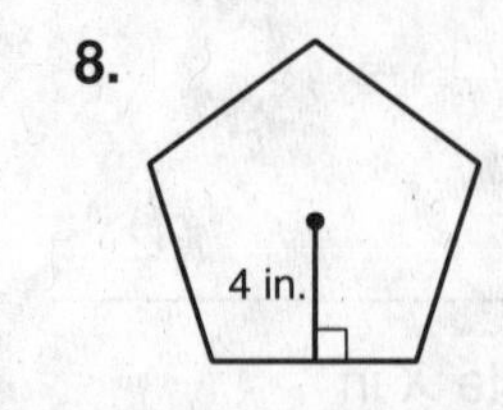

_______________ _______________

9. a regular hexagon with an apothem of 3 m

10. a regular decagon with a perimeter of 70 ft

_______________ _______________

Name ________________________ Date ____________ Class ____________

LESSON 9-3

Review for Mastery

Composite Figures

The figure at right is called a **composite figure** because it is made up of simple shapes. To find its area, first find the areas of the simple shapes and then add.

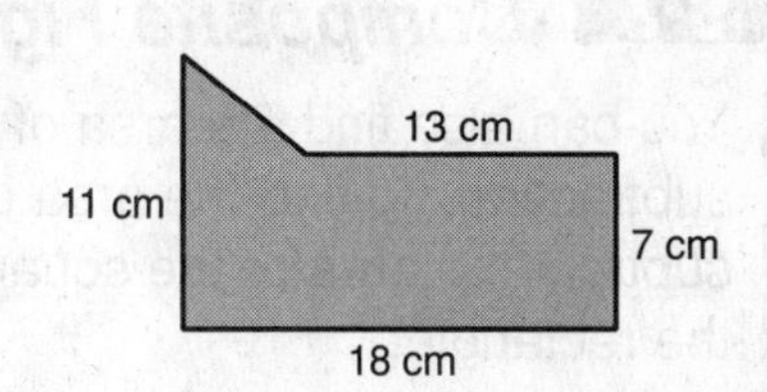

Divide the figure into a triangle and a rectangle.

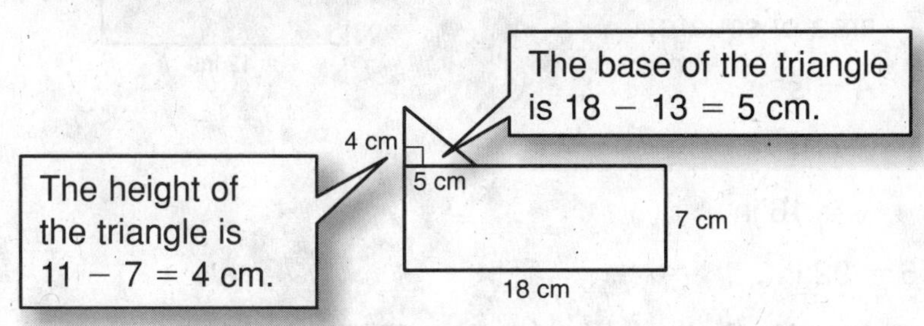

area of triangle: $A = \frac{1}{2}bh$

$= \frac{1}{2}(5)(4)$

$= 10 \text{ cm}^2$

area of rectangle: $A = bh$

$= 18(7)$

$= 126 \text{ cm}^2$

The area of the whole figure is $10 + 126 = 136 \text{ cm}^2$.

Find the shaded area. Round to the nearest tenth if necessary.

1. 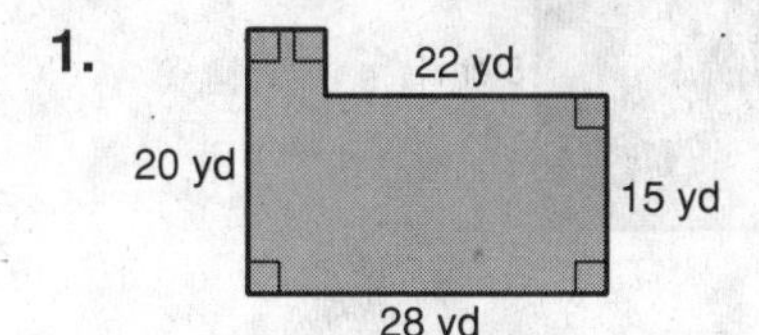

____

2. 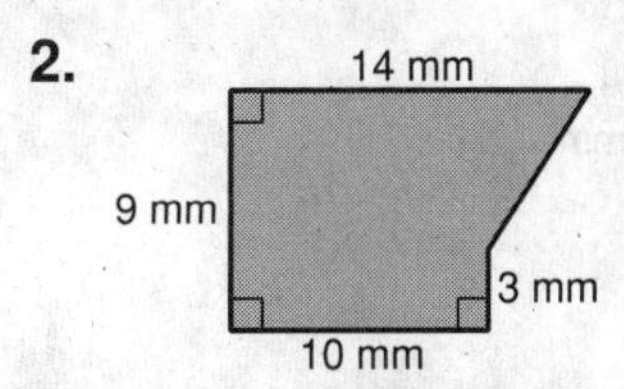

____

3. 16 ft
16 ft
16 ft

____

4. 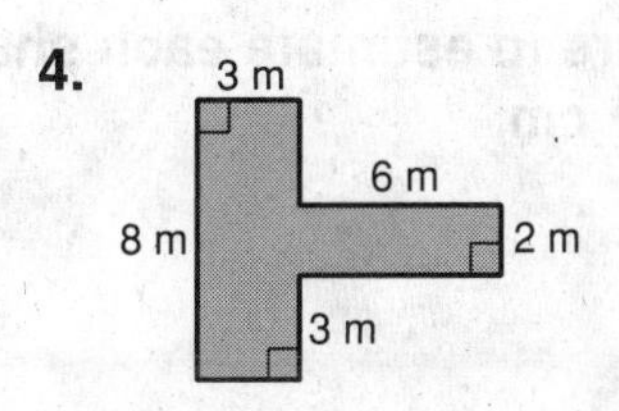

____

Name ______________________ Date __________ Class __________

Review for Mastery

Composite Figures continued

You can also find the area of composite figures by using subtraction. To find the area of the figure at right, subtract the area of the square from the area of the rectangle.

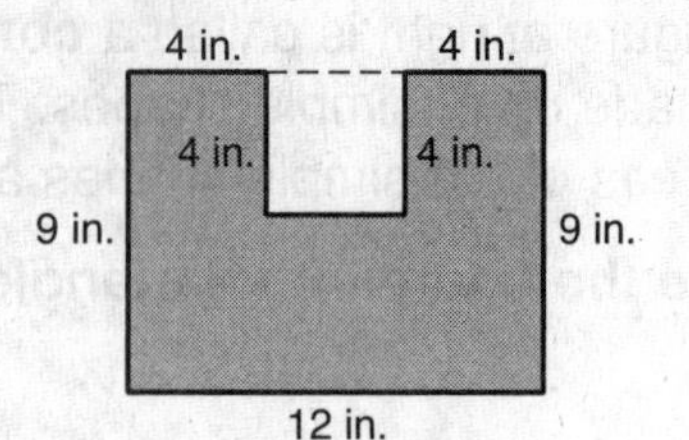

area of rectangle:

$A = bh$

$= 12(9)$

$= 108 \text{ in}^2$

area of square:

$A = s^2$

$= 4^2$

$= 16 \text{ in}^2$

The shaded area is $108 - 16 = 92 \text{ in}^2$.

You can use composite figures to estimate the area of an irregular shape like the one shown at right. The grid has squares with side lengths of 1 cm.

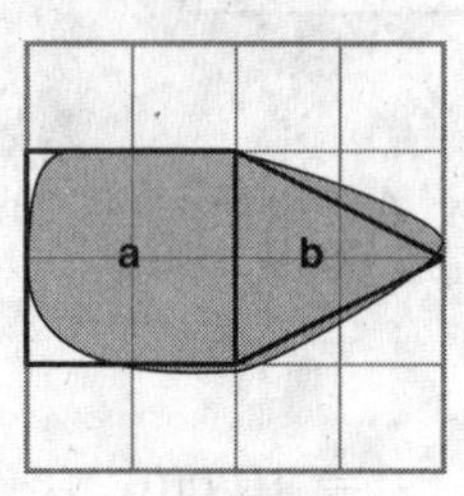

area of square *a*: $A = 2 \cdot 2 = 4 \text{ cm}^2$

area of triangle *b*: $A = \frac{1}{2}(2)(2) = 2 \text{ cm}^2$

The shaded area is about $4 + 2$ or 6 cm^2.

Find the shaded area. Round to the nearest tenth if necessary.

5.

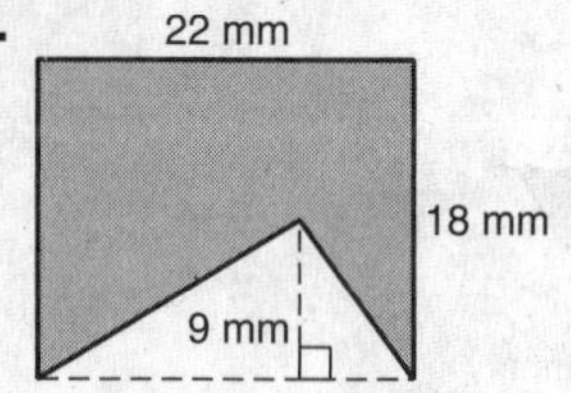

____

6.

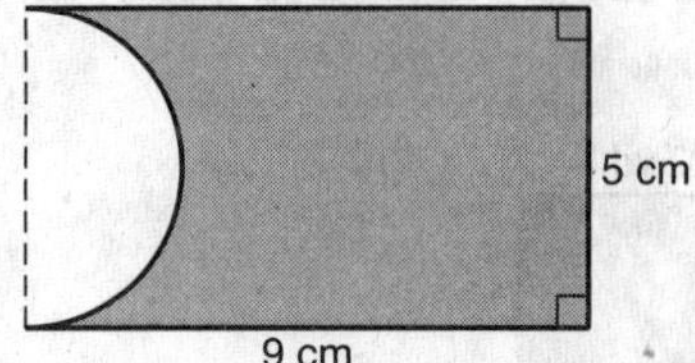

____

Use a composite figure to estimate each shaded area. The grid has squares with side lengths of 1 cm.

7.

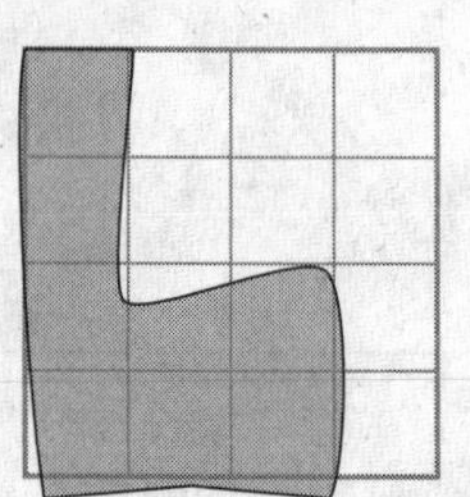

____

8.

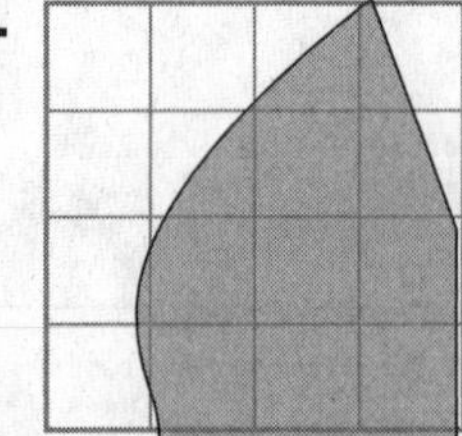

____

Name ________________________ Date ____________ Class ____________

LESSON 9-4

Review for Mastery

Perimeter and Area in the Coordinate Plane

One way to estimate the area of irregular shapes in the coordinate plane is to count the squares on the grid. You can estimate the number of whole squares and the number of half squares and then add.

The polygon with vertices $A(-3, -1)$, $B(-3, 3)$, $C(2, 3)$, and $D(4, -1)$ is drawn in the coordinate plane. The figure is a trapezoid. Use the Distance Formula to find the length of $\overline{CD}$.

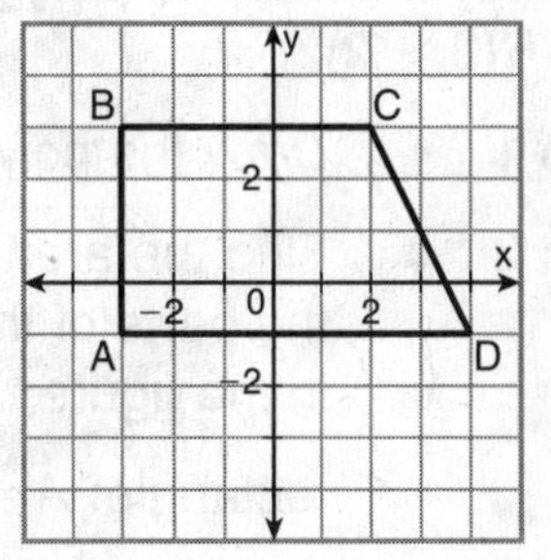

$CD = \sqrt{(4-2)^2 + (-1-3)^2} = \sqrt{20} = 2\sqrt{5}$

perimeter of $ABCD$: $P = AB + BC + CD + DA$

$= 4 + 5 + 2\sqrt{5} + 7$

≈ 20.5 units

area of $ABCD$: $A = \frac{1}{2}(b_1 + b_2)(h)$

$= \frac{1}{2}(5 + 7)(4) = 24 \text{ units}^2$

Estimate the area of each irregular shape.

1.

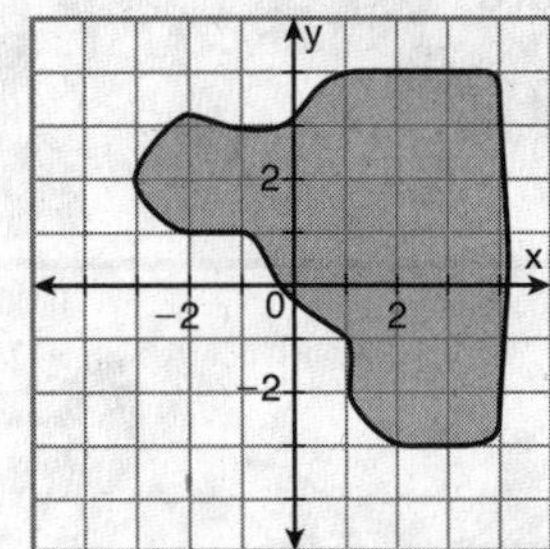

2.

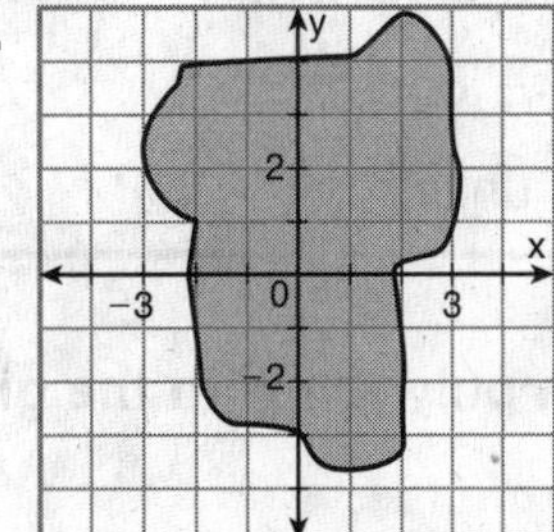

________________________ ________________________

Draw and classify each polygon with the given vertices. Find the perimeter and area of each polygon.

3. $F(-2, -3)$, $G(-2, 3)$, $H(2, 0)$

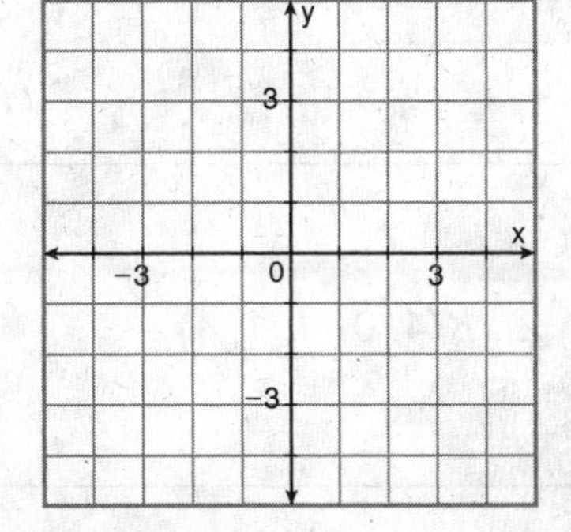

4. $Q(-4, 0)$, $R(-2, 4)$, $S(2, 2)$, $T(0, -2)$

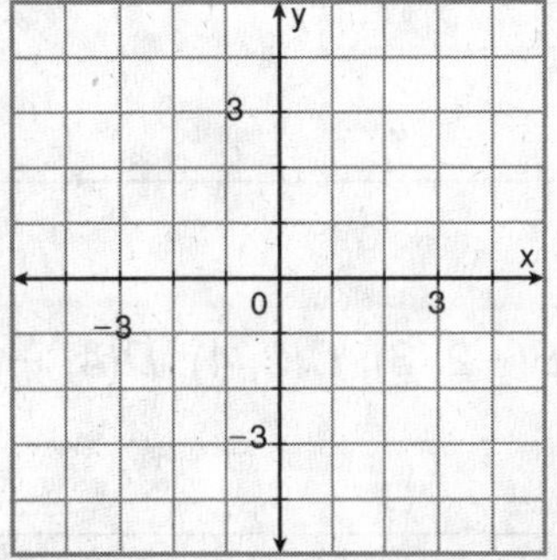

________________________ ________________________

________________________ ________________________

Name ______________________ Date __________ Class __________

LESSON 9-4

Review for Mastery

Perimeter and Area in the Coordinate Plane continued

When a figure in a coordinate plane does not have an area formula, another method can be used to find its area.

Find the area of the polygon with vertices $N(-4, -1)$, $P(-1, 3)$, $Q(4, 3)$, and $R(2, -2)$.

Step 1 Draw the polygon and enclose it in a rectangle.

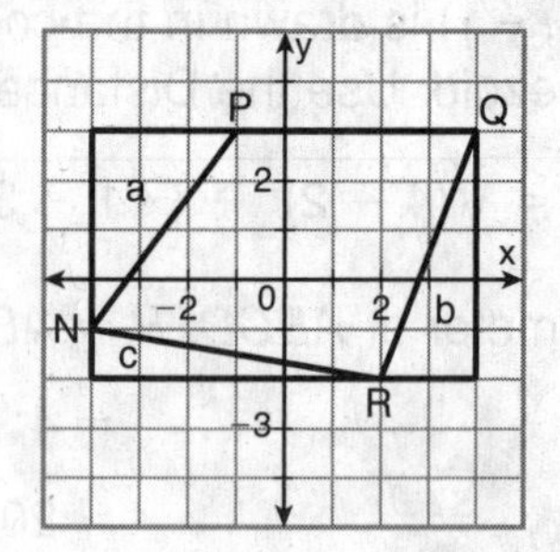

Step 2 Find the area of the rectangle and the areas of the parts of the rectangle that are not included in the figure.

rectangle: $A = bh = 8 \cdot 5 = 40$ units^2

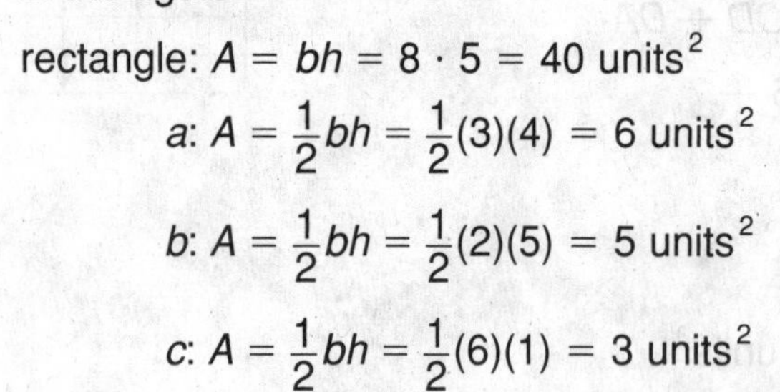

a: $A = \frac{1}{2}bh = \frac{1}{2}(3)(4) = 6$ units^2

b: $A = \frac{1}{2}bh = \frac{1}{2}(2)(5) = 5$ units^2

c: $A = \frac{1}{2}bh = \frac{1}{2}(6)(1) = 3$ units^2

Step 3 Subtract to find the area of polygon *NPQR*.

A = area of rectangle − area of parts not included in figure

$= 40 - 6 - 5 - 3$

$= 26$ units^2

Find the area of each polygon with the given vertices.

5.

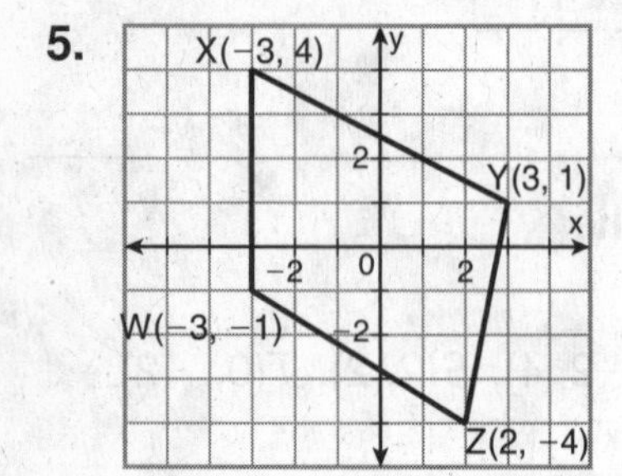

____

6.

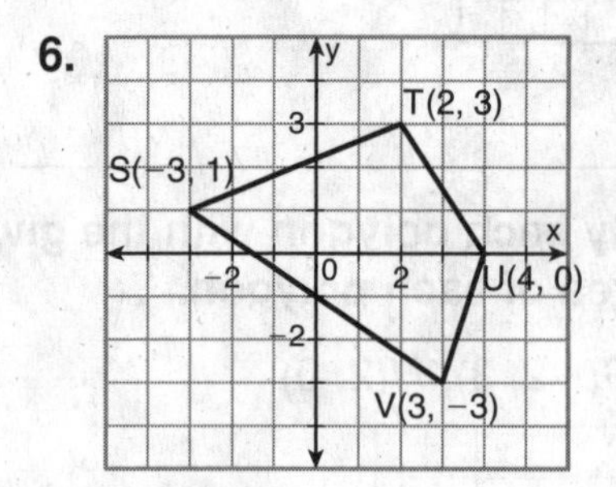

____

7. $A(-1, -1)$, $B(-2, 3)$, $C(2, 4)$, $D(4, -1)$

____

8. $H(3, 7)$, $J(7, 2)$, $K(4, 0)$, $L(1, 1)$

____

Name ________________________ Date ____________ Class ____________

LESSON 9-5

Review for Mastery

Effects of Changing Dimensions Proportionally

What happens to the area of the parallelogram if the base is tripled?

original dimensions:

$A = bh$

$= 4(5)$

$= 20 \text{ cm}^2$

triple the base:

$A = bh$

$= 12(5)$

$= 60 \text{ cm}^2$

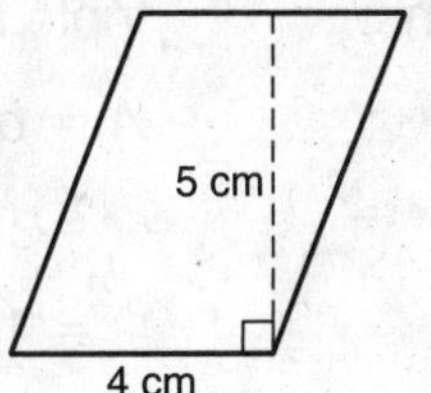

Notice that $60 = 3(20)$. If the base is multiplied by 3, the area is also multiplied by 3.

Describe the effect of each change on the area of the given figure.

1. The length of the rectangle is doubled.

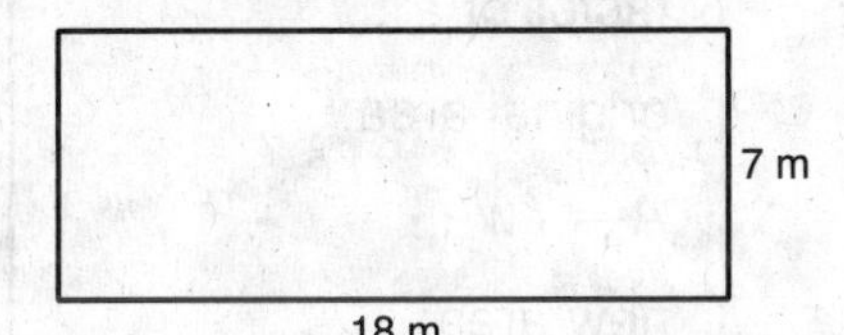

____

2. The base of the triangle is multiplied by 4.

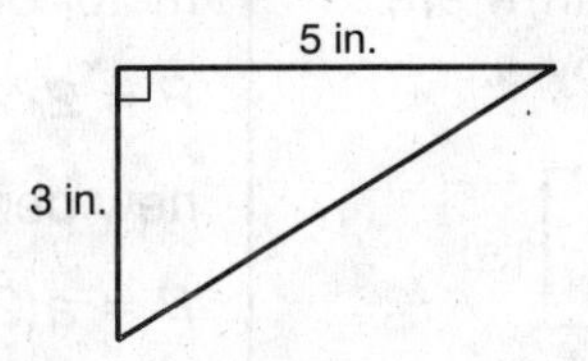

____

3. The height of the parallelogram is multiplied by 5.

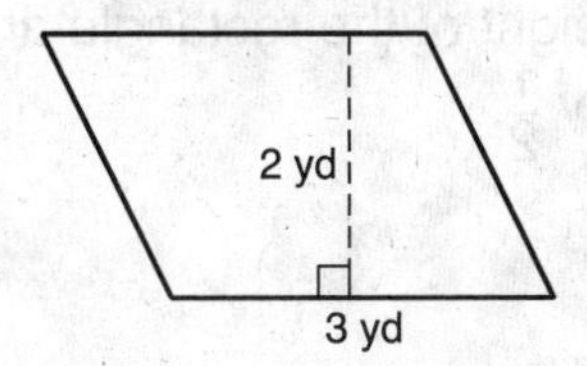

____

4. The width of the rectangle is multiplied by $\frac{1}{2}$.

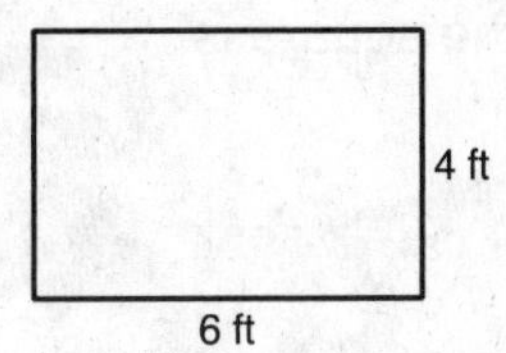

____

5. The height of the trapezoid is multiplied by 3.

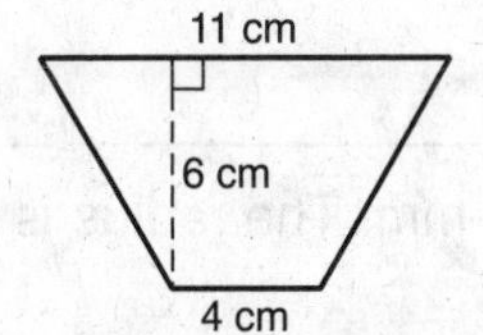

____

6. The radius of the circle is multiplied by $\frac{1}{2}$.

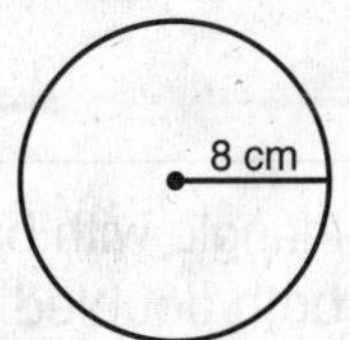

____

Name ______________________ Date ____________ Class ____________

LESSON 9-5

Review for Mastery

Effects of Changing Dimensions Proportionally continued

What happens if both the base and height of the parallelogram are tripled?

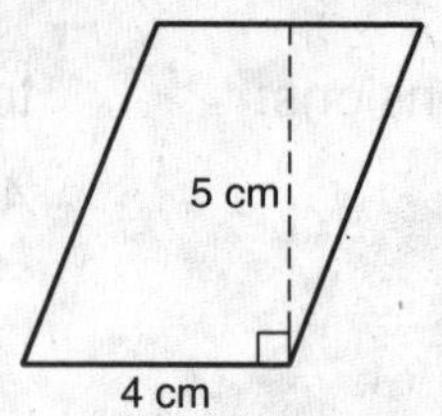

original dimensions:

$A = bh$

$= 4(5)$

$= 20 \text{ cm}^2$

triple the base and height:

$A = bh$

$= 12(15)$

$= 180 \text{ cm}^2$

When just the base is multiplied by 3, the area is also multiplied by 3. When both the base and height are multiplied by 3, the area is multiplied by 3^2, or 9.

Effects of Changing Dimensions Proportionally		
Change in Dimensions	**Perimeter or Circumference**	**Area**
Consider a rectangle whose length ℓ and width w are each multiplied by a. [rectangle with sides a(w) and a(ℓ)]	The perimeter changes by a factor of a. $P = 2\ell + 2w$ new perimeter: $P = a(2\ell + 2w)$	The area changes by a factor of a^2. original area: $A = \ell w$ new area: $A = a^2(\ell w)$

Describe the effect of each change on the perimeter or circumference and the area of the given figure.

7. The side length of the square is multiplied by 6.

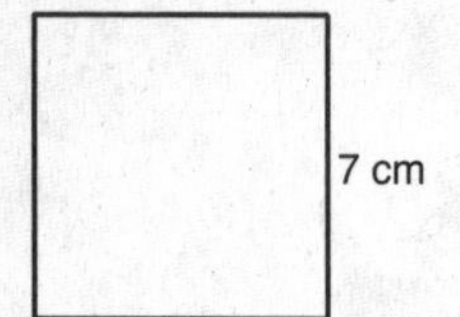

____

____

8. The base and height of the rectangle are both multiplied by $\frac{1}{2}$.

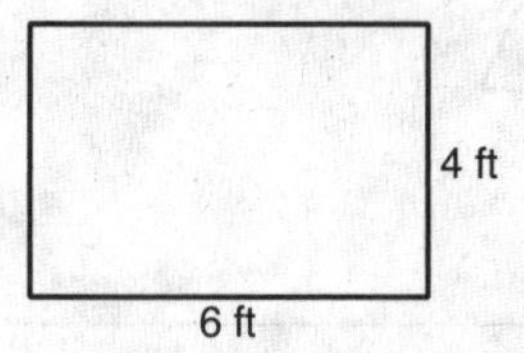

____

____

9. The base and height of a triangle with base 7 in. and height 3 in. are both doubled.

____

____

10. A circle has radius 5 mm. The radius is multiplied by 4.

____

____

Name ________________________ Date __________ Class __________

LESSON 9-6

Review for Mastery

Geometric Probability

The theoretical probability of an event occurring is

$$P = \frac{\text{number of outcomes in the event}}{\text{number of outcomes in the sample space}}.$$

The **geometric probability** of an event occurring is found by determining a ratio of geometric measures such as length or area. Geometric probability is used when an experiment has an infinite number of outcomes.

Finding Geometric Probability

Use Length

A point is chosen randomly on $\overline{AD}$. Find the probability that the point is on $\overline{BD}$.

A — 2 — B — 4 — C — 6 — D

$$P = \frac{\text{all points on } \overline{BD}}{\text{all points on } \overline{AD}}$$

$$= \frac{BD}{AD}$$

$$= \frac{10}{12} = \frac{5}{6}$$

Use Angle Measures

Use the spinner to find the probability of the pointer landing on the 160° space.

Spinner: 80°, 160°, 120°

$$P = \frac{\text{all points in 160° region}}{\text{all points in circle}}$$

$$= \frac{160}{360}$$

$$= \frac{4}{9}$$

A point is chosen randomly on $\overline{EH}$. Find the probability of each event.

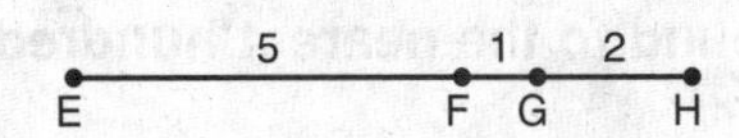

1. The point is on $\overline{FH}$. ____________________

2. The point is not on $\overline{EF}$. ____________________

3. The point is on $\overline{EF}$ or $\overline{GH}$. ____________________

4. The point is on $\overline{EG}$. ____________________

Use the spinner to find the probability of each event.

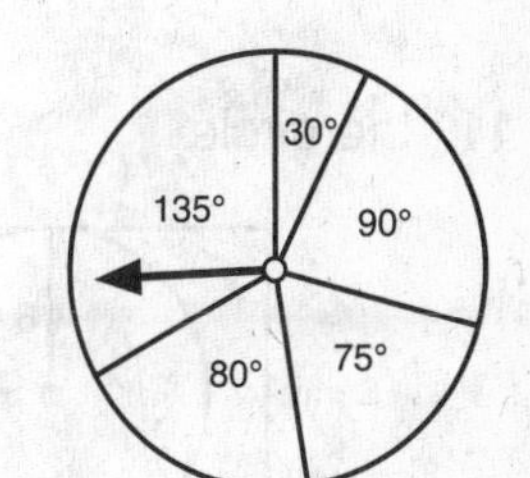

5. the pointer landing on 135° ____________________

6. the pointer landing on 75° ____________________

7. the pointer landing on 90° or 75° ____________________

8. the pointer landing on 30° ____________________

Name ______________________ Date ____________ Class ____________

LESSON 9-6

Review for Mastery

Geometric Probability continued

You can also use area to find geometric probability.

Find the probability that a point chosen randomly inside the rectangle is in the triangle.

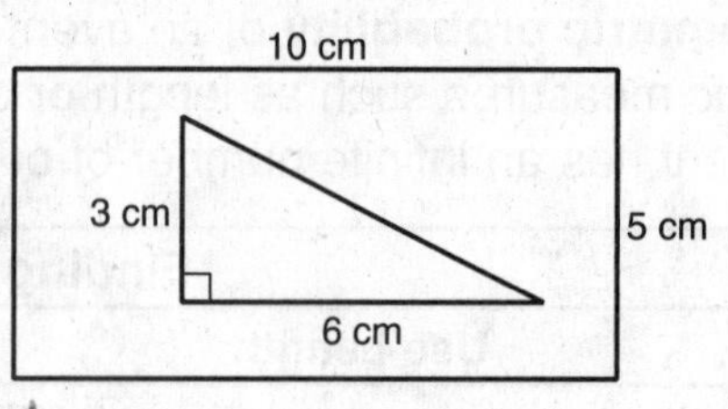

area of triangle: $A = \frac{1}{2}bh$

$= \frac{1}{2}(6)(3) = 9 \text{ cm}^2$

area of rectangle: $A = bh$

$= 10(5) = 50 \text{ cm}^2$

$P = \frac{\text{all points in triangle}}{\text{all points in rectangle}}$

$= \frac{\text{area of triangle}}{\text{area of rectangle}}$

$= \frac{9 \text{ cm}^2}{50 \text{ cm}^2}$

The probability is $P = 0.18$.

Find the probability that a point chosen randomly inside the rectangle is in each shape. Round to the nearest hundredth.

9. the square

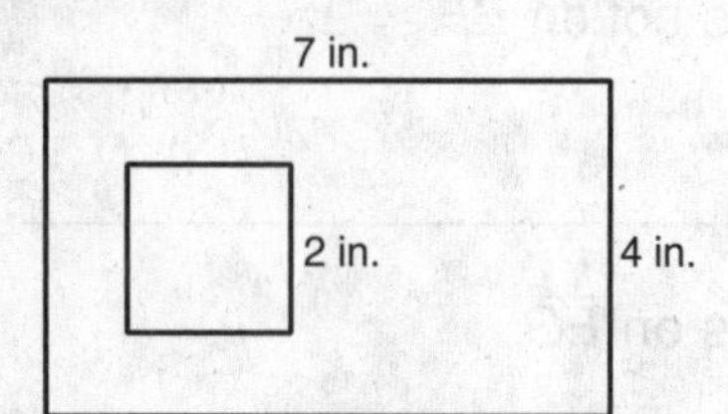

____

10. the triangle

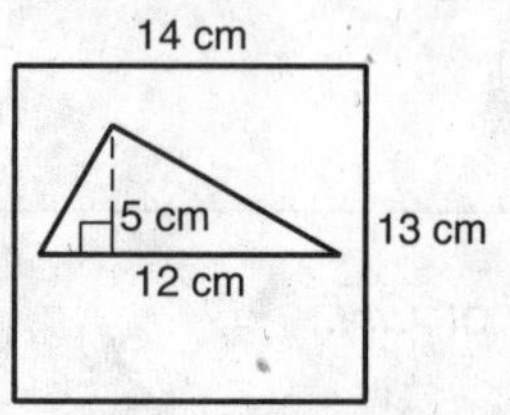

____

11. the circle

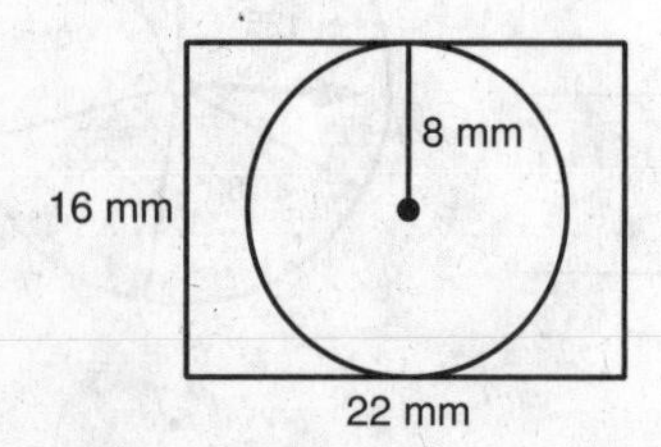

____

12. the regular pentagon

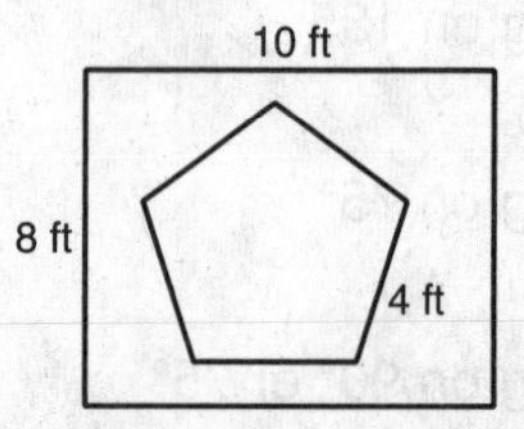

____

Name ______________________ Date ____________ Class ____________

LESSON 10-1

Review for Mastery

Solid Geometry

Three-dimensional figures, or *solids,* can have flat or curved surfaces.

Prisms and pyramids are named by the shapes of their *bases.*

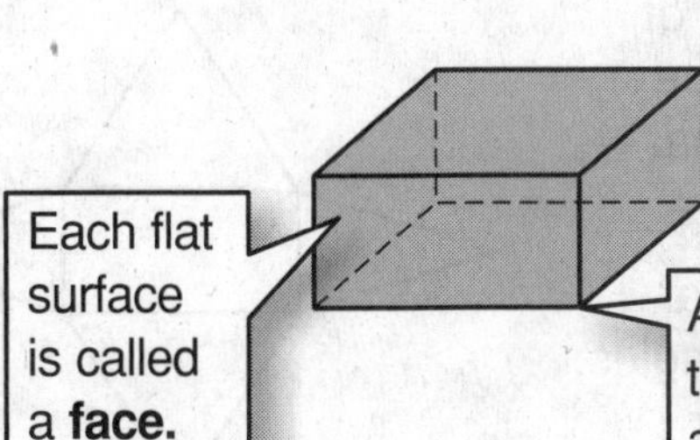

Solids			
Prisms	Pyramids	Cylinder	Cone
bases triangular prism, rectangular prism	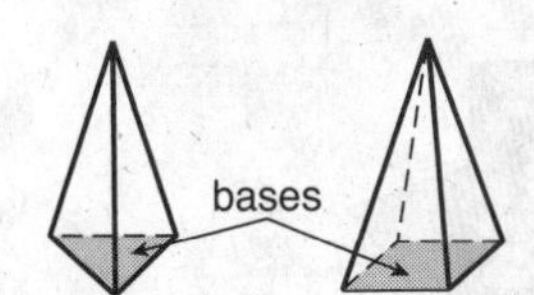 bases triangular pyramid, rectangular pyramid	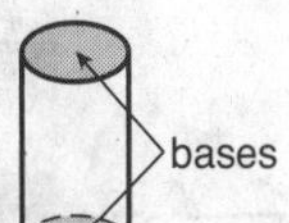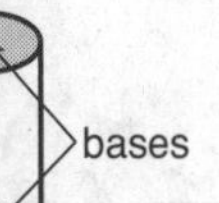 bases	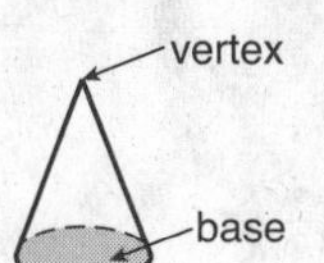 vertex base
		Neither cylinders nor cones have edges.	

Classify each figure. Name the vertices, edges, and bases.

1. 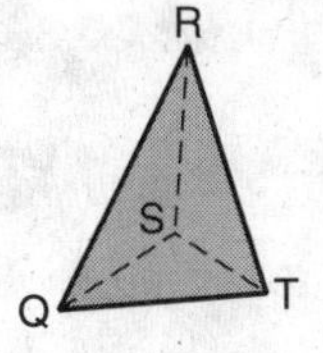

____

____

____

2. 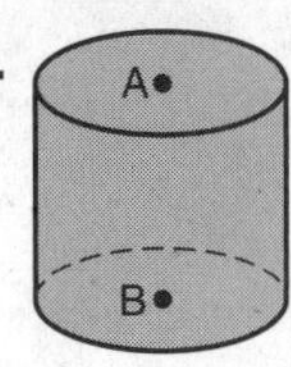

____

____

____

3. 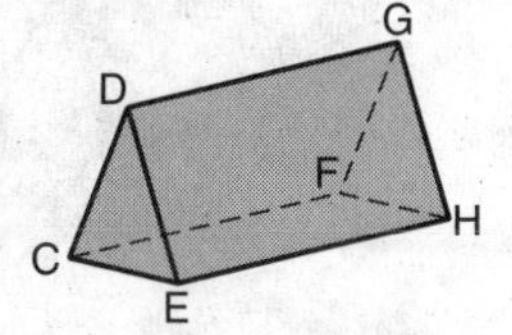

____

____

____

4. 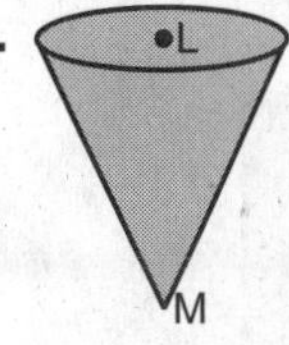

____

____

____

Name ______________________ Date ____________ Class ____________

LESSON **10-1**

Review for Mastery

Solid Geometry continued

A **net** is a diagram of the surfaces of a three-dimensional figure. It can be folded to form the three-dimensional figure.

The net at right has one rectangular face. The remaining faces are triangles, and so the net forms a rectangular pyramid.

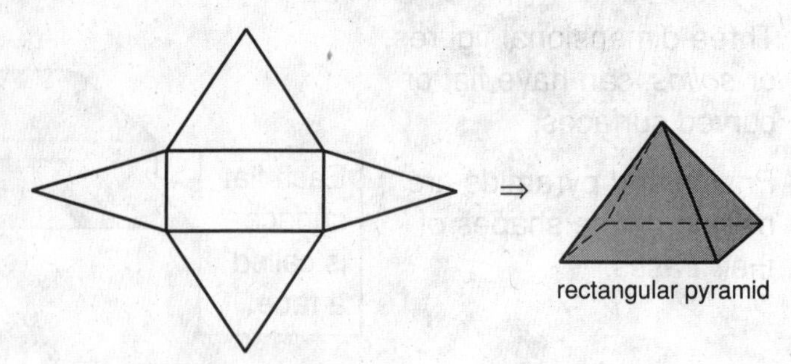

A **cross section** is the intersection of a three-dimensional figure and a plane.

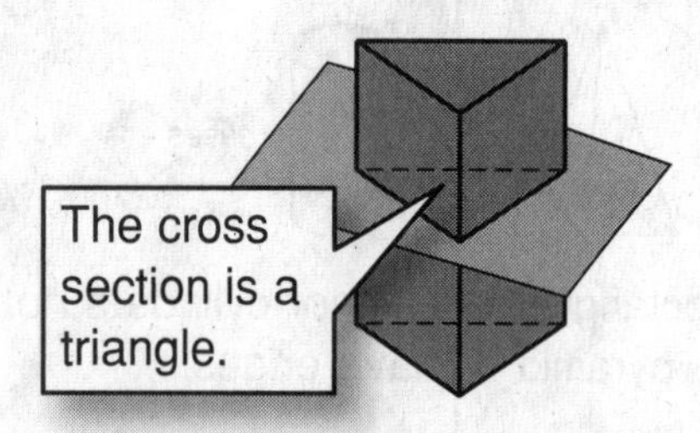

Describe the three-dimensional figure that can be made from the given net.

5.

____

6.

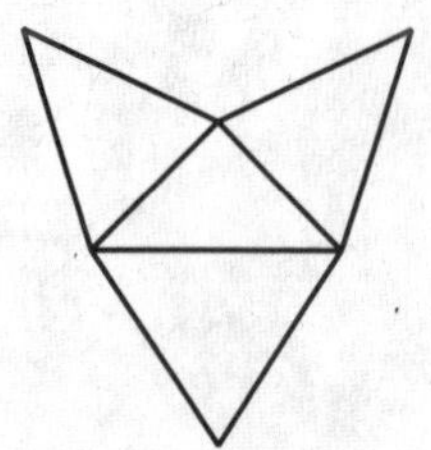

____

Describe each cross section.

7.

____

8.

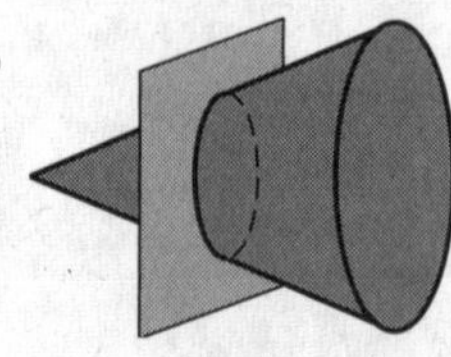

____

Name ______________________ Date __________ Class __________

LESSON 10-2

Review for Mastery

Representations of Three-Dimensional Figures

An **orthographic drawing** of a three-dimensional object shows six different views of the object. The six views of the figure at right are shown below.

Top:

Bottom:

Front:

Back:

Left:

Right:

Draw all six orthographic views of each object. Assume there are no hidden cubes.

1.

2.

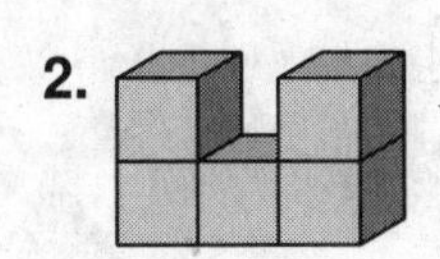

Name ______________________ Date __________ Class __________

LESSON 10-2

Review for Mastery

Representations of Three-Dimensional Figures continued

An **isometric drawing** is drawn on isometric dot paper and shows three sides of a figure from a corner view. A solid and an isometric drawing of the solid are shown.

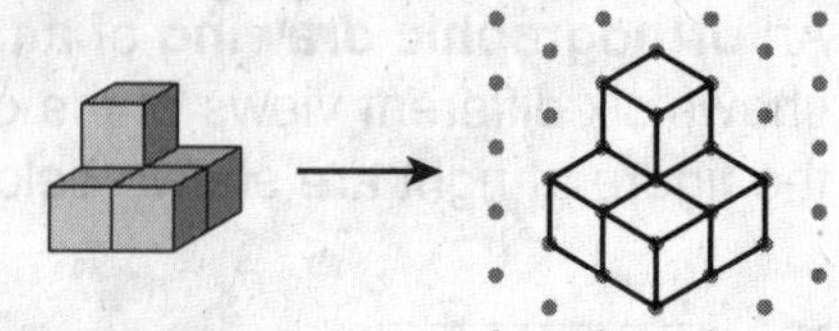

In a **one-point perspective drawing,** nonvertical lines are drawn so that they meet at a **vanishing point.** You can make a one-point perspective drawing of a triangular prism.

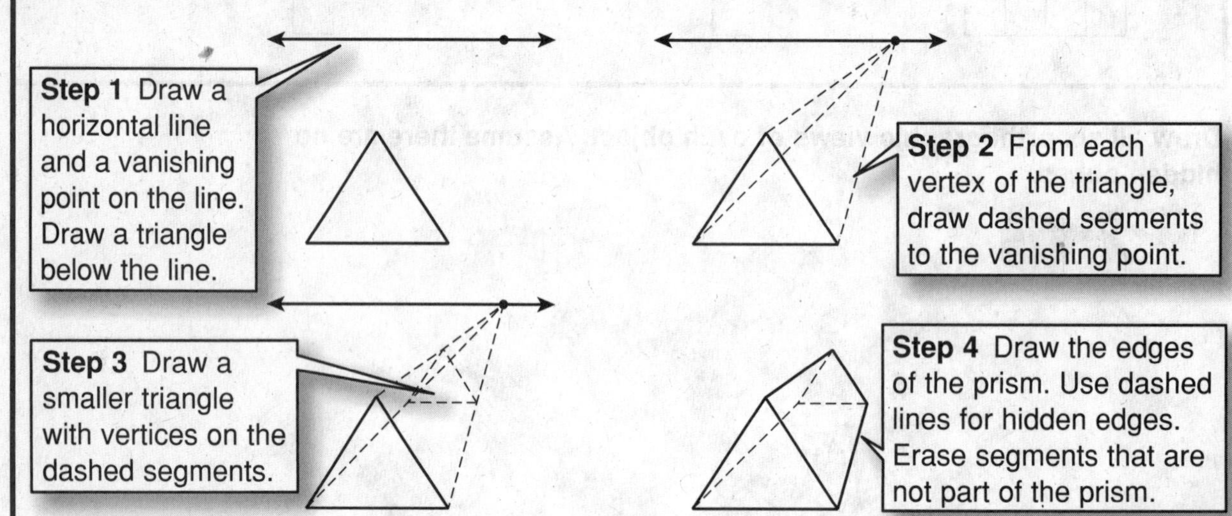

Draw an isometric view of each object. Assume there are no hidden cubes.

3.

4.

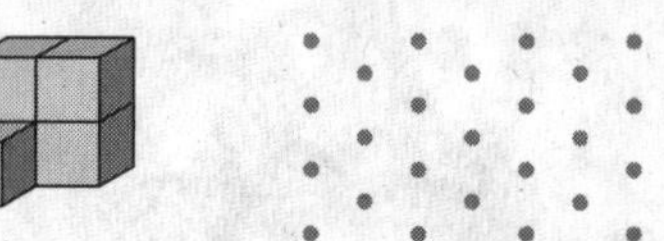

Draw each object in one-point perspective.

5. a triangular prism with bases that are obtuse triangles

6. a rectangular prism

Name ______________________ Date ____________ Class ____________

LESSON 10-3

Review for Mastery

Formulas in Three Dimensions

A **polyhedron** is a solid formed by four or more polygons that intersect only at their edges. Prisms and pyramids are polyhedrons. Cylinders and cones are not.

Euler's Formula	
For any polyhedron with V vertices, E edges, and F faces, $V - E + F = 2$.	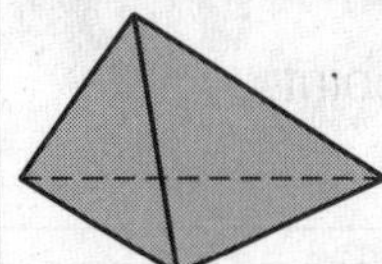 **Example** $V - E + F = 2$ Euler's Formula; $4 - 6 + 4 = 2$ $V = 4, E = 6, F = 4$; $2 = 2$; 4 vertices, 6 edges, 4 faces

Diagonal of a Right Rectangular Prism
The length of a diagonal d of a right rectangular prism with length ℓ, width w, and height h is $d = \sqrt{\ell^2 + w^2 + h^2}$. 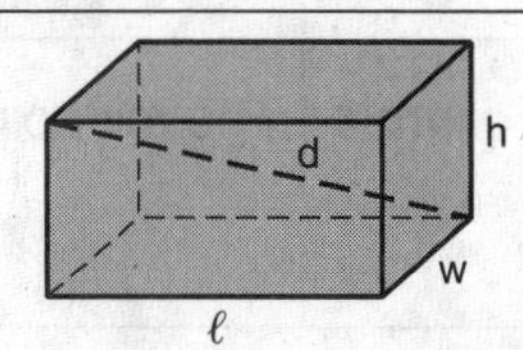

Find the height of a rectangular prism with a 4 cm by 3 cm base and a 7 cm diagonal.

$d = \sqrt{\ell^2 + w^2 + h^2}$	Formula for the diagonal of a right rectangular prism
$7 = \sqrt{4^2 + 3^2 + h^2}$	Substitute 7 for d, 4 for ℓ, and 3 for w.
$49 = 4^2 + 3^2 + h^2$	Square both sides of the equation.
$24 = h^2$	Simplify.
$4.9 \text{ cm} \approx h$	Take the square root of each side.

Find the number of vertices, edges, and faces of each polyhedron. Use your results to verify Euler's Formula.

1. 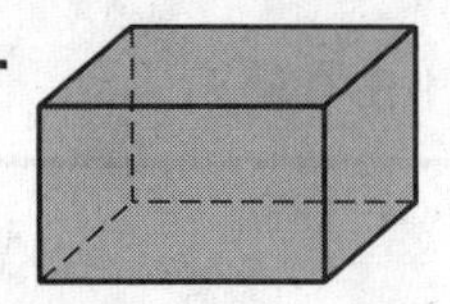

____

____

2. 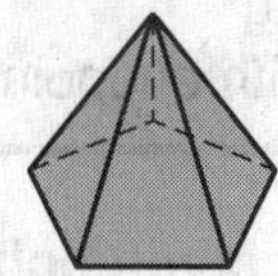

____

____

Find the unknown dimension in each figure. Round to the nearest tenth if necessary.

3. the length of the diagonal of a 6 cm by 8 cm by 11 cm rectangular prism

____

4. the height of a rectangular prism with a 4 in. by 5 in. base and a 9 in. diagonal

____

Name ______________________ Date ____________ Class ____________

LESSON 10-3

Review for Mastery

Formulas in Three Dimensions continued

A three-dimensional coordinate system has three perpendicular axes:

- *x*-axis
- *y*-axis
- *z*-axis

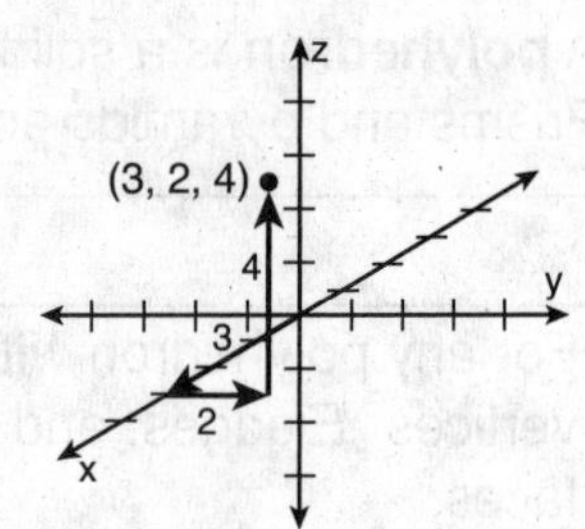

An *ordered triple* (x, y, z) is used to locate a point.
The point at (3, 2, 4) is graphed at right.

Formulas in Three Dimensions	
Distance Formula	The distance between the points (x_1, y_1, z_1) and (x_2, y_2, z_2) is $d = \sqrt{(x_2 - x_1)^2 + (y_2 - y_1)^2 + (z_2 - z_1)^2}$.
Midpoint Formula	The midpoint of the segment with endpoints (x_1, y_1, z_1) and (x_2, y_2, z_2) is $M\left(\frac{x_1 + x_2}{2}, \frac{y_1 + y_2}{2}, \frac{z_1 + z_2}{2}\right)$.

Find the distance between the points (4, 0, 1) and (2, 3, 0). Find the midpoint of the segment with the given endpoints.

$d = \sqrt{(x_2 - x_1)^2 + (y_2 - y_1)^2 + (z_2 - z_1)^2}$ Distance Formula

$= \sqrt{(2 - 4)^2 + (3 - 0)^2 + (0 - 1)^2}$ $(x_1, y_1, z_1) = (4, 0, 1), (x_2, y_2, z_2) = (2, 3, 0)$

$= \sqrt{4 + 9 + 1}$ Simplify.

$= \sqrt{14} \approx 3.7$ units Simplify.

The distance between the points (4, 0, 1) and (2, 3, 0) is about 3.7 units.

$M\left(\frac{x_1 + x_2}{2}, \frac{y_1 + y_2}{2}, \frac{z_1 + z_2}{2}\right) = M\left(\frac{4 + 2}{2}, \frac{0 + 3}{2}, \frac{1 + 0}{2}\right)$ Midpoint Formula

$= M(3, 1.5, 0.5)$ Simplify.

The midpoint of the segment with endpoints (4, 0, 1) and (2, 3, 0) is $M(3, 1.5, 0.5)$.

Find the distance between the given points. Find the midpoint of the segment with the given endpoints. Round to the nearest tenth if necessary.

5. (0, 0, 0) and (6, 8, 2)

____

6. (0, 6, 0) and (4, 8, 0)

____

7. (9, 1, 4) and (7, 0, 7)

____

8. (2, 4, 1) and (3, 3, 5)

____

Name ______________________________ Date ____________ Class ____________

LESSON 10-4

Review for Mastery

Surface Area of Prisms and Cylinders

The *lateral area* of a prism is the sum of the areas of all the *lateral faces*. A lateral face is not a base. The **surface area** is the total area of all faces.

Lateral and Surface Area of a Right Prism		
Lateral Area	The lateral area of a right prism with base perimeter P and height h is $$L = Ph.$$	h; lateral face
Surface Area	The surface area of a right prism with lateral area L and base area B is $$S = L + 2B, \text{ or } S = Ph + 2B.$$	

The lateral area of a right cylinder is the curved surface that connects the two bases. The **surface area** is the total area of the curved surface and the bases.

Lateral and Surface Area of a Right Cylinder		
Lateral Area	The lateral area of a right cylinder with radius r and height h is $$L = 2\pi rh.$$	h; lateral surface
Surface Area	The surface area of a right cylinder with lateral area L and base area B is $$S = L + 2B, \text{ or } S = 2\pi rh + 2\pi r^2.$$	

Find the lateral area and surface area of each right prism.

1.

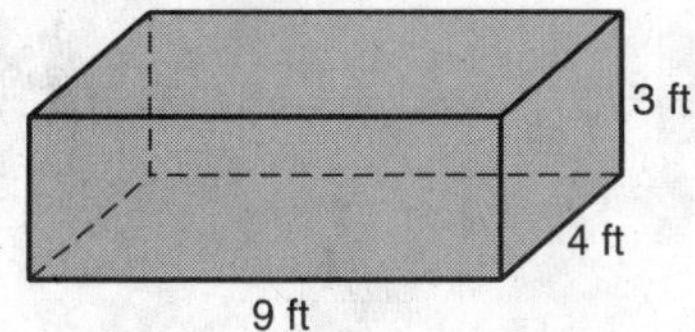

2.

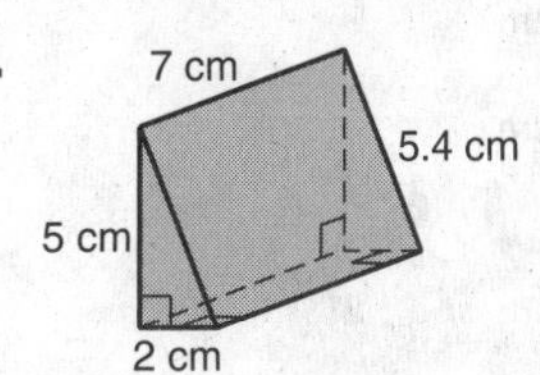

______________________________ ______________________________

Find the lateral area and surface area of each right cylinder. Give your answers in terms of π.

3.

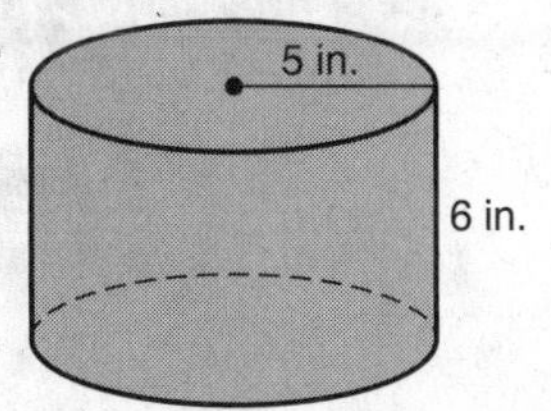

4.

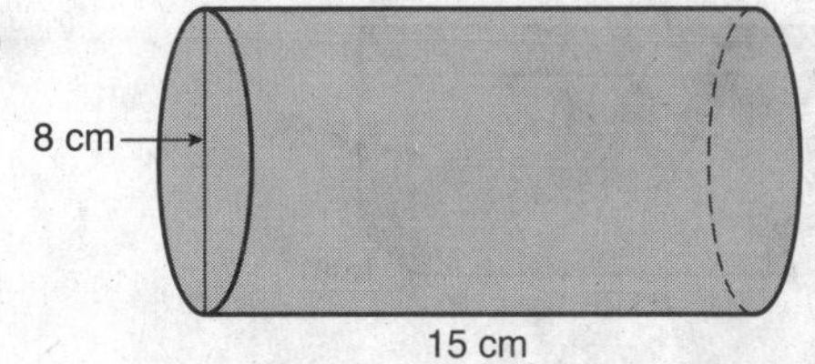

______________________________ ______________________________

Name ______________________ Date __________ Class __________

LESSON **10-4**

Review for Mastery

Surface Area of Prisms and Cylinders continued

You can find the surface area of a composite three-dimensional figure like the one shown at right.

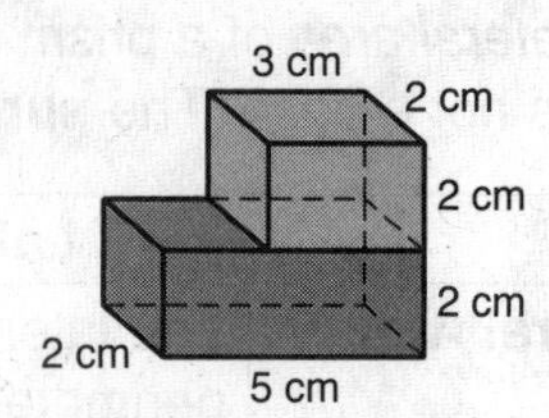

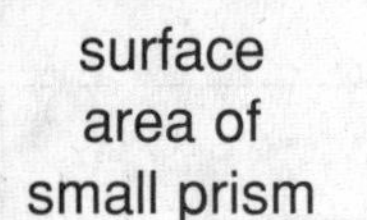

surface area of small prism + surface area of large prism − hidden surfaces

The dimensions are multiplied by 3. Describe the effect on the surface area.

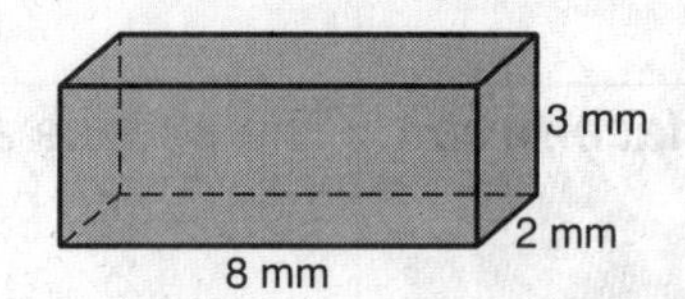

original surface area:

$S = Ph + 2B$

$= 20(3) + 2(16)$ $\quad P = 20, h = 3, B = 16$

$= 92 \text{ mm}^2$ $\quad$ Simplify.

new surface area, dimensions multiplied by 3:

$S = Ph + 2B$

$= 60(9) + 2(144)$ $\quad P = 60, h = 9, B = 144$

$= 828 \text{ mm}^2$ $\quad$ Simplify.

Notice that $92 \cdot 9 = 828$. If the dimensions are multiplied by 3, the surface area is multiplied by 3^2, or 9.

Find the surface area of each composite figure. Be sure to subtract the hidden surfaces of each part of the composite solid. Round to the nearest tenth.

5.

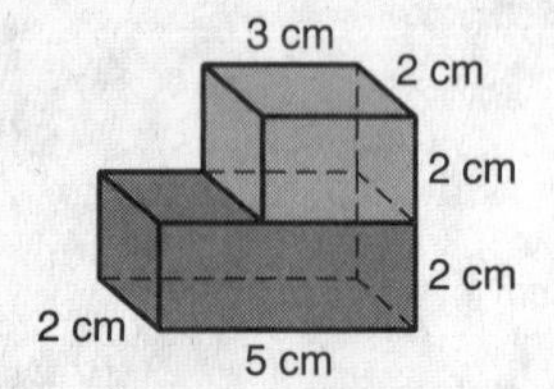

____

6.

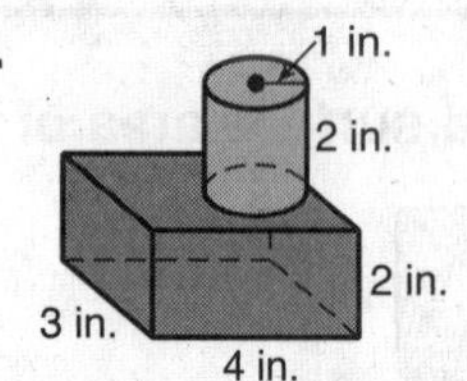

____

Describe the effect of each change on the surface area of the given figure.

7. The length, width, and height are multiplied by 2.

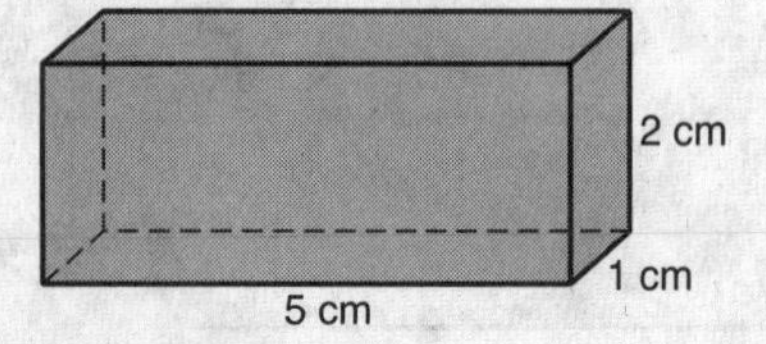

____

8. The height and radius are multiplied by $\frac{1}{2}$.

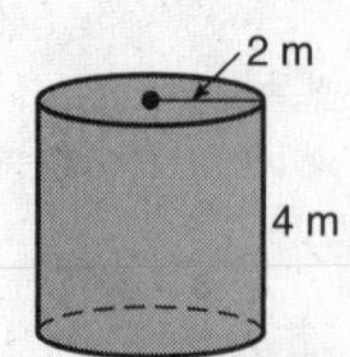

____

Name ______________________ Date __________ Class __________

LESSON 10-5

Review for Mastery

Surface Area of Pyramids and Cones

Lateral and Surface Area of a Regular Pyramid		
Lateral Area	The lateral area of a regular pyramid with perimeter P and slant height ℓ is $$L = \frac{1}{2}P\ell.$$	slant height; ℓ; base
Surface Area	The surface area of a regular pyramid with lateral area L and base area B is $$S = L + B, \text{ or } S = \frac{1}{2}P\ell + B.$$	

Lateral and Surface Area of a Right Cone		
Lateral Area	The lateral area of a right cone with radius r and slant height ℓ is $$L = \pi r\ell.$$	slant height; ℓ; r; base
Surface Area	The surface area of a right cone with lateral area L and base area B is $$S = L + B, \text{ or } S = \pi r\ell + \pi r^2.$$	

Find the lateral area and surface area of each regular pyramid. Round to the nearest tenth.

1.

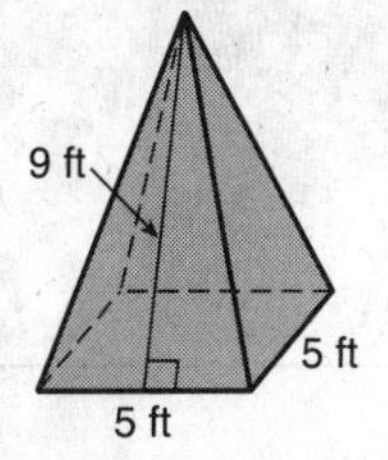

____

____

2.

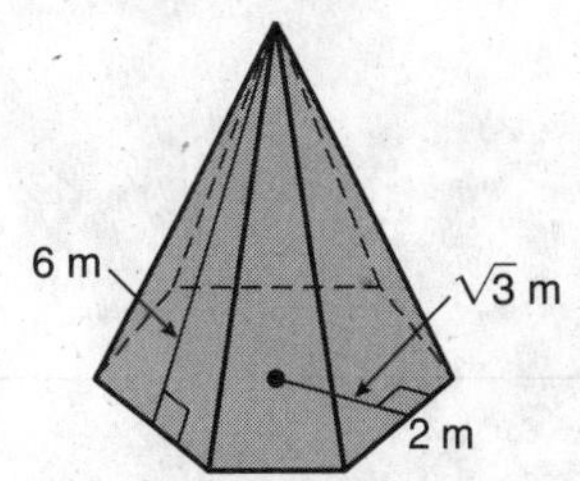

____

____

Find the lateral area and surface area of each right cone. Give your answers in terms of π.

3.

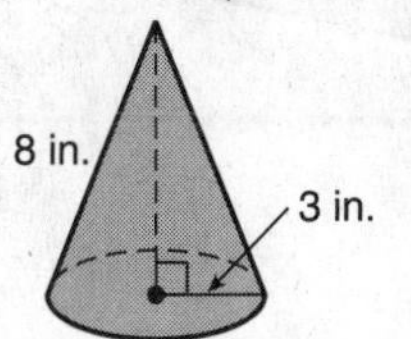

____

4.

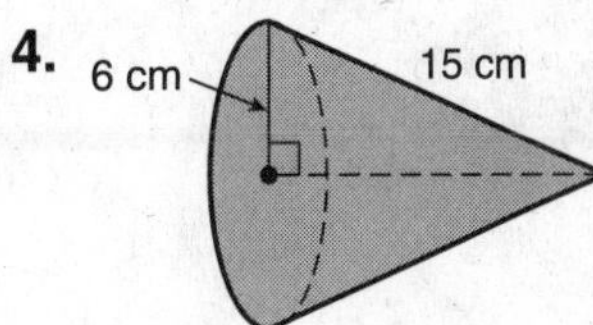

____

Name ______________________ Date __________ Class __________

LESSON 10-5

Review for Mastery

Surface Area of Pyramids and Cones continued

The radius and slant height of the cone at right are doubled. Describe the effect on the surface area.

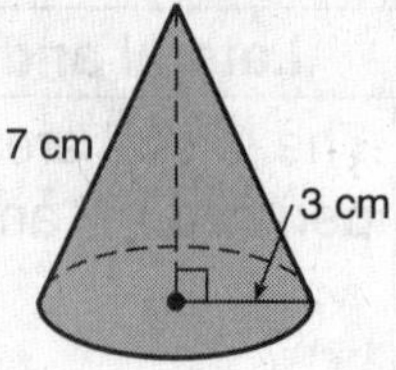

original surface area:		new surface area, dimensions doubled:	
$S = \pi r\ell + \pi r^2$		$S = \pi r\ell + \pi r^2$	
$= \pi(3)(7) + \pi(3)^2$	$r = 3, \ell = 7$	$= \pi(6)(14) + \pi(6)^2$	$r = 6, \ell = 14$
$= 30\pi \text{ cm}^2$	Simplify.	$= 120\pi \text{ cm}^2$	Simplify.

If the dimensions are doubled, then the surface area is multiplied by 2^2, or 4.

Describe the effect of each change on the surface area of the given figure.

5. The dimensions are tripled.

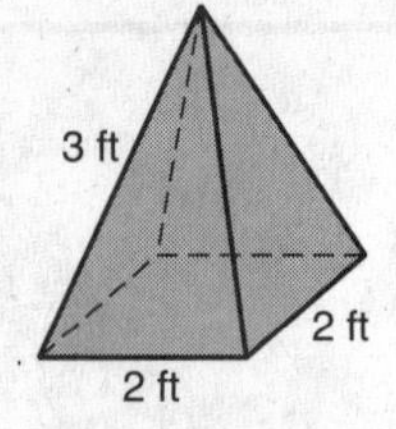

____

6. The dimensions are multiplied by $\frac{1}{2}$.

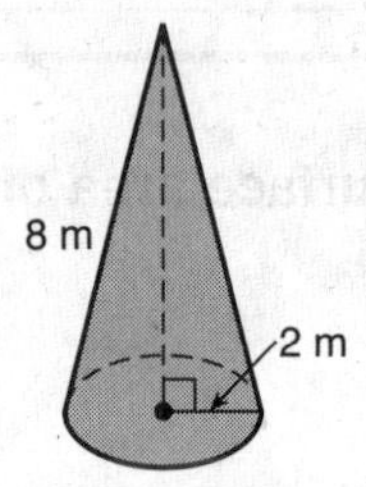

____

Find the surface area of each composite figure.

7. *Hint:* Do not include the base area of the pyramid or the upper surface area of the rectangular prism.

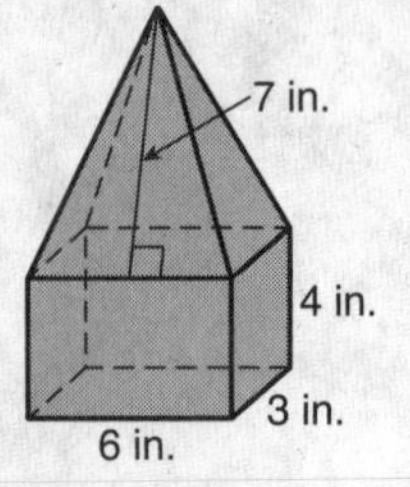

____

8. *Hint:* Add the lateral areas of the cones.

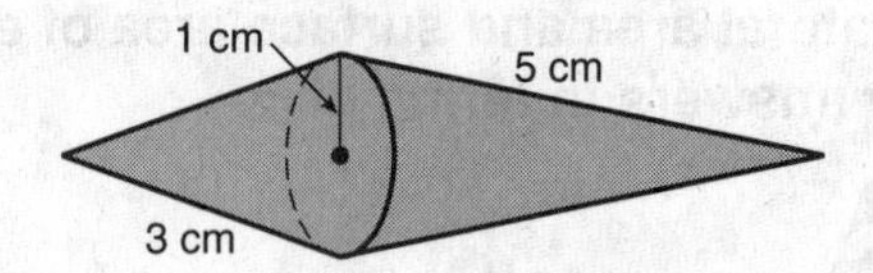

____

Name ______________________ Date __________ Class __________

LESSON 10-6

Review for Mastery

Volume of Prisms and Cylinders

Volume of Prisms		
Prism	The volume of a prism with base area B and height h is $V = Bh$.	
Right Rectangular Prism	The volume of a right rectangular prism with length ℓ, width w, and height h is $V = \ell wh$.	
Cube	The volume of a cube with edge length s is $V = s^3$.	

Volume of a Cylinder

The volume of a cylinder with base area B, radius r, and height h is

$$V = Bh, \text{ or } V = \pi r^2 h.$$

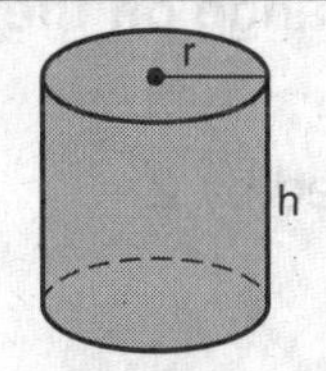

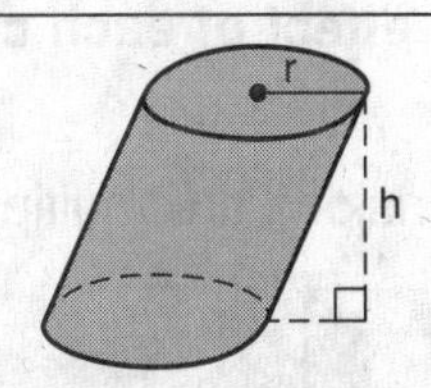

Find the volume of each prism.

1.

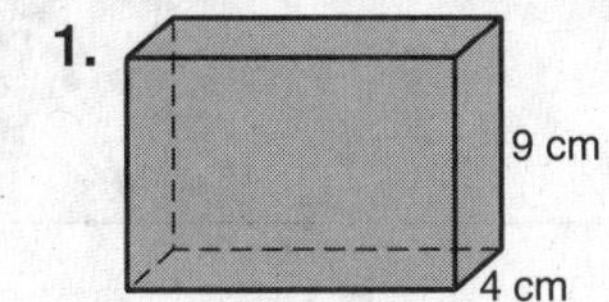

2.

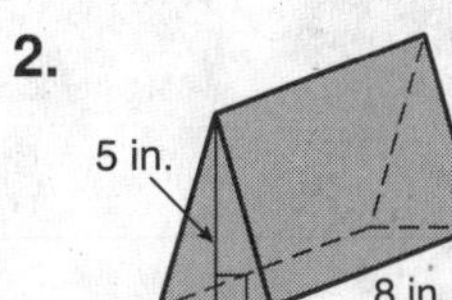

______________________ ______________________

Find the volume of each cylinder. Give your answers both in terms of π and rounded to the nearest tenth.

3.

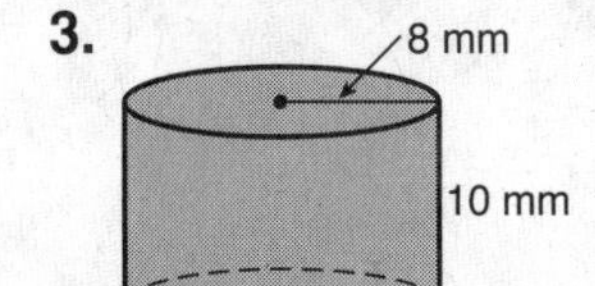

4.

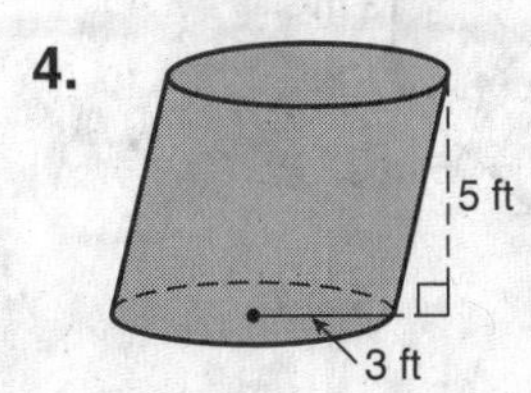

______________________ ______________________

Name ______________________ Date __________ Class __________

LESSON 10-6

Review for Mastery

Volume of Prisms and Cylinders continued

The dimensions of the prism are multiplied by $\frac{1}{3}$. Describe the effect on the volume.

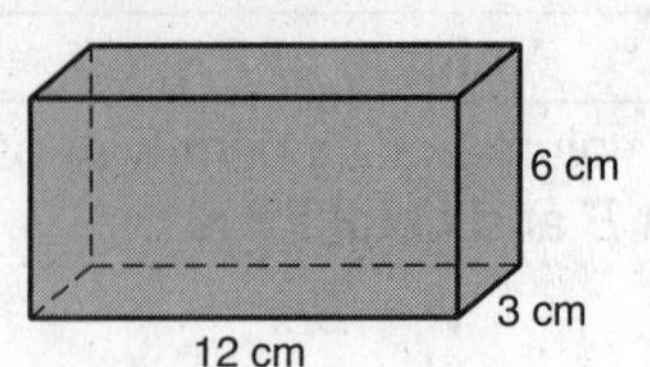

original volume:

$V = \ell wh$

$= (12)(3)(6)$ $\quad \ell = 12, w = 3, h = 6$

$= 216 \text{ cm}^3$ $\quad$ Simplify.

new volume, dimensions multiplied by $\frac{1}{3}$:

$V = \ell wh$

$= (4)(1)(2)$ $\quad \ell = 4, w = 1, h = 2$

$= 8 \text{ cm}^3$ $\quad$ Simplify.

Notice that $216 \cdot \frac{1}{27} = 8$. If the dimensions are multiplied by $\frac{1}{3}$, the volume is multiplied by $\left(\frac{1}{3}\right)^3$, or $\frac{1}{27}$.

Describe the effect of each change on the volume of the given figure.

5. The dimensions are multiplied by 2.

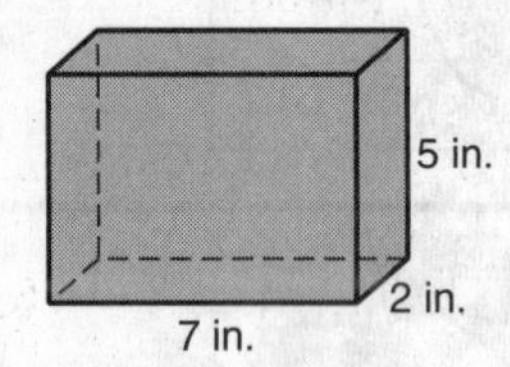

____

6. The dimensions are multiplied by $\frac{1}{4}$.

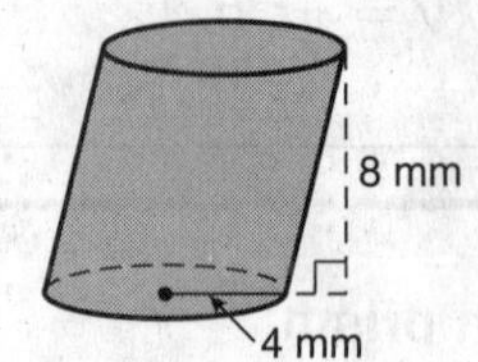

____

Find the volume of each composite figure. Round to the nearest tenth.

7.

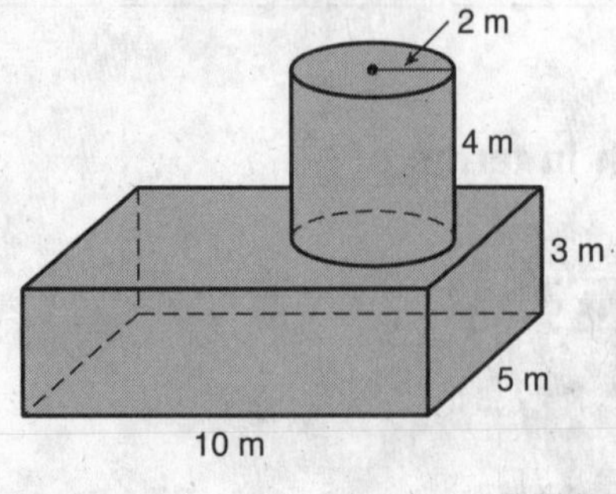

____

8.

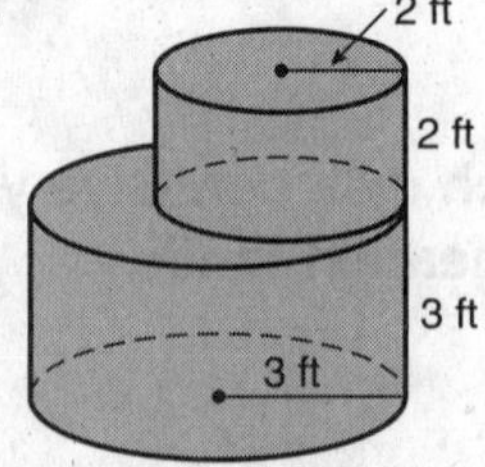

____

Name ______________________ Date ____________ Class ____________

LESSON 10-7

Review for Mastery

Volume of Pyramids and Cones

Volume of a Pyramid
The volume of a pyramid with base area B and height h is $V = \frac{1}{3}Bh$.

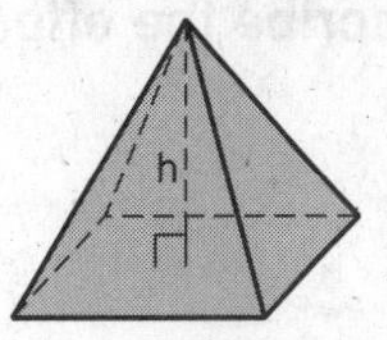

Volume of a Cone
The volume of a cone with base area B, radius r, and height h is $V = \frac{1}{3}Bh$, or $V = \frac{1}{3}\pi r^2 h$.

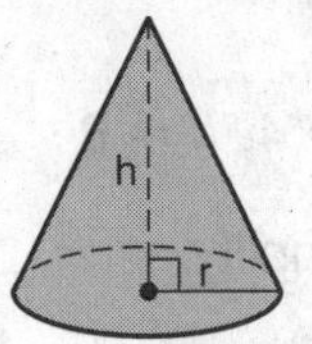

Find the volume of each pyramid. Round to the nearest tenth if necessary.

1.

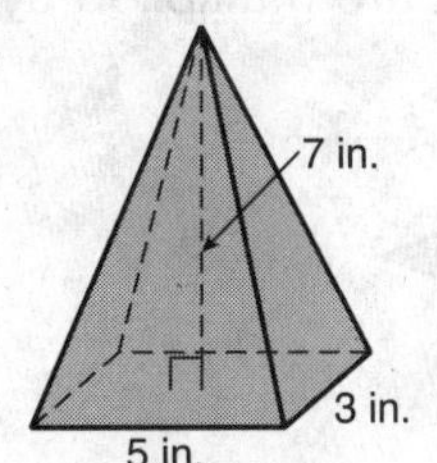

2.

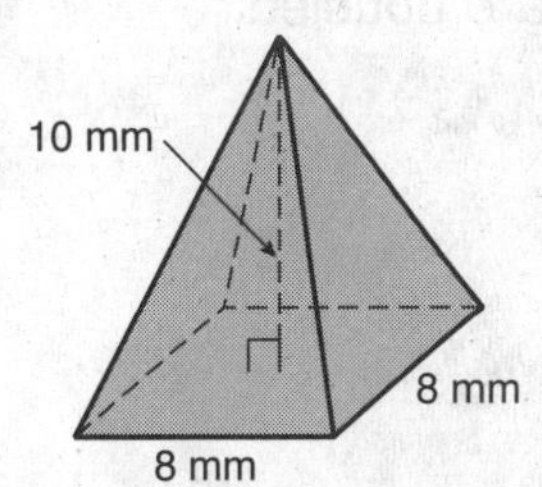

______________________ ______________________

Find the volume of each cone. Give your answers both in terms of π and rounded to the nearest tenth.

3.

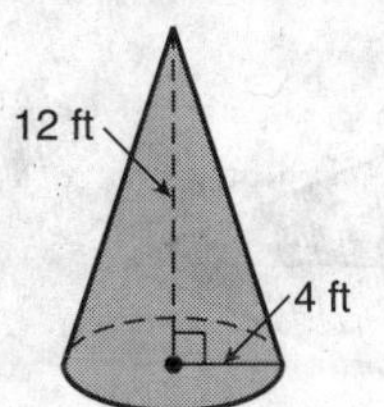

4.

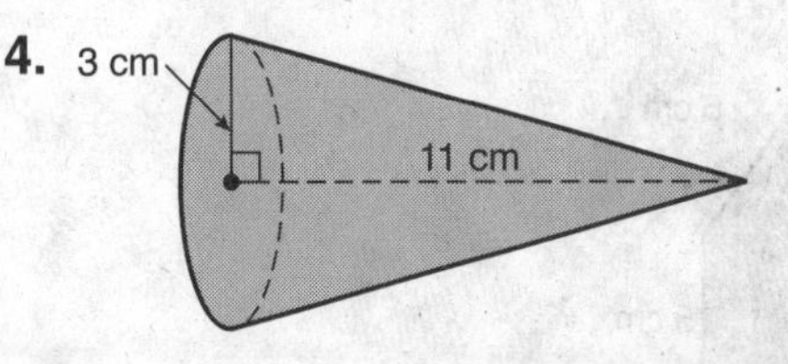

______________________ ______________________

Name ________________________ Date __________ Class __________

LESSON 10-7

Review for Mastery

Volume of Pyramids and Cones continued

The radius and height of the cone are multiplied by $\frac{1}{2}$. Describe the effect on the volume.

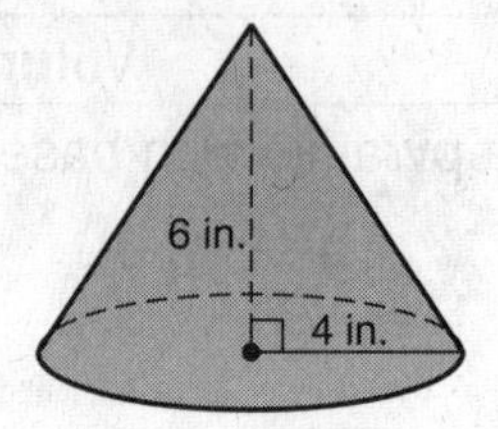

original volume:

$V = \frac{1}{3}\pi r^2 h$

$= \frac{1}{3}\pi(4)^2(6)$ $\quad r = 4, h = 6$

$= 32\pi \text{ in}^3$ $\quad$ Simplify.

new volume, dimensions multiplied by $\frac{1}{2}$:

$V = \frac{1}{3}\pi r^2 h$

$= \frac{1}{3}\pi(2)^2(3)$ $\quad r = 2, h = 3$

$= 4\pi \text{ in}^3$ $\quad$ Simplify.

If the dimensions are multiplied by $\frac{1}{2}$, then the volume is multiplied by $\left(\frac{1}{2}\right)^3$, or $\frac{1}{8}$.

Describe the effect of each change on the volume of the given figure.

5. The dimensions are doubled.

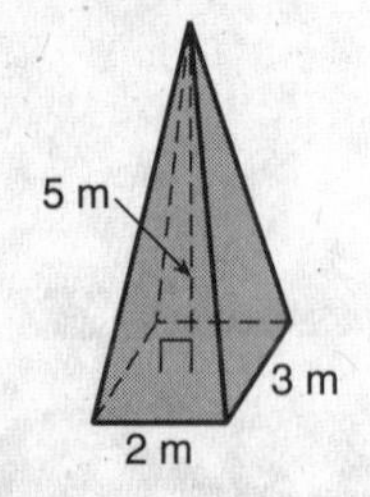

____

6. The radius and height are multiplied by $\frac{1}{3}$.

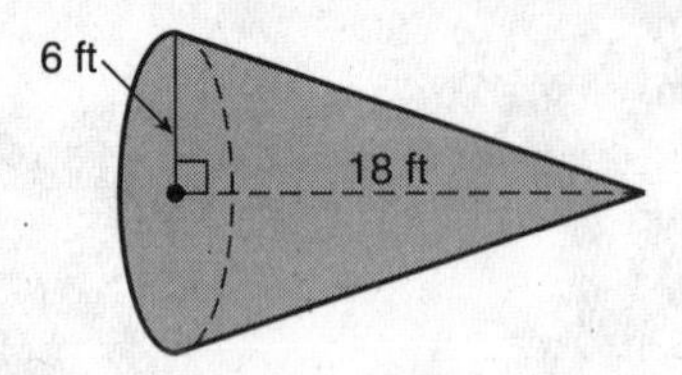

____

Find the volume of each composite figure. Round to the nearest tenth if necessary.

7.

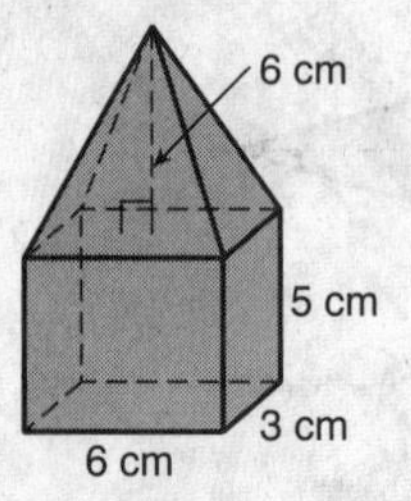

____

8.

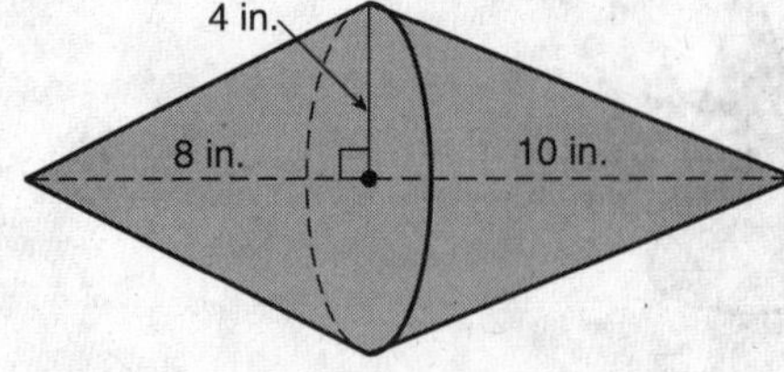

____

Name ______________________ Date ____________ Class ____________

LESSON 10-8

Review for Mastery

Spheres

Volume and Surface Area of a Sphere		
Volume	The volume of a sphere with radius r is $V = \frac{4}{3}\pi r^3$.	
Surface Area	The surface area of a sphere with radius r is $S = 4\pi r^2$.	

Find each measurement. Give your answer in terms of π.

1. the volume of the sphere

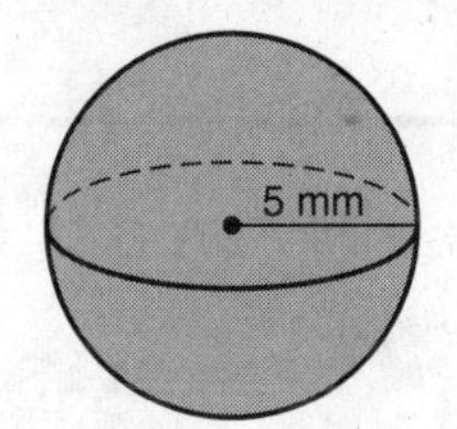

____

2. the volume of the sphere

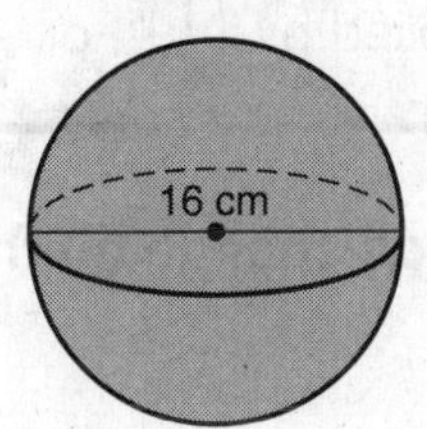

____

3. the volume of the hemisphere

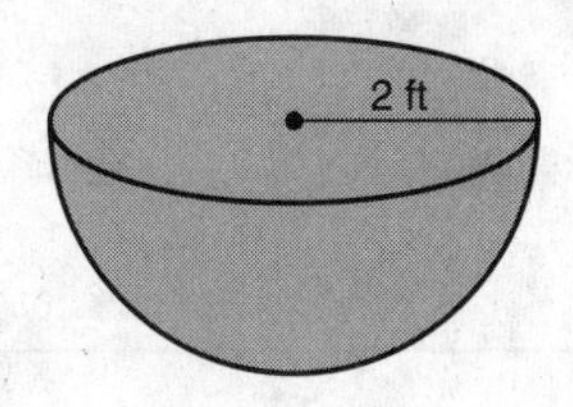

____

4. the radius of a sphere with volume 7776π in^3

____

5. the surface area of the sphere

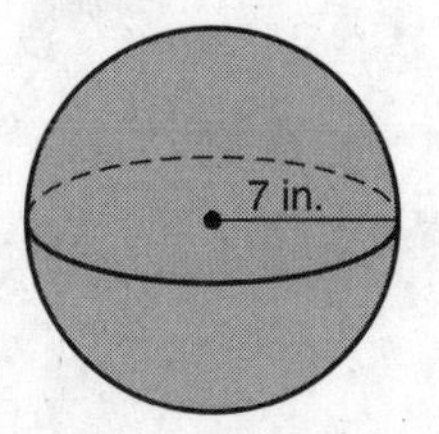

____

6. the surface area of the sphere

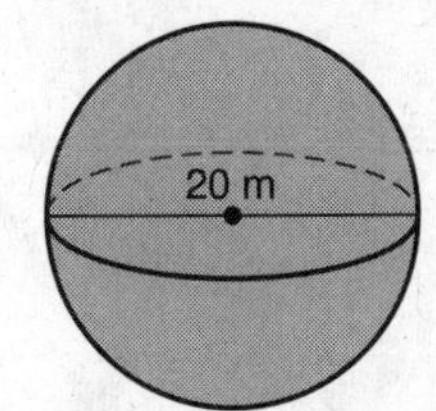

____

Name ______________________ Date __________ Class __________

LESSON 10-8

Review for Mastery

Spheres continued

The radius of the sphere is multiplied by $\frac{1}{4}$. Describe the effect on the surface area.

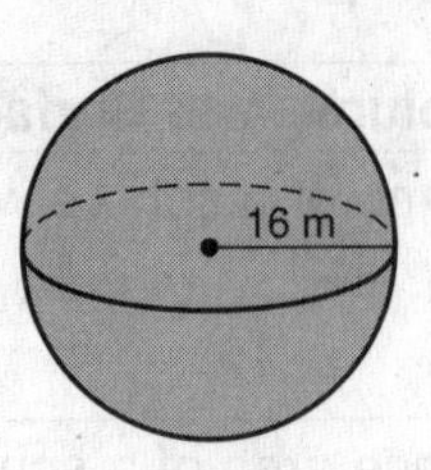

original surface area:

$S = 4\pi r^2$

$= 4\pi(16)^2$ $\quad r = 16$

$= 1024\pi \text{ m}^2$ $\quad$ Simplify.

new surface area, radius multiplied by $\frac{1}{4}$:

$S = 4\pi r^2$

$= 4\pi(4)^2$ $\quad r = 4$

$= 64\pi \text{ m}^2$ $\quad$ Simplify.

Notice that $1024 \cdot \frac{1}{16} = 64$. If the dimensions are multiplied by $\frac{1}{4}$, the surface area is multiplied by $\left(\frac{1}{4}\right)^2$, or $\frac{1}{16}$.

Describe the effect of each change on the given measurement of the figure.

7. surface area
The radius is multiplied by 4.

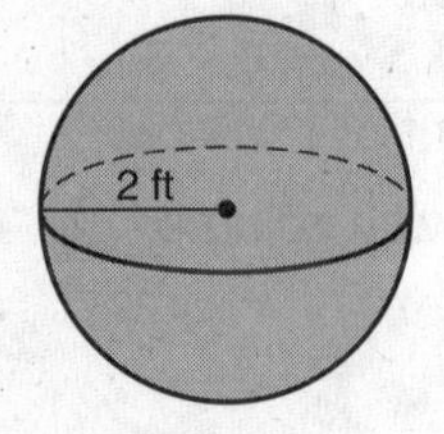

____

8. volume
The dimensions are multiplied by $\frac{1}{2}$.

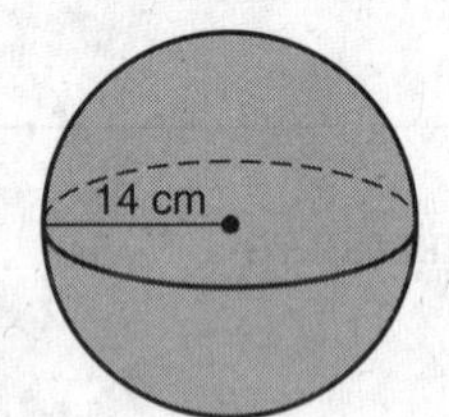

____

Find the surface area and volume of each composite figure. Round to the nearest tenth.

9. *Hint:* To find the surface area, add the lateral area of the cylinder, the area of one base, and the surface area of the hemisphere.

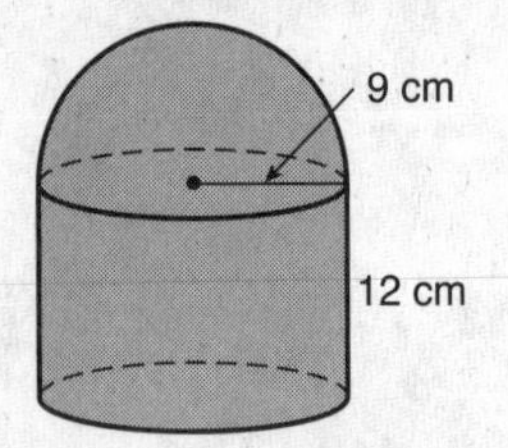

____

10. *Hint:* To find the volume, subtract the volume of the hemisphere from the volume of the cylinder.

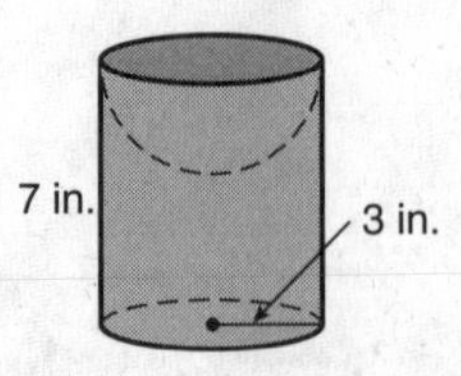

____

Name ______________________ Date __________ Class __________

LESSON 11-1

Review for Mastery

Lines That Intersect Circles

Lines and Segments That Intersect Circles

- A **chord** is a segment whose endpoints lie on a circle.
- A **secant** is a line that intersects a circle at two points.
- A **tangent** is a line in the same plane as a circle that intersects the circle at exactly one point, called the **point of tangency.**
- Radii and diameters also intersect circles.

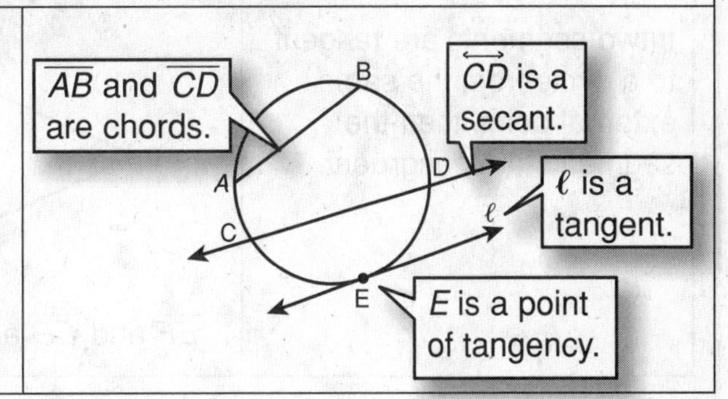

Tangent Circles

Two coplanar circles that intersect at exactly one point are called **tangent circles.**

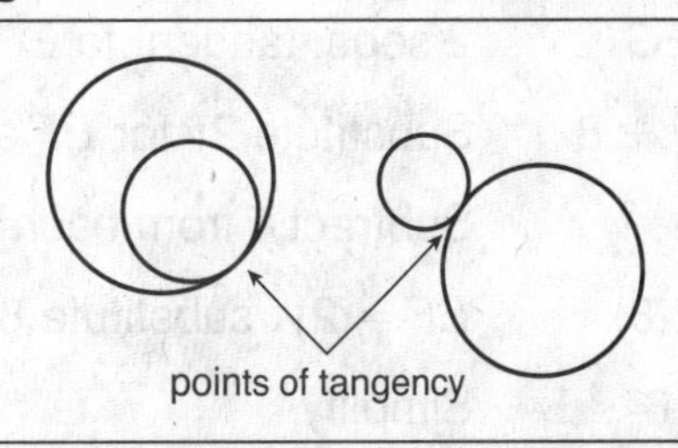

Identify each line or segment that intersects each circle.

1.

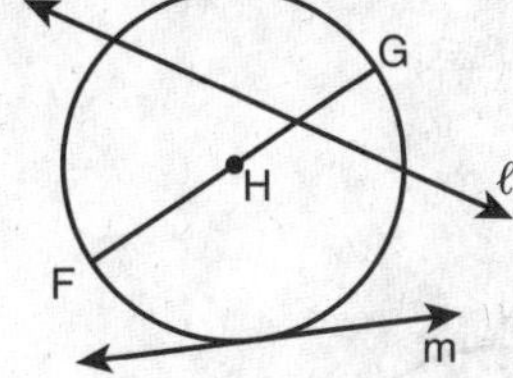

____

____

2.

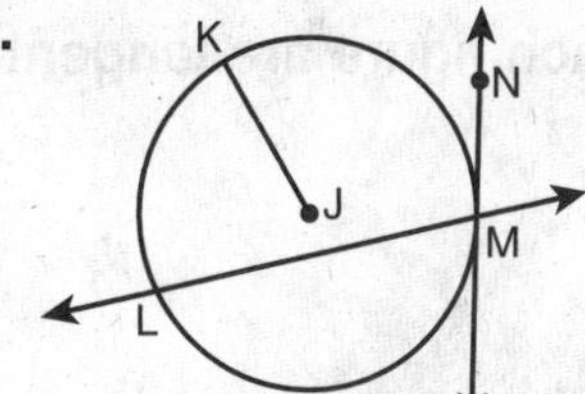

____

____

Find the length of each radius. Identify the point of tangency and write the equation of the tangent line at that point.

3.

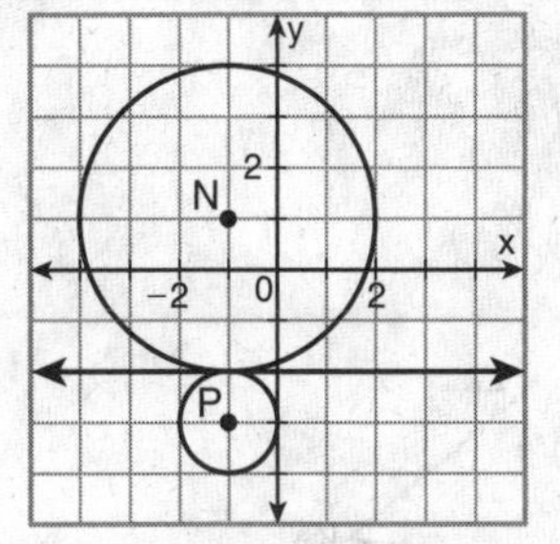

____

____

____

4.

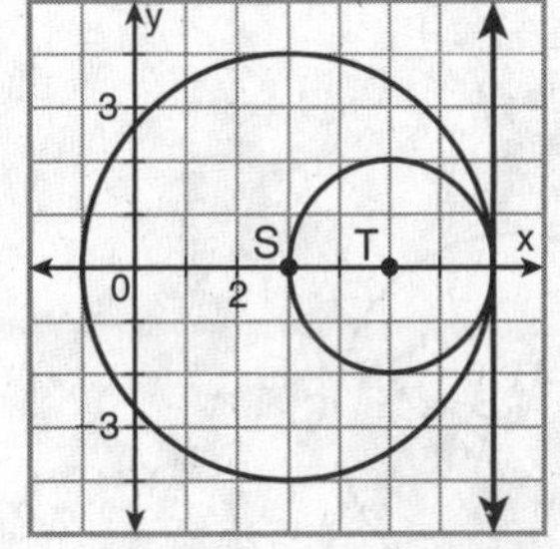

____

____

____

Name ______________________ Date ____________ Class ____________

LESSON 11-1

Review for Mastery

Lines That Intersect Circles continued

Theorem	Hypothesis	Conclusion
If two segments are tangent to a circle from the same external point, then the segments are congruent.	$\overline{EF}$ and $\overline{EG}$ are tangent to $\odot C$.	$\overline{EF} \cong \overline{EG}$

In the figure above, $EF = 2y$ and $EG = y + 8$. Find EF.

$EF = EG$	2 segs. tangent to ⊙ from same ext. pt. → segs. ≅.
$2y = y + 8$	Substitute $2y$ for EF and $y + 8$ for EG.
$y = 8$	Subtract y from each side.
$EF = 2(8)$	$EF = 2y$; substitute 8 for y.
$= 16$	Simplify.

The segments in each figure are tangent to the circle. Find each length.

5. BC

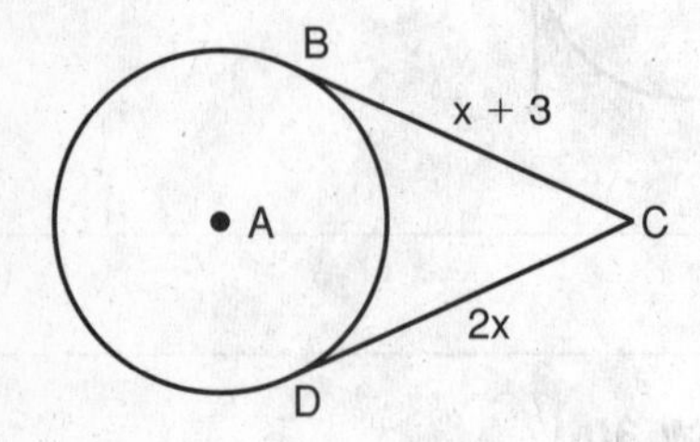

____

6. LM

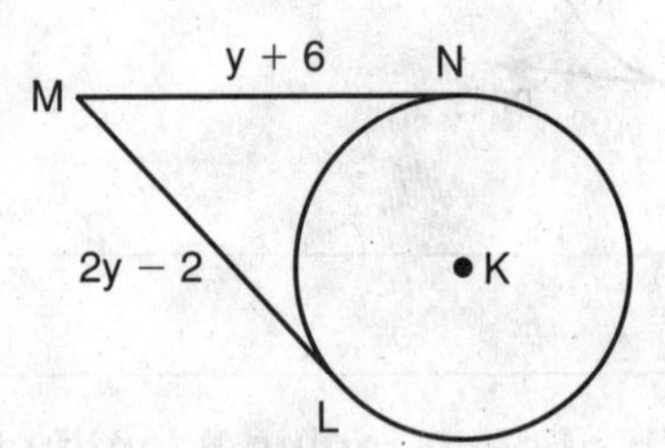

____

7. RS

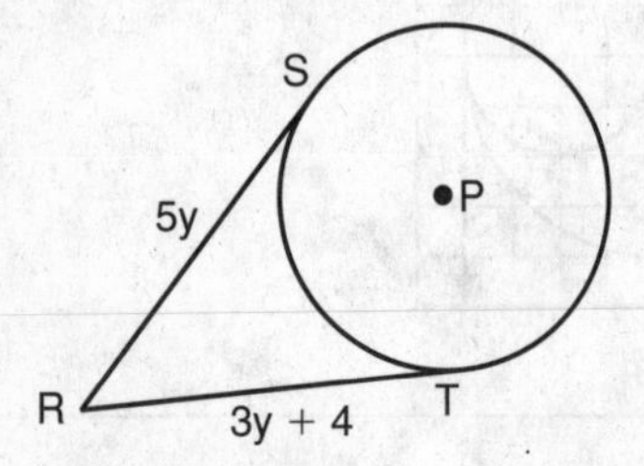

____

8. JK

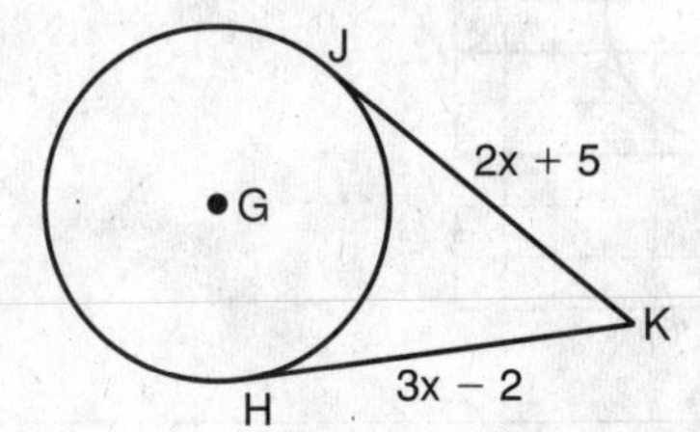

____

Name ______________________ Date ___________ Class ___________

LESSON 11-2

Review for Mastery

Arcs and Chords

Arcs and Their Measure

- A **central angle** is an angle whose vertex is the center of a circle.
- An **arc** is an unbroken part of a circle consisting of two points on a circle and all the points on the circle between them.

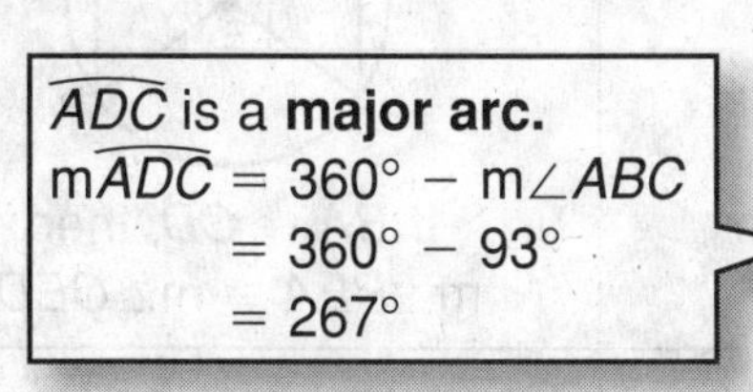

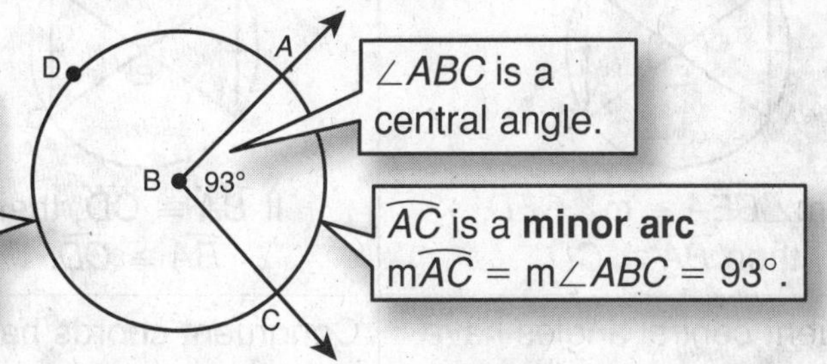

- If the endpoints of an arc lie on a diameter, the arc is a semicircle and its measure is 180°.

Arc Addition Postulate

The measure of an arc formed by two adjacent arcs is the sum of the measures of the two arcs.

$m\overset{\frown}{ABC} = m\overset{\frown}{AB} + m\overset{\frown}{BC}$

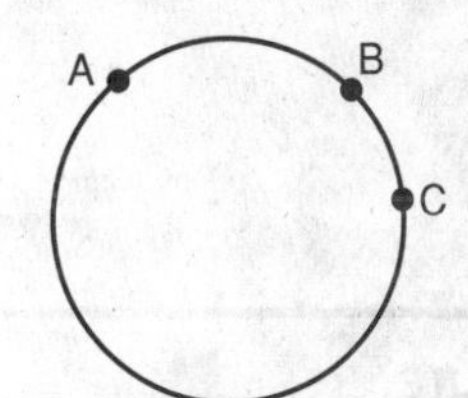

Find each measure.

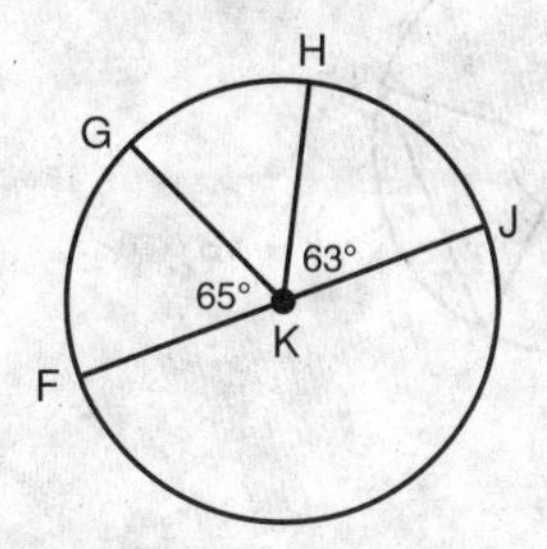

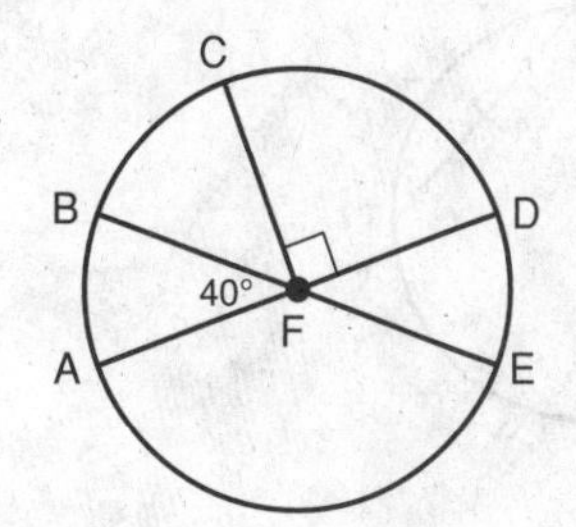

1. $m\overset{\frown}{HJ}$ ______________________

3. $m\overset{\frown}{CDE}$ ______________________

2. $m\overset{\frown}{FGH}$ ______________________

4. $m\overset{\frown}{BCD}$ ______________________

5. $m\overset{\frown}{LMN}$ ______________________

6. $m\overset{\frown}{LNP}$ ______________________

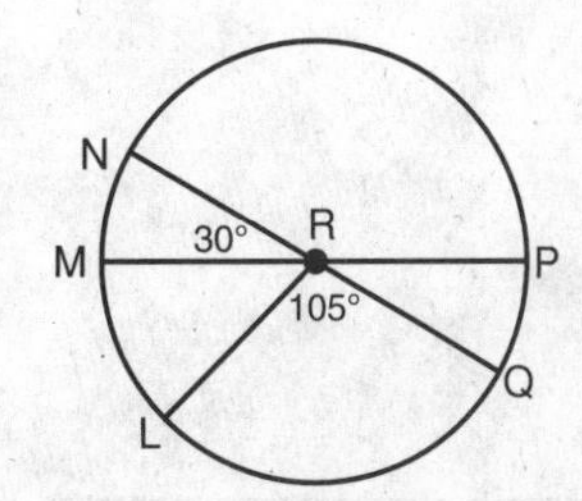

Name ______________________ Date ____________ Class ____________

LESSON 11-2

Review for Mastery

Arcs and Chords continued

Congruent arcs are arcs that have the same measure.

Congruent Arcs, Chords, and Central Angles		
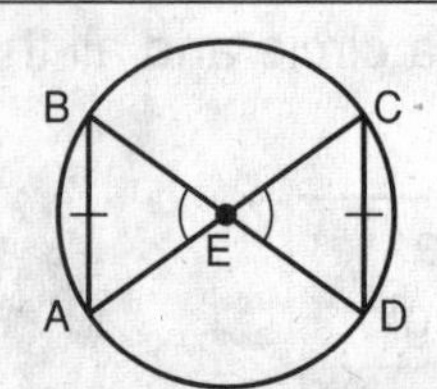 If $m\angle BEA \cong m\angle CED$, then $\overline{BA} \cong \overline{CD}$.	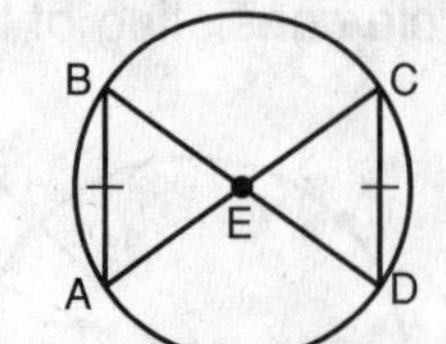 If $\overline{BA} \cong \overline{CD}$, then $\overset{\frown}{BA} \cong \overset{\frown}{CD}$.	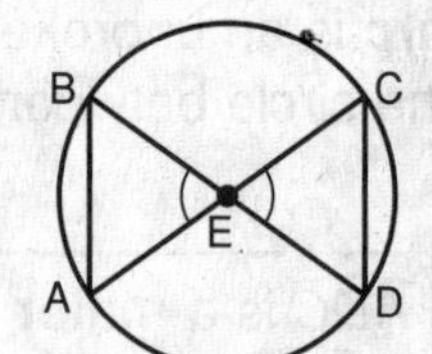 If $\overset{\frown}{BA} \cong \overset{\frown}{CD}$, then $m\angle BEA \cong m\angle CED$.
Congruent central angles have congruent chords.	Congruent chords have congruent arcs.	Congruent arcs have congruent central angles.

In a circle, if a radius or diameter is perpendicular to a chord, then it bisects the chord and its arc.

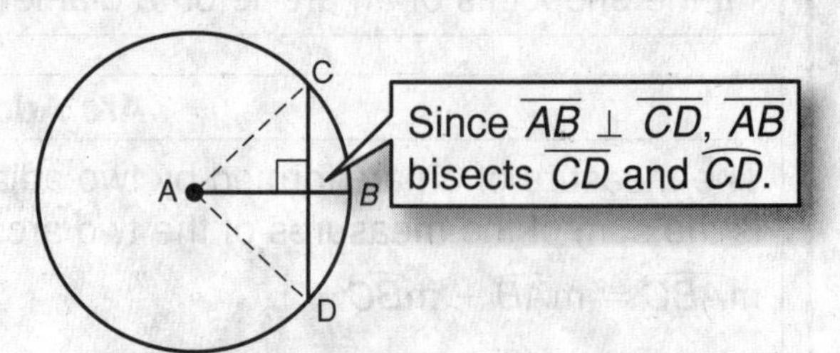

Find each measure.

7. $\overline{QR} \cong \overline{ST}$. Find $m\overset{\frown}{QR}$.

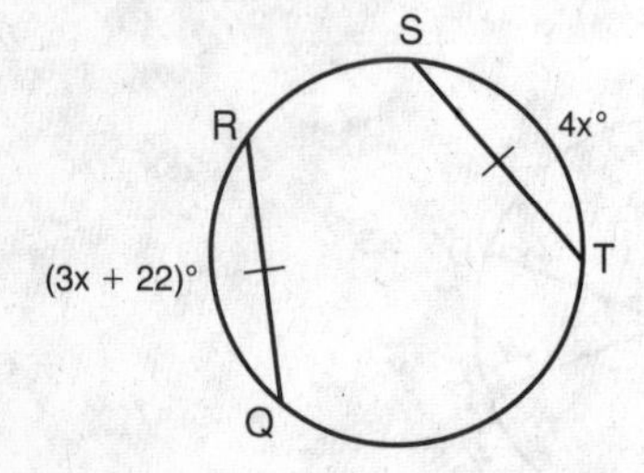

____

8. $\angle HLG \cong \angle KLJ$. Find GH.

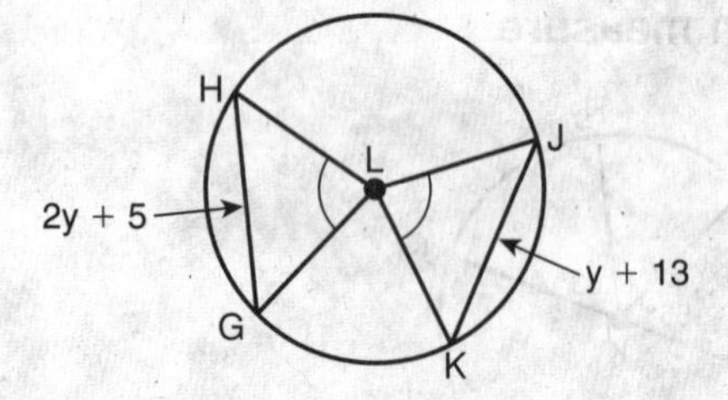

____

Find each length to the nearest tenth.

9. NP

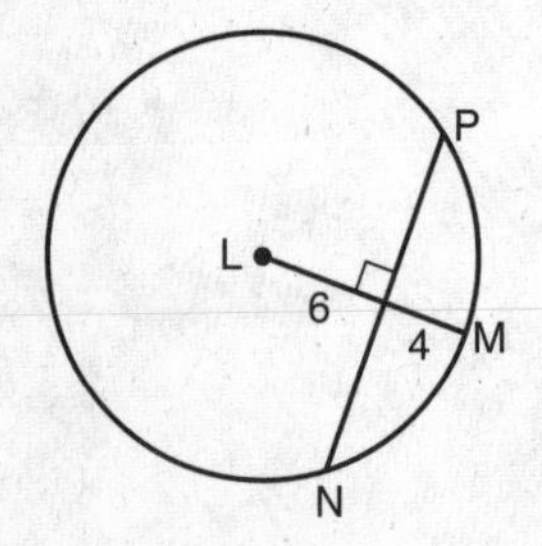

____

10. EF

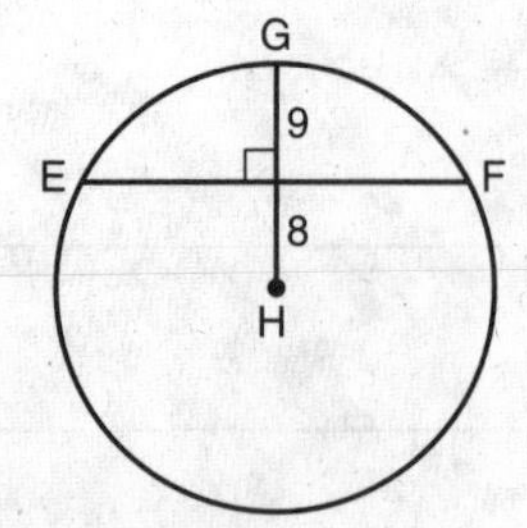

____

Name ______________________ Date ____________ Class ____________

LESSON 11-3

Review for Mastery

Sector Area and Arc Length

Sector of a Circle

A **sector of a circle** is a region bounded by two radii of the circle and their intercepted arc.

The area of a sector of a circle is given by the formula $A = \pi r^2\left(\frac{m^\circ}{360^\circ}\right)$.

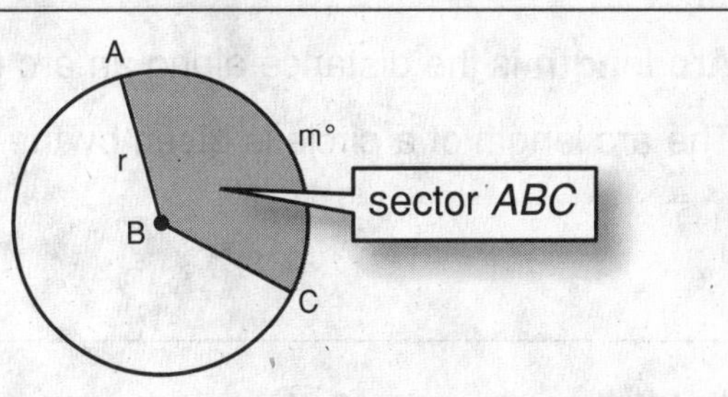

Segment of a Circle

A **segment of a circle** is a region bounded by an arc and its chord.

area of segment ABC = area of sector ABC − area of $\triangle ABC$

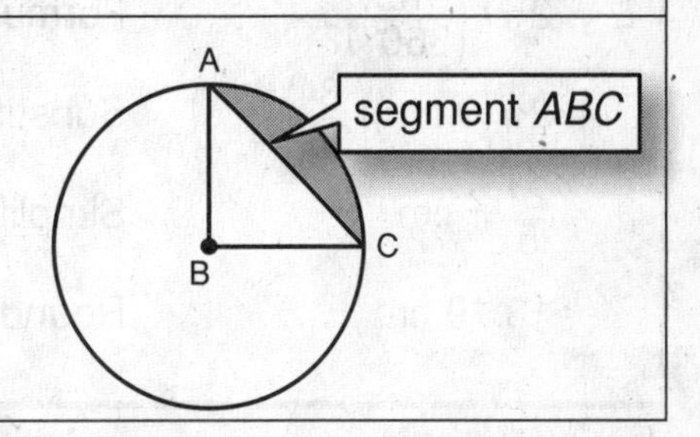

Find the area of each sector. Give your answer in terms of π and rounded to the nearest hundredth.

1. sector CDE

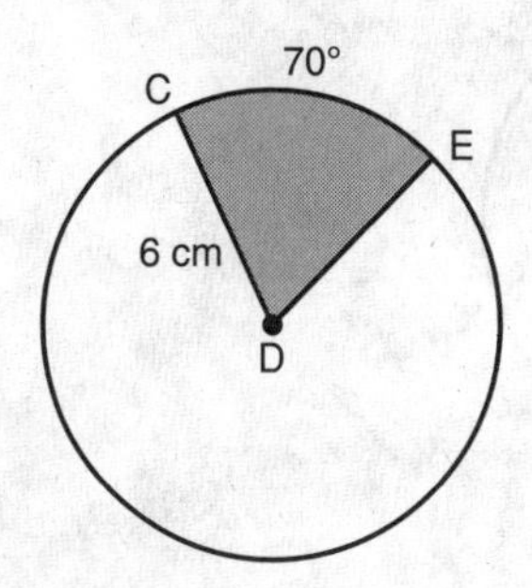

____

2. sector QRS

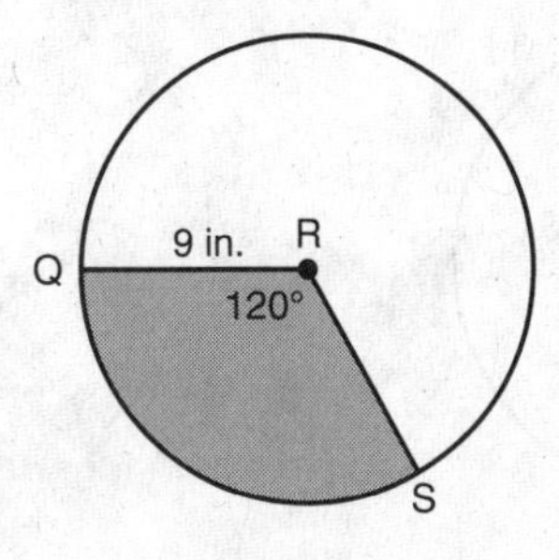

____

Find the area of each segment to the nearest hundredth.

3.

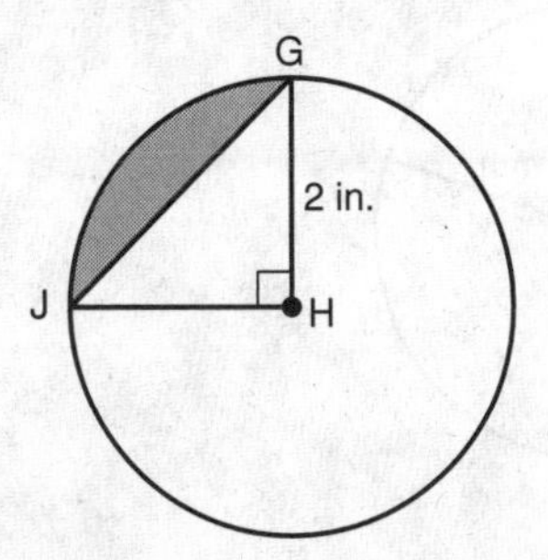

____

4.

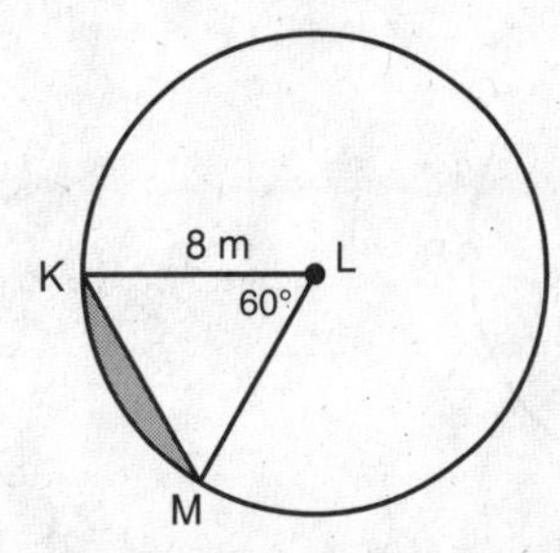

____

Name ________________________ Date ____________ Class ____________

LESSON 11-3

Review for Mastery

Sector Area and Arc Length continued

Arc Length

Arc length is the distance along an arc measured in linear units.

The arc length of a circle is given by the formula $L = 2\pi r\left(\frac{m^\circ}{360^\circ}\right)$.

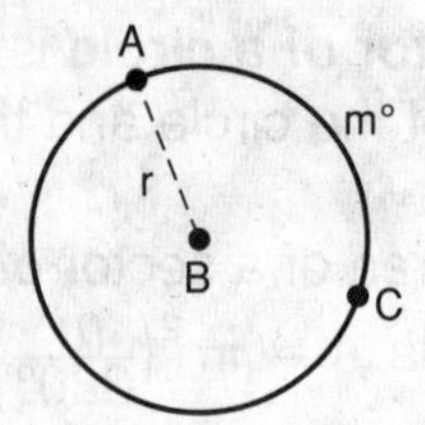

Find the arc length of $\overset{\frown}{JK}$.

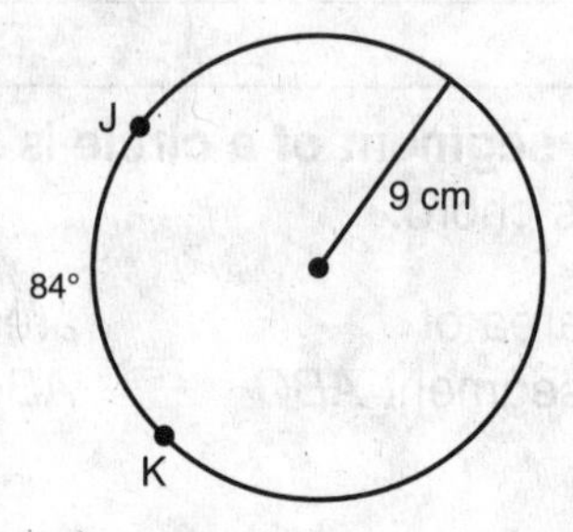

$L = 2\pi r\left(\frac{m^\circ}{360^\circ}\right)$	Formula for arc length
$= 2\pi(9 \text{ cm})\left(\frac{84^\circ}{360^\circ}\right)$	Substitute 9 cm for r and 84° for $m°$.
$= \frac{21}{5}\pi \text{ cm}$	Simplify.
$\approx 13.19 \text{ cm}$	Round to the nearest hundredth.

Find each arc length. Give your answer in terms of π and rounded to the nearest hundredth.

5. $\overset{\frown}{AB}$

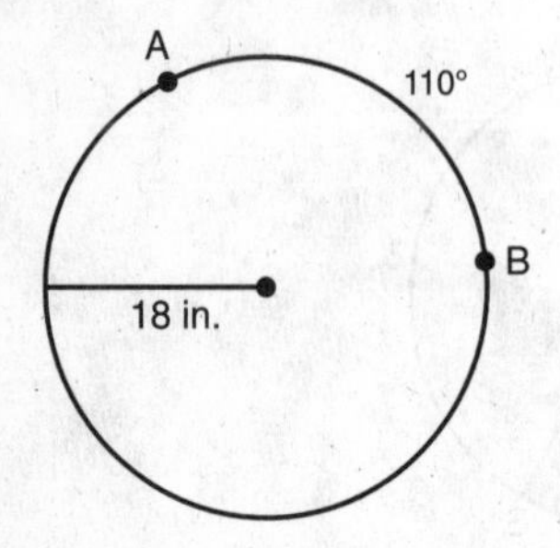

____

6. $\overset{\frown}{WX}$

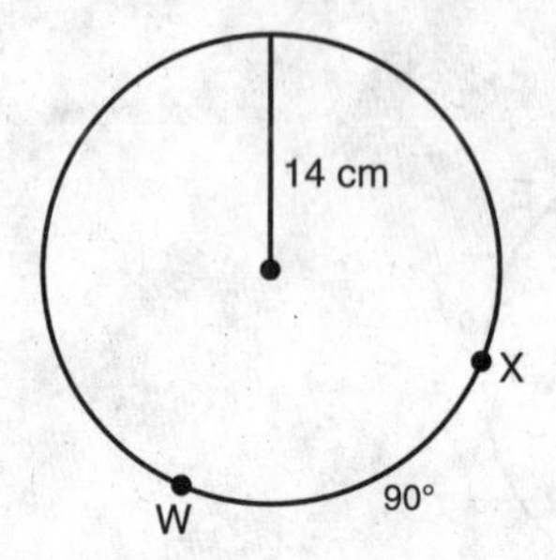

____

7. $\overset{\frown}{QR}$

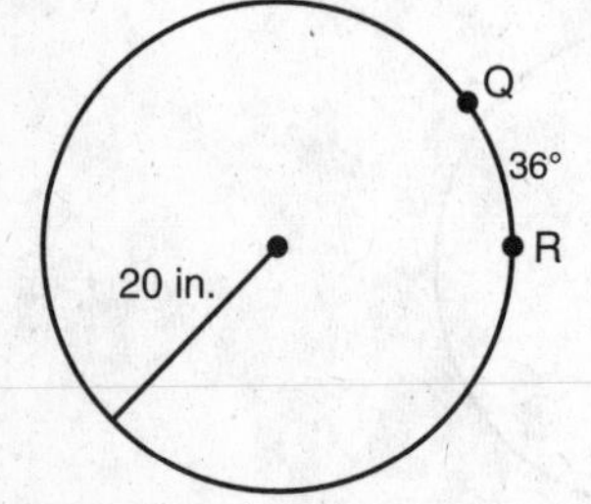

____

8. $\overset{\frown}{ST}$

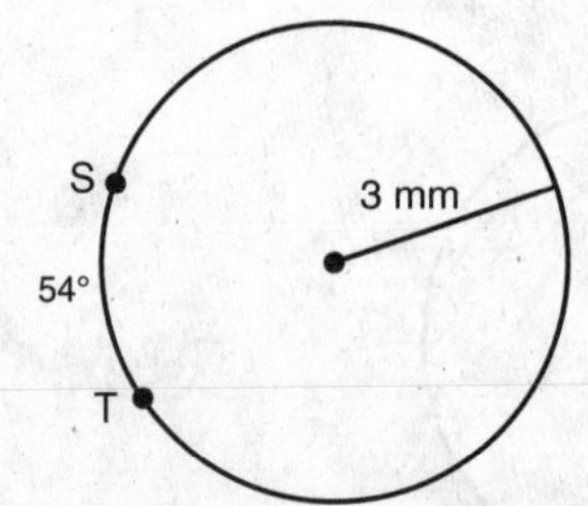

____

Name ______________________ Date __________ Class __________

LESSON 11-4
Review for Mastery
Inscribed Angles

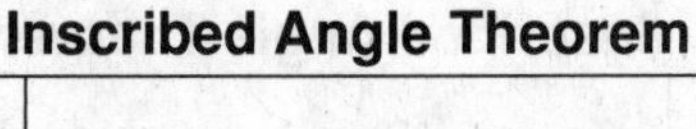

Inscribed Angle Theorem

The measure of an inscribed angle is half the measure of its intercepted arc.

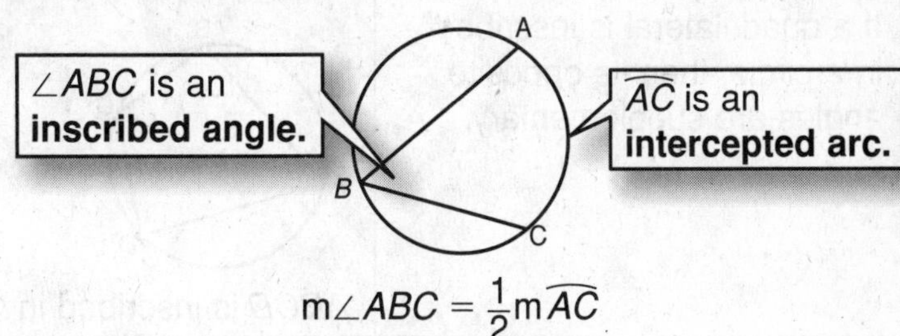

$$m\angle ABC = \frac{1}{2}m\overset{\frown}{AC}$$

Inscribed Angles

If inscribed angles of a circle intercept the same arc, then the angles are congruent.

$\angle ABC$ and $\angle ADC$ intercept $\overset{\frown}{AC}$, so $\angle ABC \cong \angle ADC$.

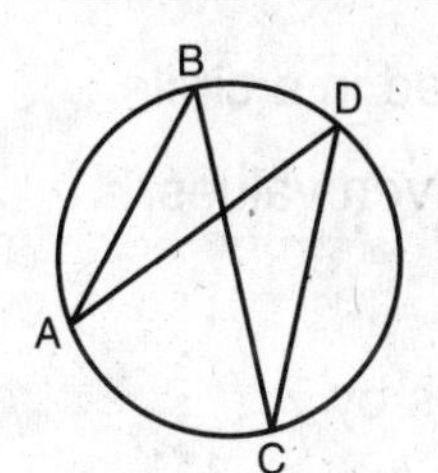

An inscribed angle subtends a semicircle if and only if the angle is a right angle.

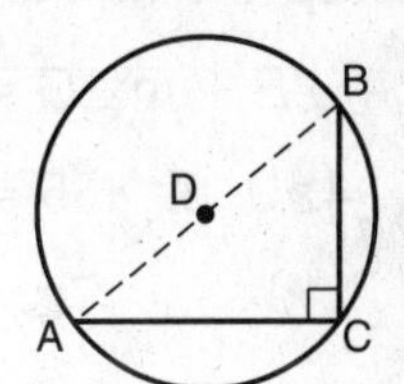

Find each measure.

1. $m\angle LMP$ and $m\overset{\frown}{MN}$

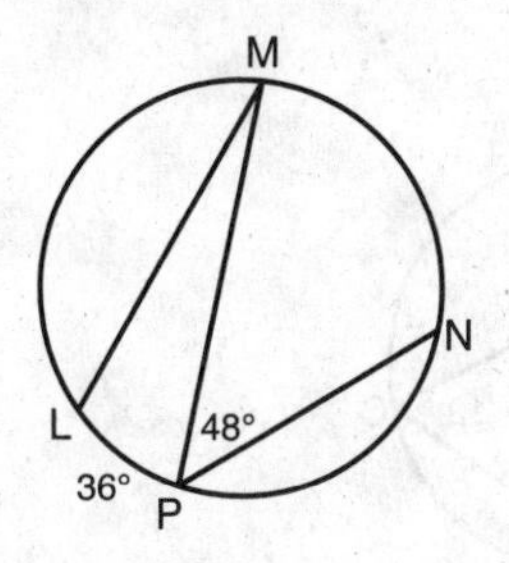

____

2. $m\angle GFJ$ and $m\overset{\frown}{FH}$

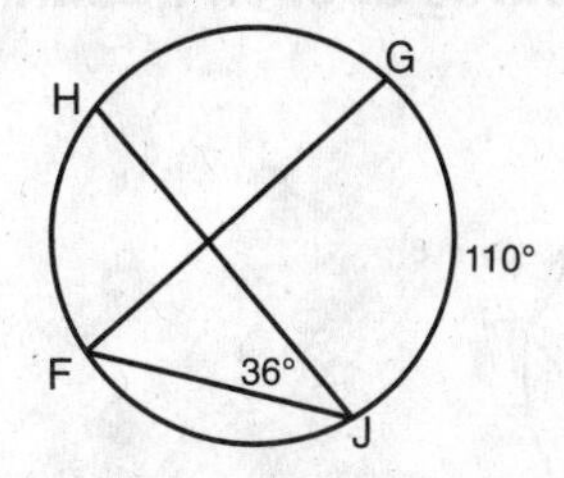

____

Find each value.

3. x

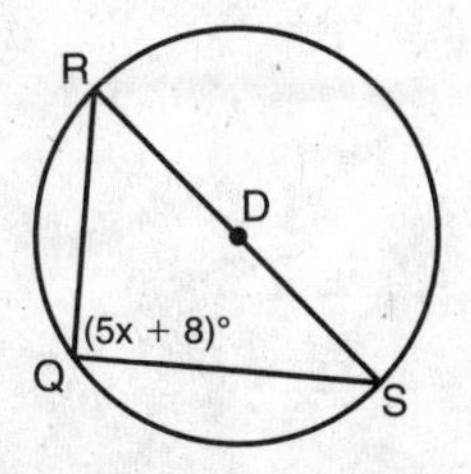

____

4. $m\angle FJH$

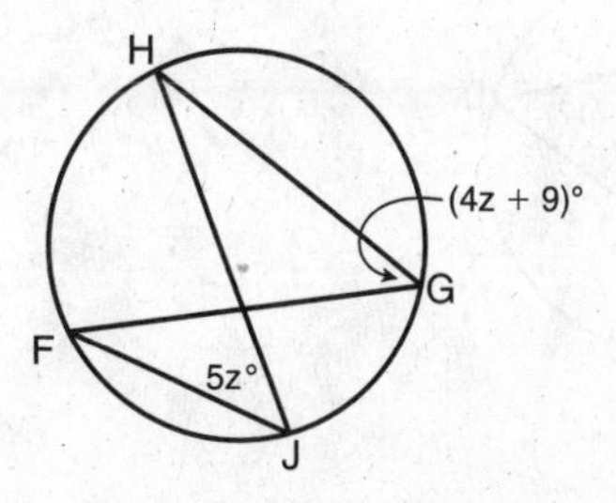

____

Name ______________________ Date __________ Class __________

LESSON 11-4

Review for Mastery

Inscribed Angles continued

Inscribed Angle Theorem		
If a quadrilateral is inscribed in a circle, then its opposite angles are supplementary.	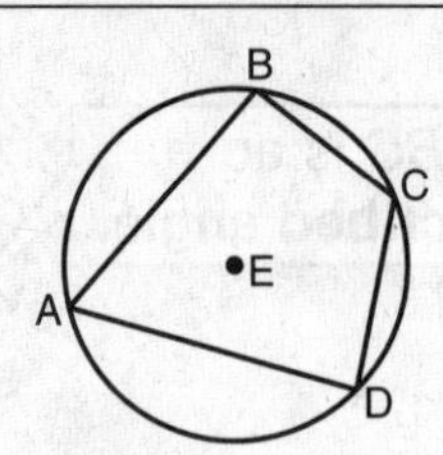 $ABCD$ is inscribed in $\odot E$.	$\angle A$ and $\angle C$ are supplementary. $\angle B$ and $\angle D$ are supplementary.

Find m$\angle G$.

Step 1 Find the value of z.

$m\angle E + m\angle G = 180°$ — *EFGH* is inscribed in a circle.

$4z + 3z + 5 = 180$ — Substitute the given values.

$7z = 175$ — Simplify.

$z = 25$ — Divide both sides by 7.

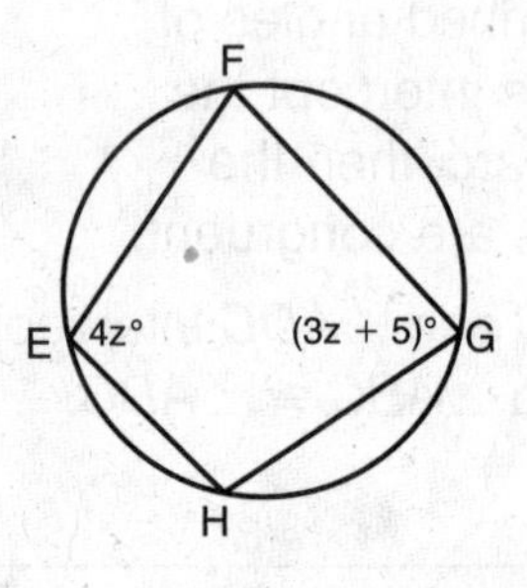

Step 2 Find the measure of $\angle G$.

$m\angle G = 3z + 5$

$= 3(25) + 5 = 80°$ — Substitute 25 for z.

Find the angle measures of each quadrilateral.

5. *RSTV*

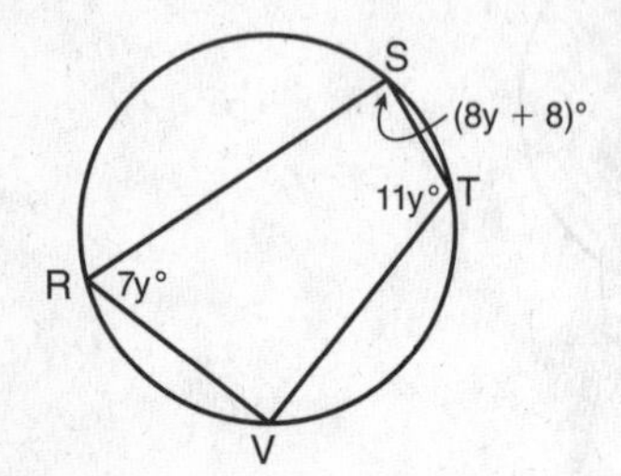

____

6. *ABCD*

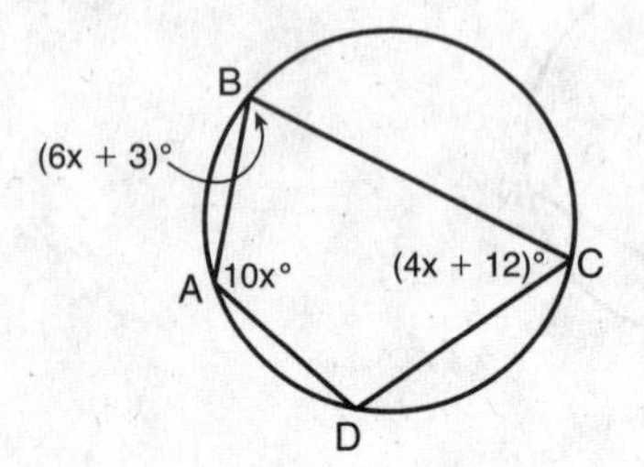

____

7. *JKLM*

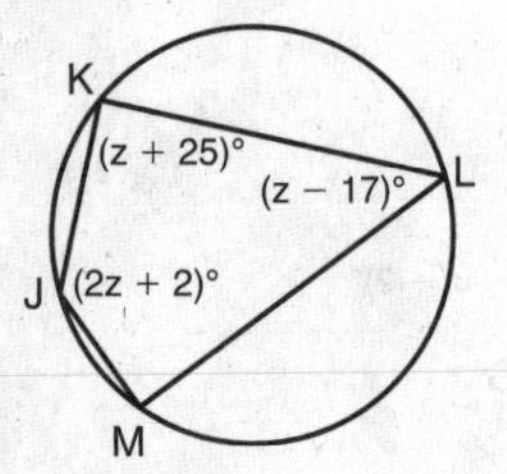

____

8. *MNPQ*

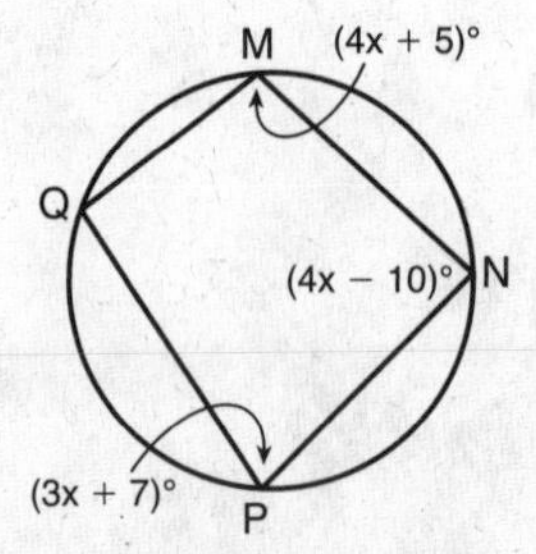

____

Name ______________________ Date ____________ Class ____________

LESSON 11-5

Review for Mastery

Angle Relationships in Circles

If a tangent and a secant (or chord) intersect on a circle at the point of tangency, then the measure of the angle formed is half the measure of its intercepted arc.	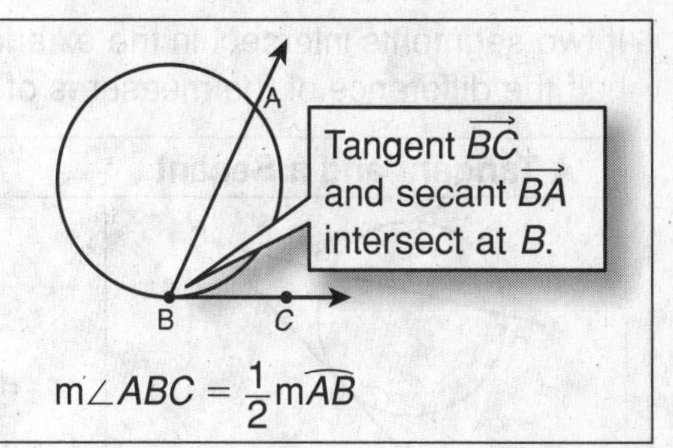 $m\angle ABC = \frac{1}{2}m\overset{\frown}{AB}$

If two secants or chords intersect in the interior of a circle, then the measure of the angle formed is half the sum of the measures of its intercepted arcs.	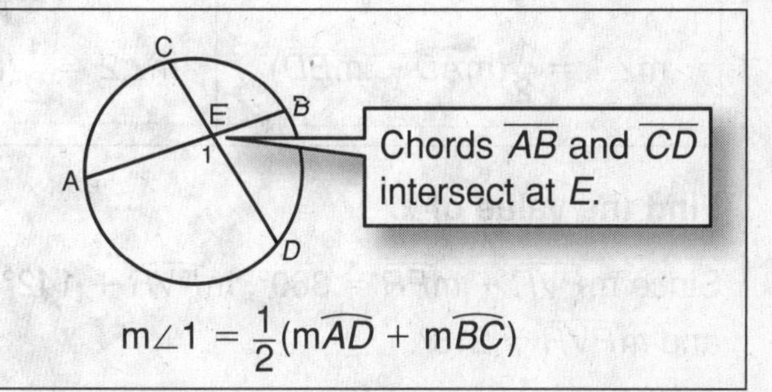 $m\angle 1 = \frac{1}{2}(m\overset{\frown}{AD} + m\overset{\frown}{BC})$

Find each measure.

1. $m\angle FGH$

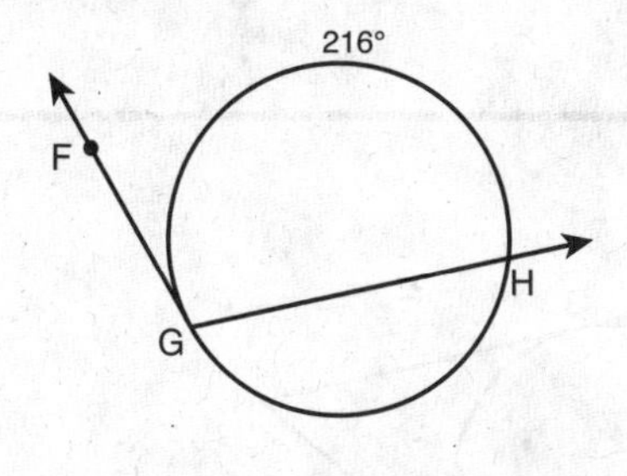

____

2. $m\overset{\frown}{LM}$

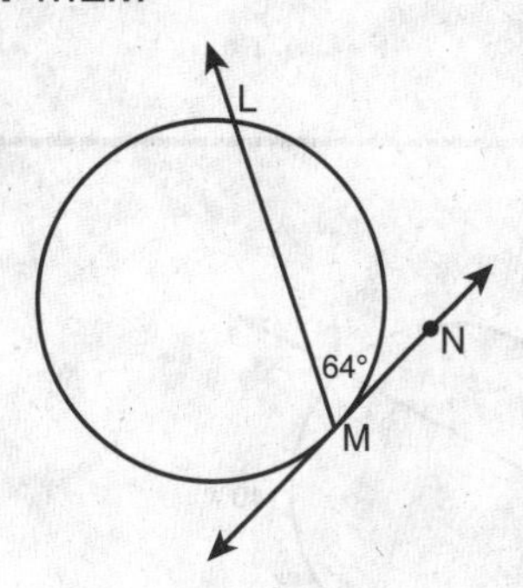

____

3. $m\angle JML$

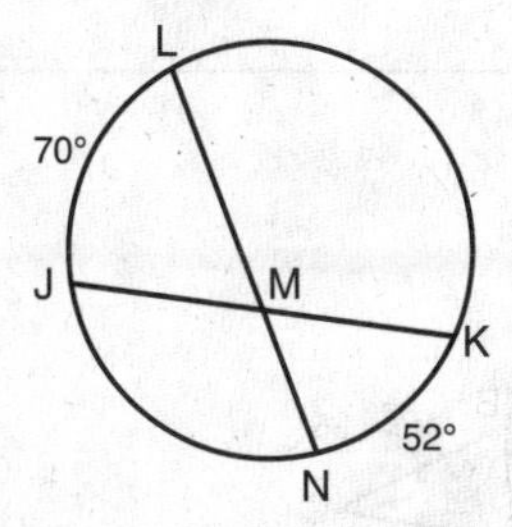

____

4. $m\angle STR$

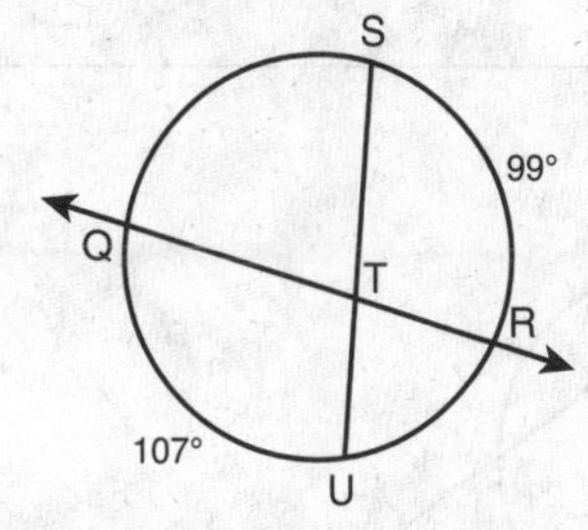

____

Name ______________________ Date ____________ Class ____________

LESSON 11-5

Review for Mastery

Angle Relationships in Circles continued

If two segments intersect in the exterior of a circle, then the measure of the angle formed is half the difference of the measures of its intercepted arcs.

A Tangent and a Secant	Two Tangents	Two Secants
$m\angle 1 = \frac{1}{2}(m\overset{\frown}{AD} - m\overset{\frown}{BD})$	$m\angle 2 = \frac{1}{2}(m\overset{\frown}{EHG} - m\overset{\frown}{EG})$	$m\angle 3 = \frac{1}{2}(m\overset{\frown}{JN} - m\overset{\frown}{KM})$

Find the value of *x*.

Since $m\overset{\frown}{PVR} + m\overset{\frown}{PR} = 360°$, $m\overset{\frown}{PVR} + 142° = 360°$, and $m\overset{\frown}{PVR} = 218°$.

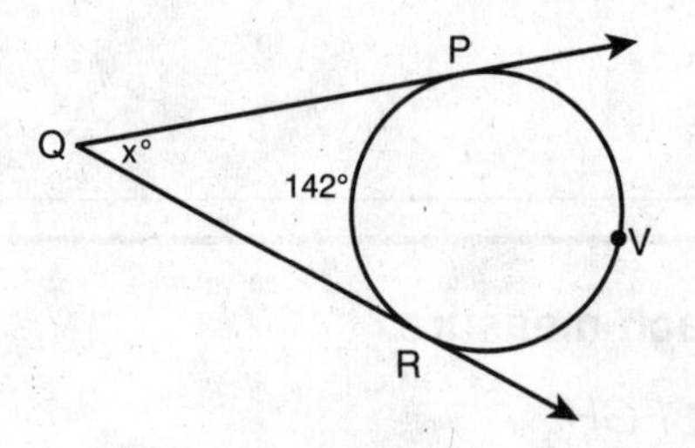

$x° = \frac{1}{2}(m\overset{\frown}{PVR} - m\overset{\frown}{PR})$

$= \frac{1}{2}(218° - 142°)$

$x° = 38°$

$x = 38$

Find the value of *x*.

5.

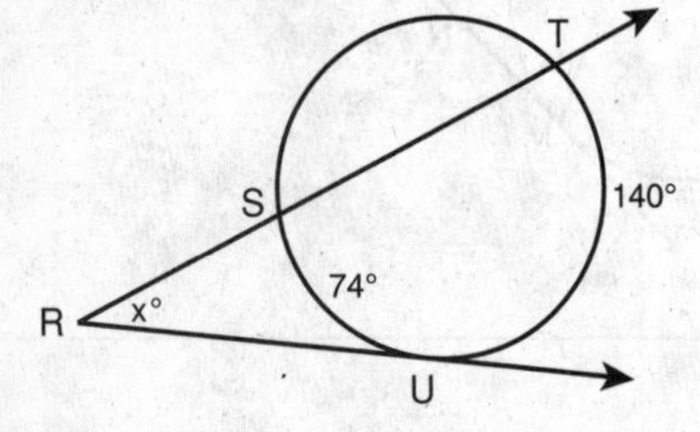

____

6.

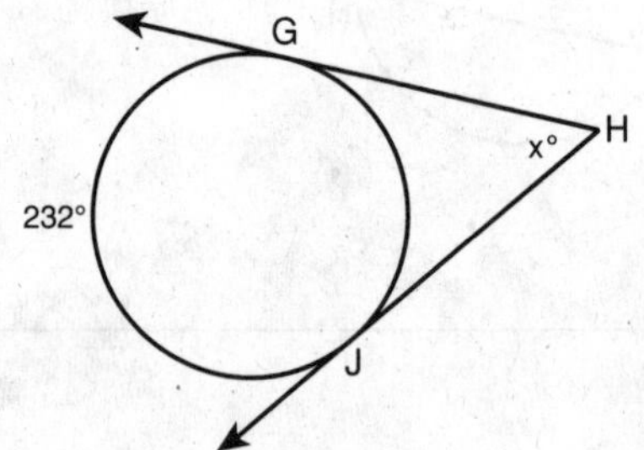

____

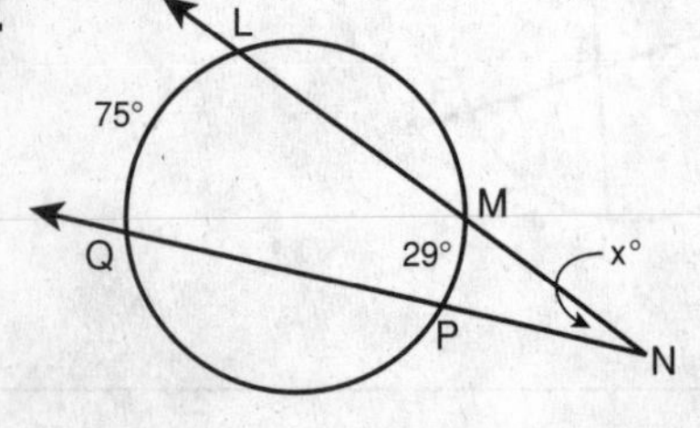

____

8.

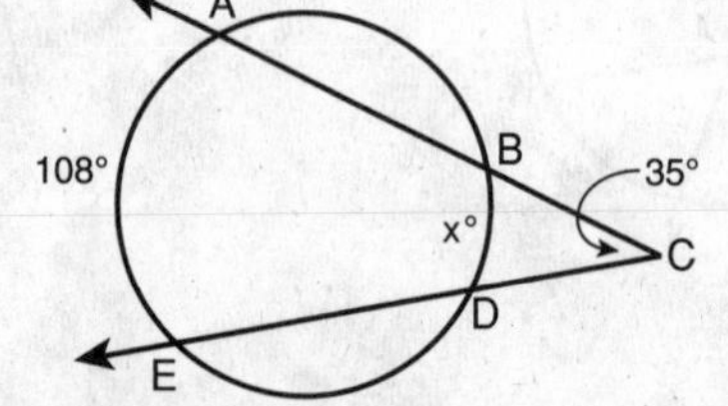

____

Name ______________________ Date __________ Class __________

LESSON 11-6

Review for Mastery

Segment Relationships in Circles

Chord-Chord Product Theorem	
If two chords intersect in the interior of a circle, then the products of the lengths of the segments of the chords are equal.	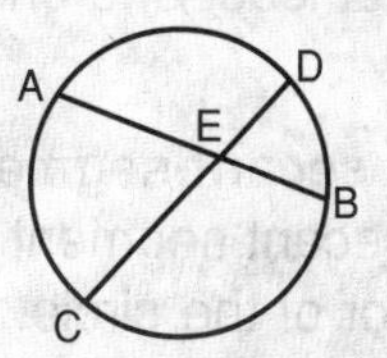 $AE \cdot EB = CE \cdot ED$

Find the value of *x* and the length of each chord.

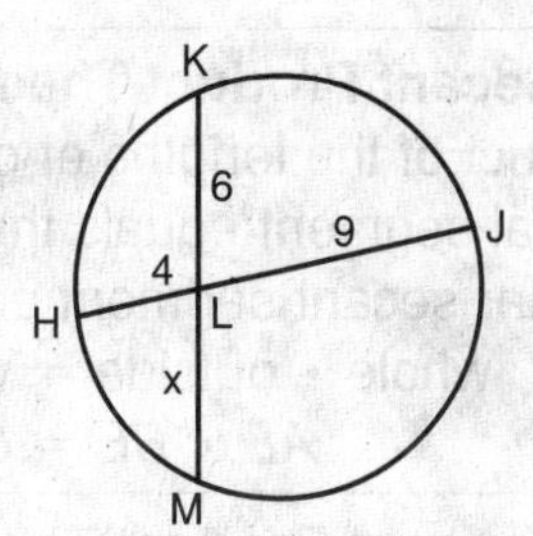

$HL \cdot LJ = KL \cdot LM$	Chord-Chord Product Thm.
$4 \cdot 9 = 6 \cdot x$	$HL = 4$, $LJ = 9$, $KL = 6$, $LM = x$
$36 = 6x$	Simplify.
$6 = x$	Divide each side by 6.

$HJ = 4 + 9 = 13$

$KM = 6 + x$

$= 6 + 6 = 12$

Find the value of the variable and the length of each chord.

1.

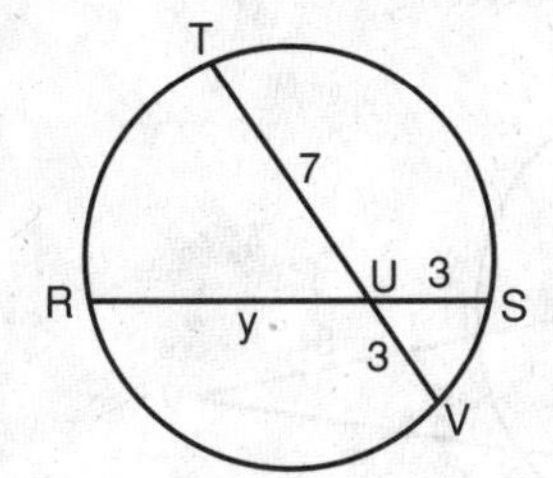

____

2.

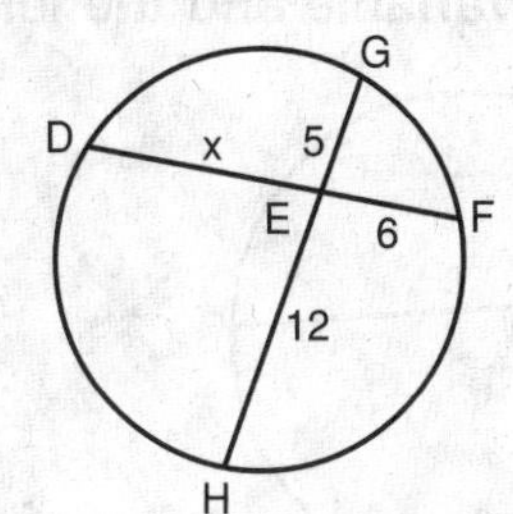

____

3.

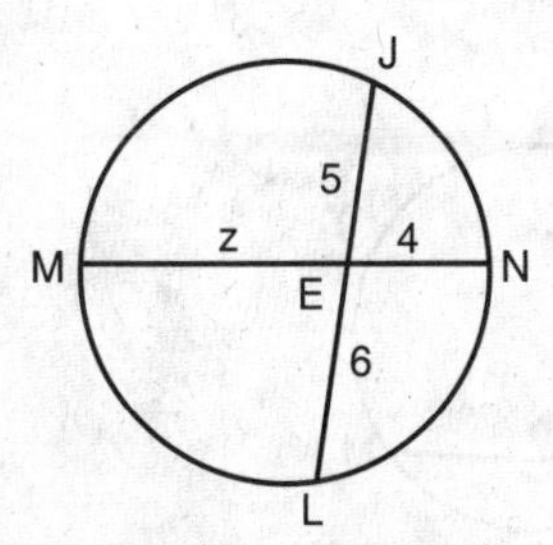

____

4.

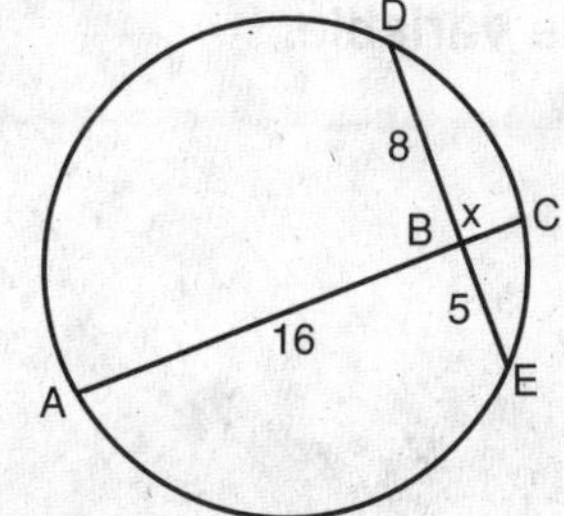

____

Name ______________________ Date __________ Class __________

LESSON 11-6

Review for Mastery

Segment Relationships in Circles continued

- A **secant segment** is a segment of a secant with at least one endpoint on the circle.
- An **external secant segment** is the part of the secant segment that lies in the exterior of the circle.
- A **tangent segment** is a segment of a tangent with one endpoint on the circle.

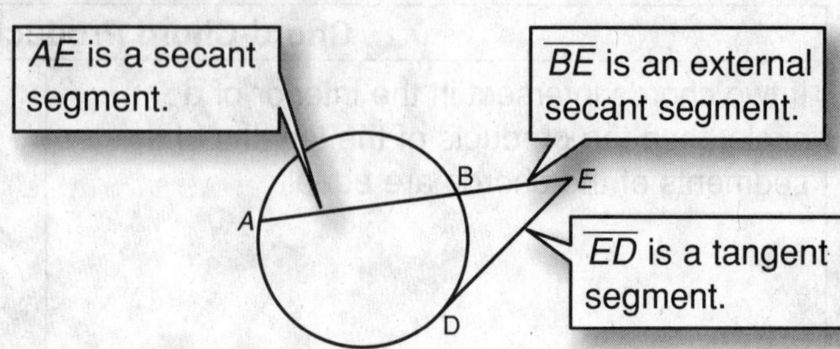

If two segments intersect outside a circle, the following theorems are true.

Secant-Secant Product Theorem The product of the lengths of one secant segment and its external segment equals the product of the lengths of the other secant segment and its external segment. whole • outside = whole • outside $AE \cdot BE = CE \cdot DE$	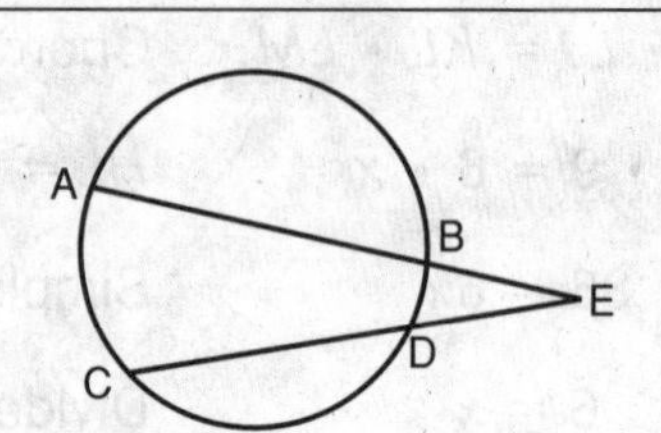
Secant-Tangent Product Theorem The product of the lengths of the secant segment and its external segment equals the length of the tangent segment squared. whole • outside = tangent^2 $AE \cdot BE = DE^2$	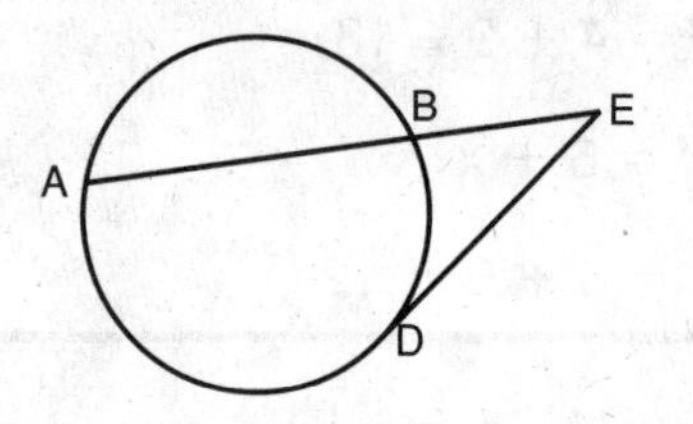

Find the value of the variable and the length of each secant segment.

5.

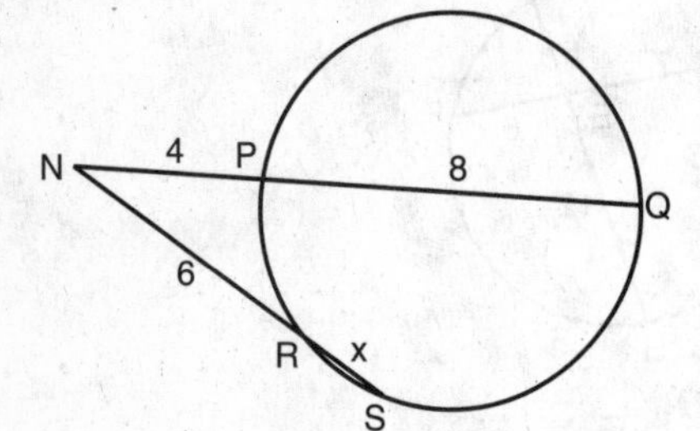

____

6.

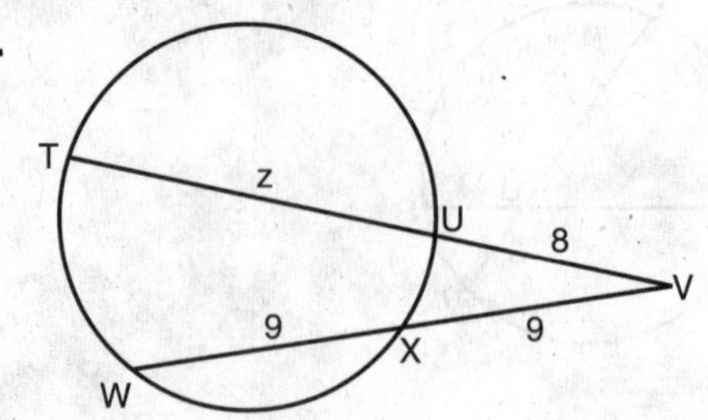

____

Find the value of the variable.

7.

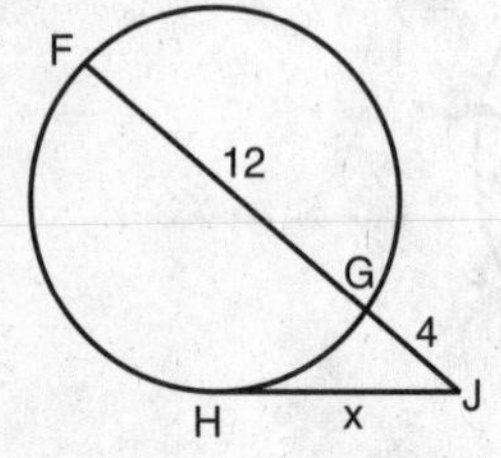

____

8.

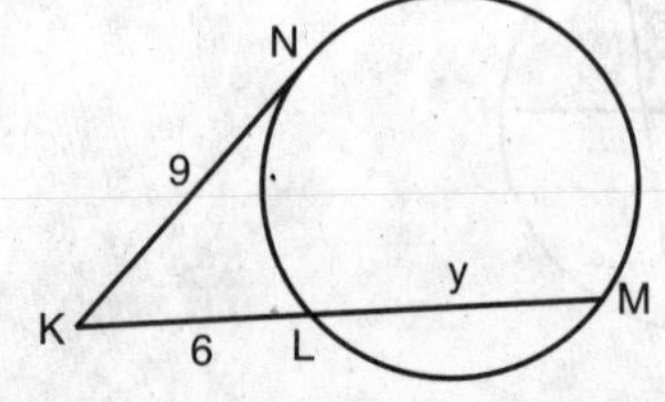

____

Name ______________________ Date ____________ Class ____________

LESSON 11-7

Review for Mastery

Circles in the Coordinate Plane

Equation of a Circle

The equation of a circle with center (h, k) and radius r is $(x - h)^2 + (y - k)^2 = r^2$.

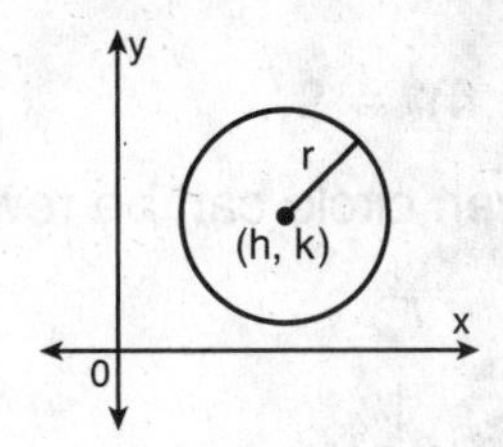

Write the equation of ⊙C with center $C(2, -1)$ and radius 6.

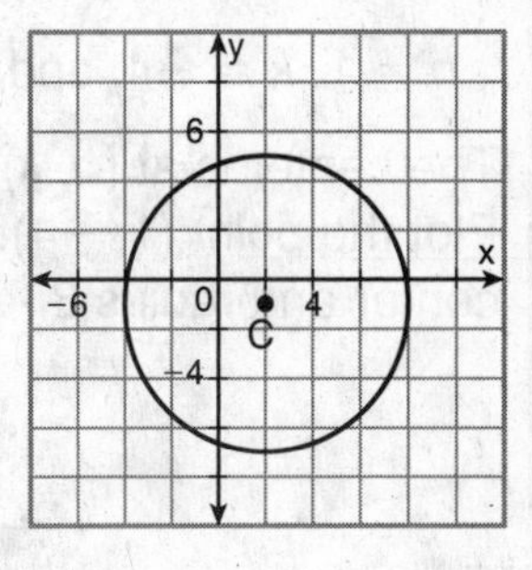

$(x - h)^2 + (y - k)^2 = r^2$	Equation of a circle
$(x - 2)^2 + (y - (-1))^2 = 6^2$	Substitute 2 for h, -1 for k, and 6 for r.
$(x - 2)^2 + (y + 1)^2 = 36$	Simplify.

You can also write the equation of a circle if you know the center and one point on the circle.

Write the equation of ⊙L that has center $L(3, 7)$ and passes through $(1, 7)$.

Step 1 Find the radius.

$r = \sqrt{(x_2 - x_1)^2 + (y_2 - y_1)^2}$	Distance Formula
$r = \sqrt{(1 - 3)^2 + (7 - 7)^2}$	Substitution
$r = \sqrt{4} = 2$	Simplify.

Step 2 Use the equation of a circle.

$(x - h)^2 + (y - k)^2 = r^2$	Equation of a circle
$(x - 3)^2 + (y - 7)^2 = 2^2$	$(h, k) = (3, 7)$
$(x - 3)^2 + (y - 7)^2 = 4$	Simplify.

Write the equation of each circle.

1. 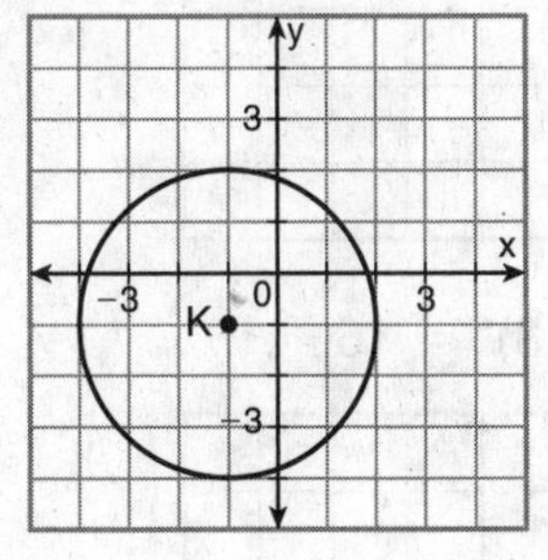

____

2. 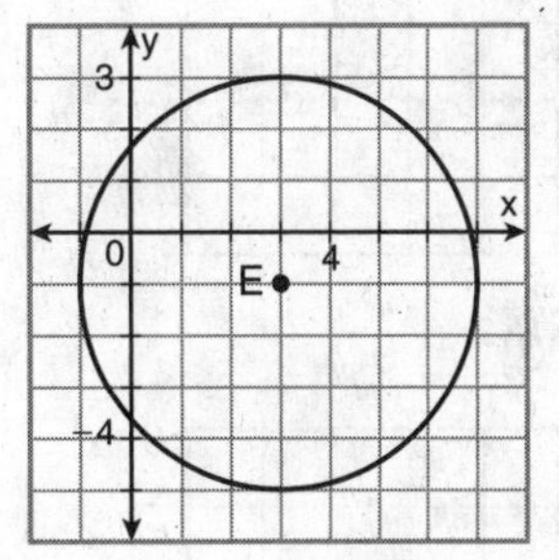

____

3. ⊙T with center $T(4, 5)$ and radius 8

____

4. ⊙B that passes through $(3, 6)$ and has center $B(-2, 6)$

____

Name ______________________________ Date ____________ Class ____________

LESSON 11-7

Review for Mastery

Circles in the Coordinate Plane continued

You can use an equation to graph a circle by making a table or by identifying its center and radius.

Graph $(x - 1)^2 + (y + 4)^2 = 9$.

The equation of the given circle can be rewritten.

$$(x - h)^2 + (y - k)^2 = r^2$$

$$\downarrow \quad \downarrow \quad \downarrow$$

$$(x - 1)^2 + (y - (-4))^2 = 3^2$$

$h = 1$, $k = -4$, and $r = 3$

The center is at (h, k) or $(1, -4)$, and the radius is 3. Plot the point $(1, -4)$. Then graph a circle having this center and radius 3.

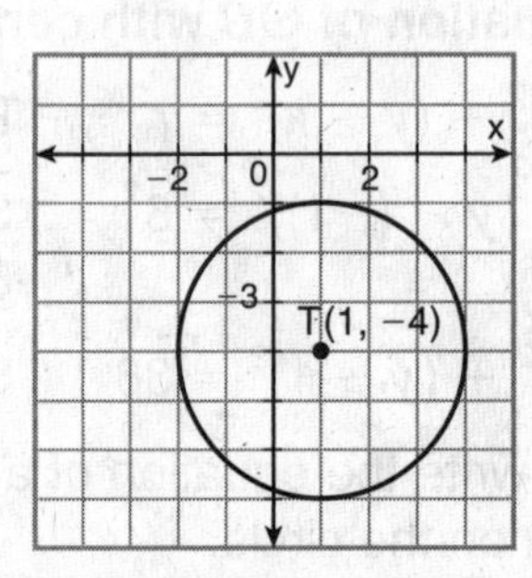

Graph each equation.

5. $(x - 1)^2 + (y - 2)^2 = 9$

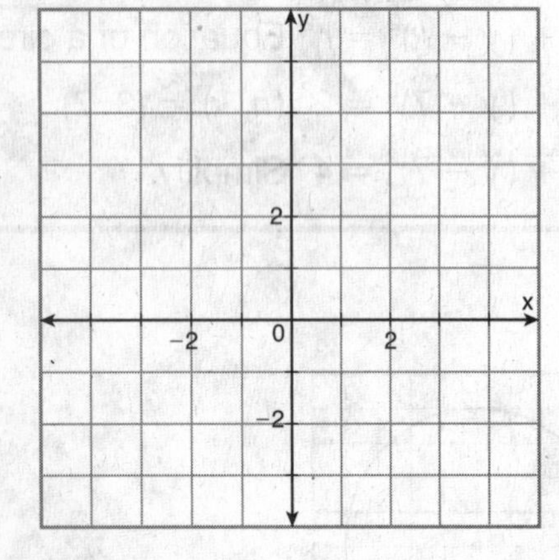

6. $(x - 3)^2 + (y + 1)^2 = 4$

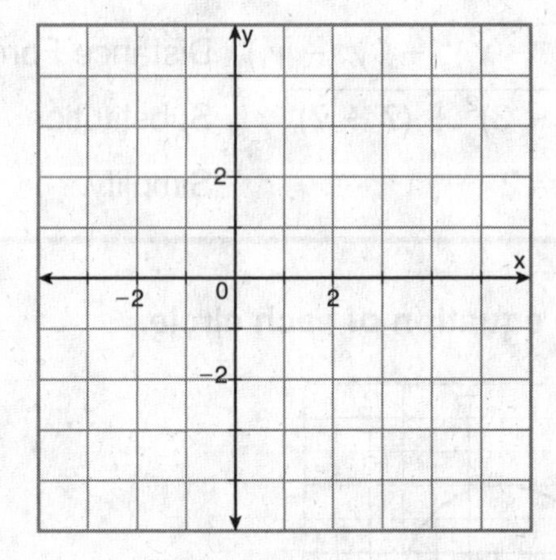

7. $(x + 2)^2 + (y - 2)^2 = 9$

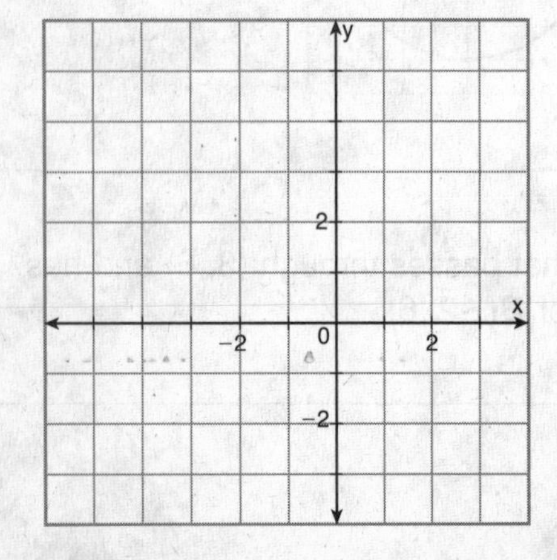

8. $(x + 1)^2 + (y + 3)^2 = 16$

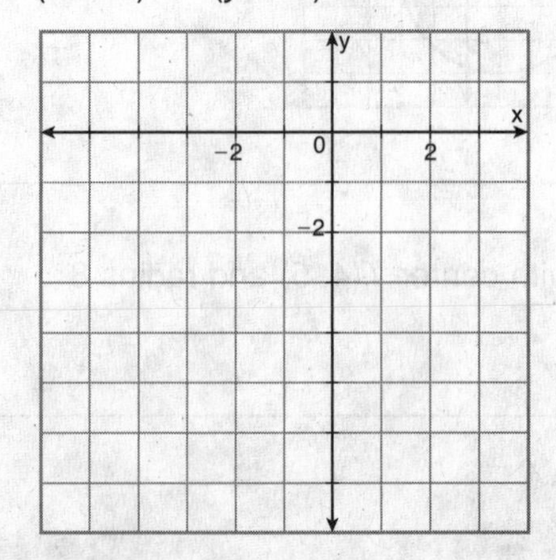

Name ______________________ Date ____________ Class ____________

LESSON 12-1

Review for Mastery

Reflections

An **isometry** is a transformation that does not change the shape or size of a figure. Reflections, translations, and rotations are all isometries.

A reflection is a transformation that flips a figure across a line.

Reflection

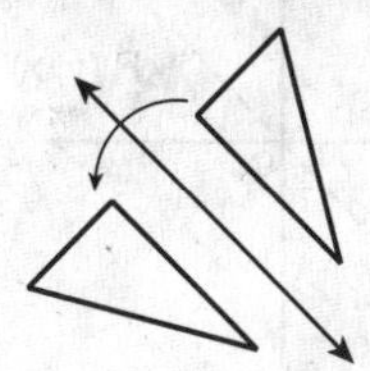

Not a Reflection

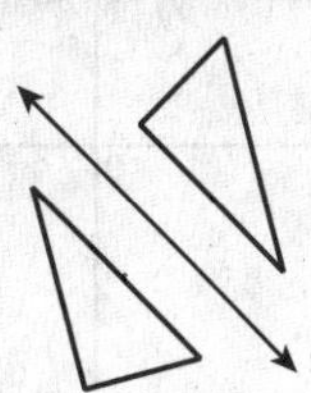

The line of reflection is the perpendicular bisector of each segment joining each point and its image.

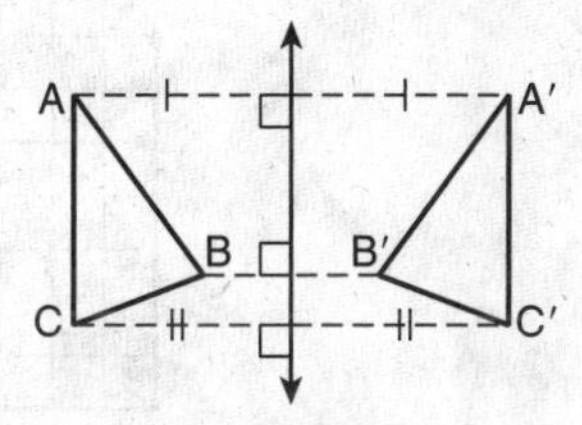

Tell whether each transformation appears to be a reflection.

1.

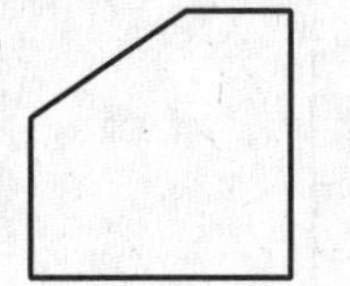

____

2.

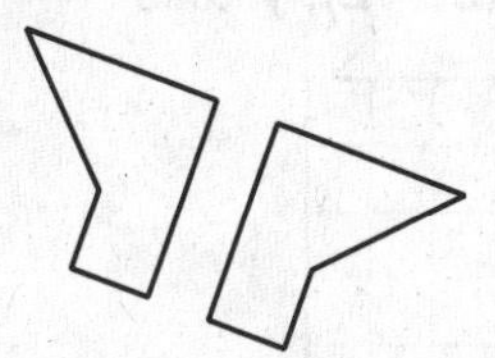

____

Copy each figure and the line of reflection. Draw the reflection of the figure across the line.

3.

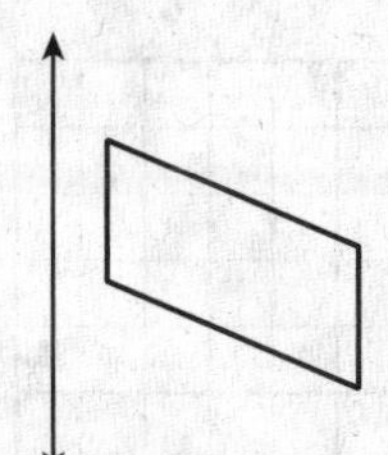

4.

Name ______________________ Date ____________ Class ____________

LESSON 12-1

Review for Mastery

Reflections continued

Reflections in the Coordinate Plane		
Across the *x*-axis	**Across the *y*-axis**	**Across the line *y* = *x***
R(x, y); R′(x, −y)	R′(−x, y); R(x, y)	R(x, y); R′(y, x); y = x
$(x, y) \rightarrow (x, -y)$	$(x, y) \rightarrow (-x, y)$	$(x, y) \rightarrow (y, x)$

Reflect $\triangle FGH$ with vertices $F(-1, 4)$, $G(2, 4)$, and $H(4, 1)$ across the *x*-axis.

The reflection of (x, y) is $(x, -y)$.

$F(-1, 4) \rightarrow F'(-1, -4)$

$G(2, 4) \rightarrow G'(2, -4)$

$H(4, 1) \rightarrow H'(4, -1)$

Graph the preimage and image.

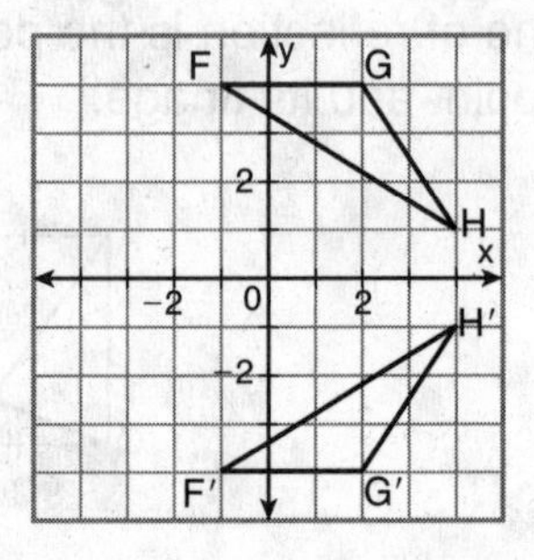

Reflect the figure with the given vertices across the line.

5. $M(2, 4)$, $N(4, 2)$, $P(3, -2)$; *y*-axis

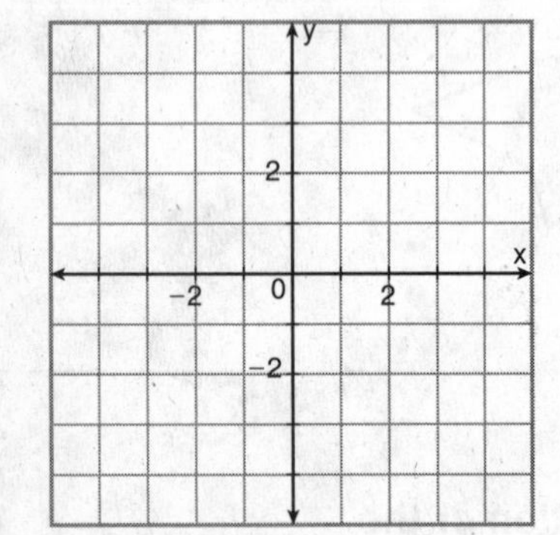

6. $T(-4, 1)$, $U(-3, 4)$, $V(2, 3)$, $W(0, 1)$; *x*-axis

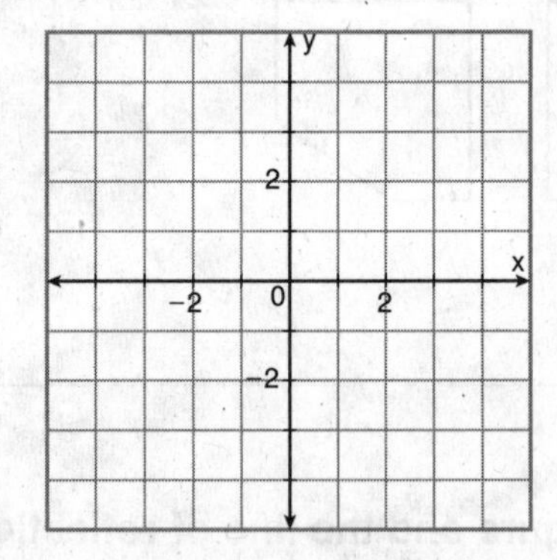

7. $Q(-3, -1)$, $R(2, 4)$, $S(2, 1)$; *x*-axis

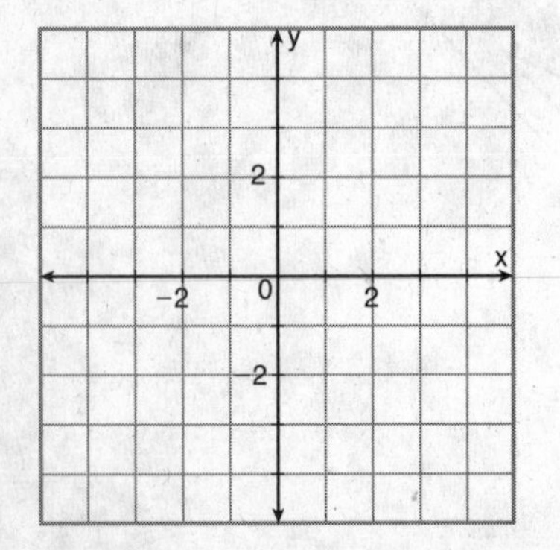

8. $A(-2, 4)$, $B(1, 1)$, $C(-5, -1)$; $y = x$

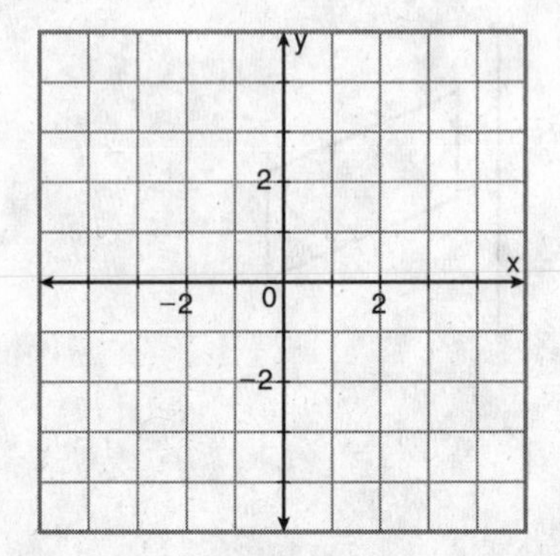

Name ______________________ Date __________ Class __________

LESSON 12-2

Review for Mastery

Translations

A translation is a transformation in which all the points of a figure are moved the same distance in the same direction.

Translation **Not a Translation**

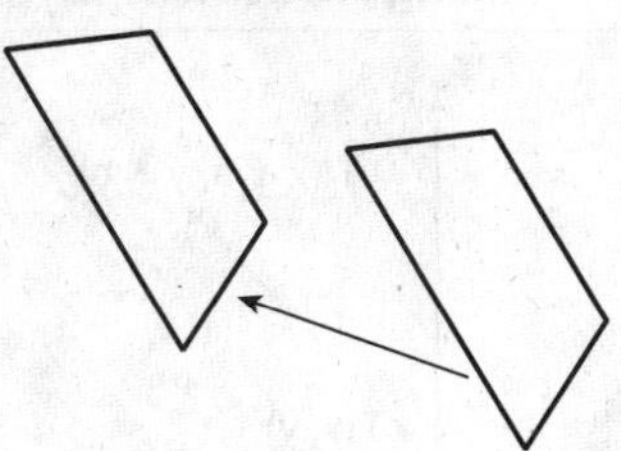

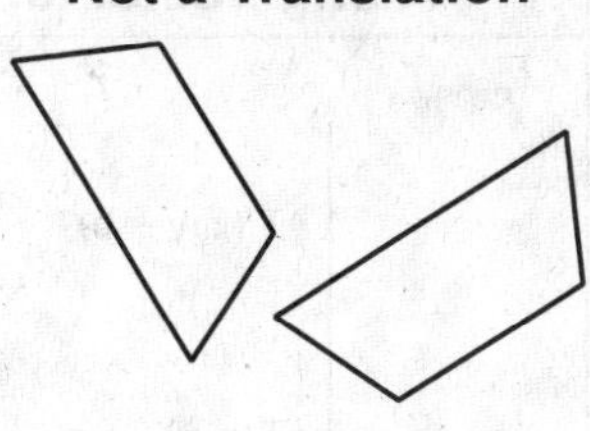

A translation is a transformation along a vector such that each segment joining a point and its image has the same length as the vector and is parallel to the vector.

$\overline{AA'}$, $\overline{BB'}$, and $\overline{CC'}$ have the same length as $\vec{v}$ and are parallel to $\vec{v}$.

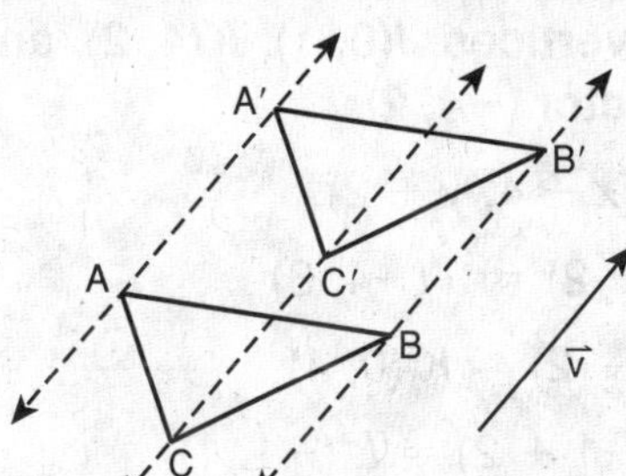

Tell whether each transformation appears to be a translation.

1.

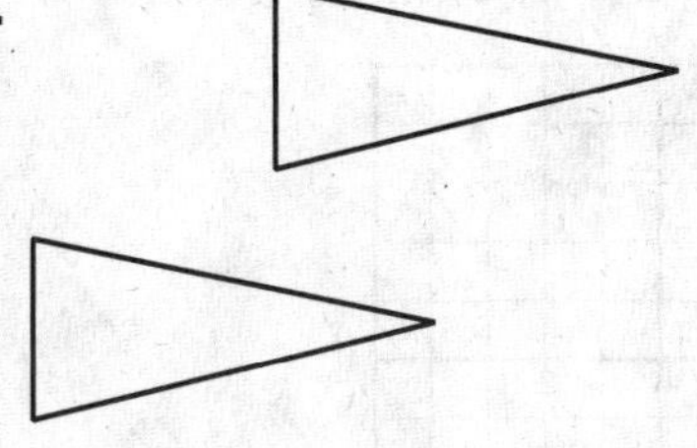

____

2.

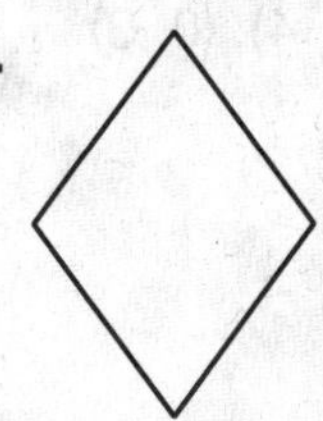

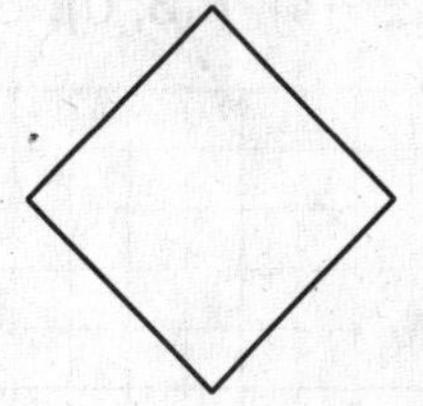

____

Copy each figure and the translation vector. Draw the translation of the figure along the given vector.

3.

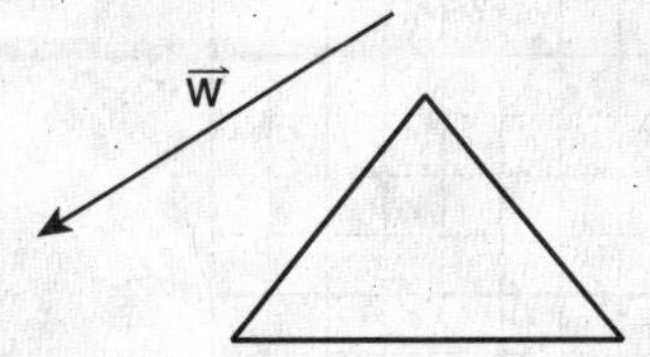

4.

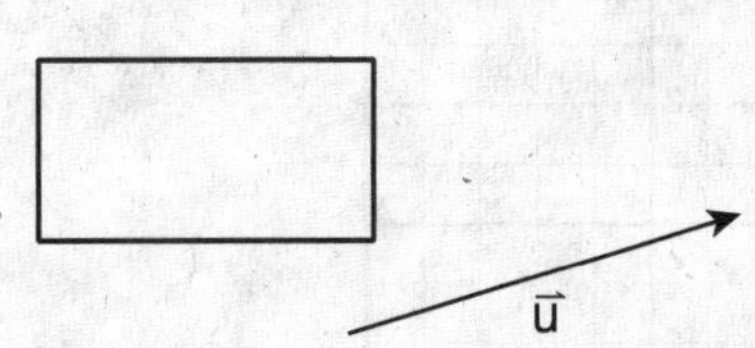

Name ______________________________ Date ____________ Class ____________

LESSON 12-2

Review for Mastery

Translations continued

Translations in the Coordinate Plane		
Horizontal Translation Along Vector $\langle a, 0\rangle$	**Horizontal Translation Along Vector $\langle 0, b\rangle$**	**Horizontal Translation Along Vector $\langle a, b\rangle$**
y; T(x, y) T′(x + a, y); 0; x	y; T′(x, y + b); T(x, y); 0; x	y; T′(x + a, y + b); T(x, y); 0; x
$(x, y) \rightarrow (x + a, y)$	$(x, y) \rightarrow (x, y + b)$	$(x, y) \rightarrow (x + a, y + b)$

Translate $\triangle JKL$ with vertices $J(0, 1)$, $K(4, 2)$, and $L(3, -1)$ along the vector $\langle -4, 2\rangle$.

The image of (x, y) is $(x - 4, y + 2)$.

$J(0, 1) \rightarrow J'(0 - 4, 1 + 2) = J'(-4, 3)$

$K(4, 2) \rightarrow K'(4 - 4, 2 + 2) = K'(0, 4)$

$L(3, -1) \rightarrow L'(3 - 4, -1 + 2) = L'(-1, 1)$

Graph the preimage and image.

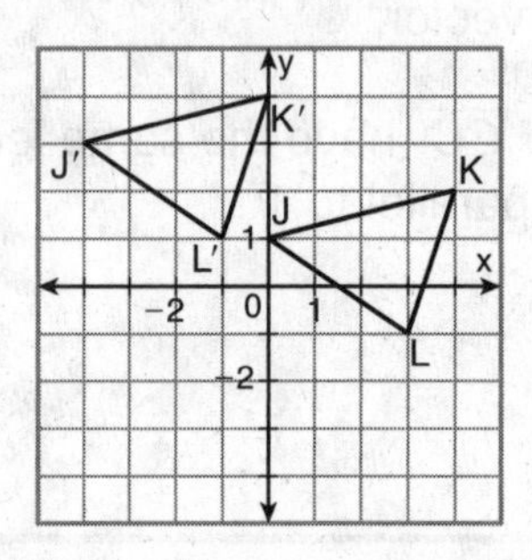

Translate the figure with the given vertices along the given vector.

5. $E(-2, -4)$, $F(3, 0)$, $G(3, -4)$; $\langle 0, 3\rangle$

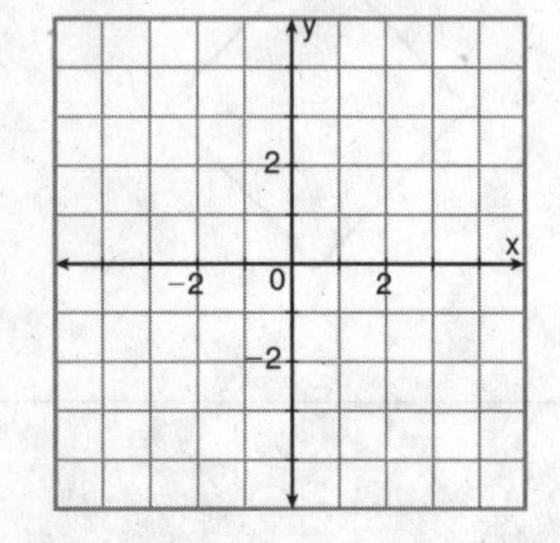

6. $P(-4, -1)$, $Q(-1, 3)$, $R(0, -4)$; $\langle 4, 1\rangle$

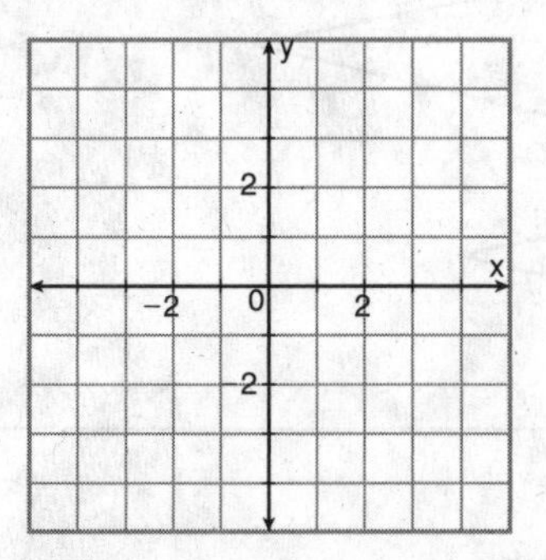

7. $A(1, -2)$, $B(1, 0)$, $C(3, 1)$, $D(4, -3)$; $\langle -5, 3\rangle$

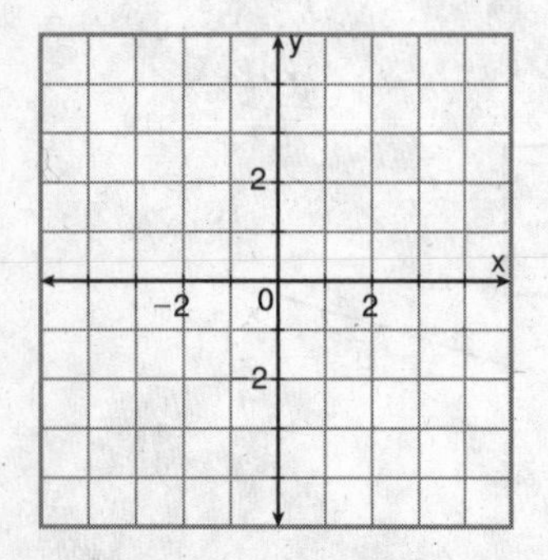

8. $G(-3, 4)$, $H(4, 3)$, $J(1, 2)$; $\langle -1, -6\rangle$

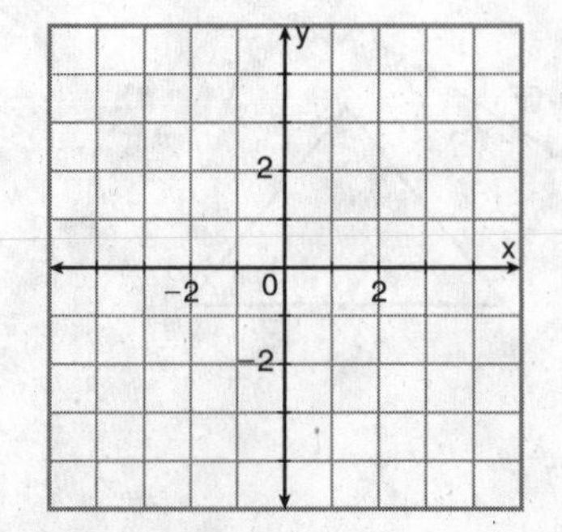

Name ______________________ Date __________ Class __________

LESSON 12-3

Review for Mastery

Rotations

A rotation is a transformation that turns a figure around a fixed point, called the center of rotation.

Rotation **Not a Rotation**

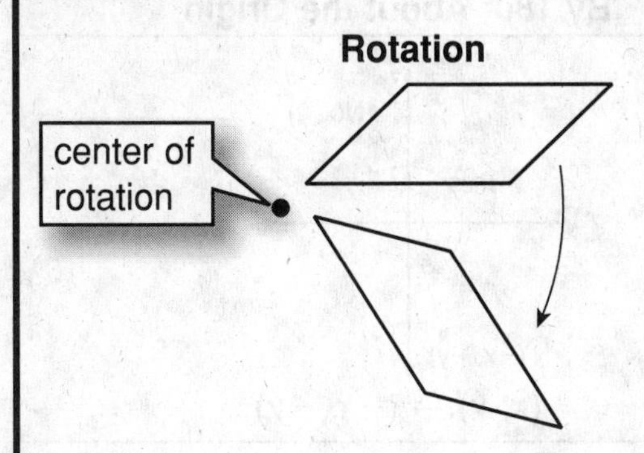

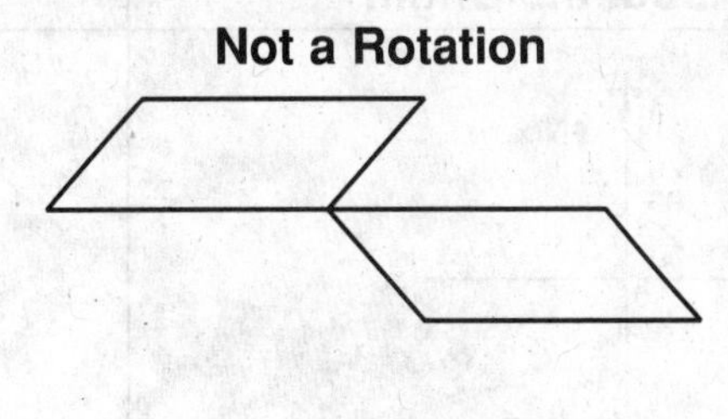

A rotation is a transformation about a point *P* such that each point and its image are the same distance from *P*.

$PQ = PQ'$

$PR = PR'$

$PS = PS'$

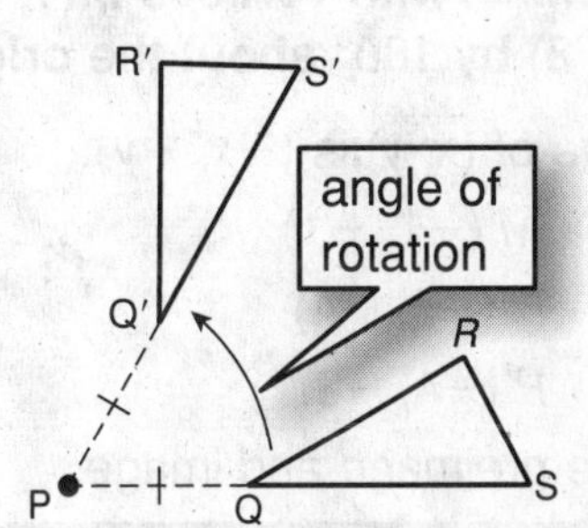

Tell whether each transformation appears to be a rotation.

1.

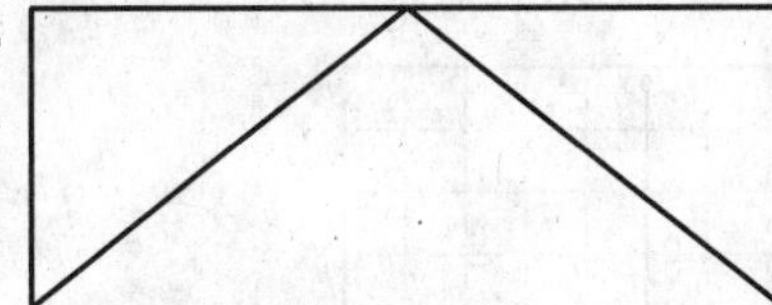

____

2.

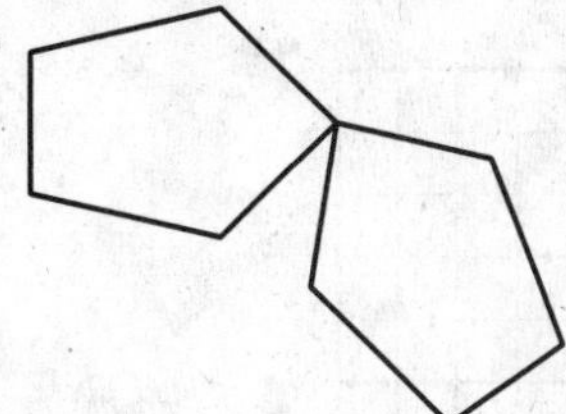

____

Copy each figure and the angle of rotation. Draw the rotation of the figure about point *P* by m∠*A*.

3.

4.

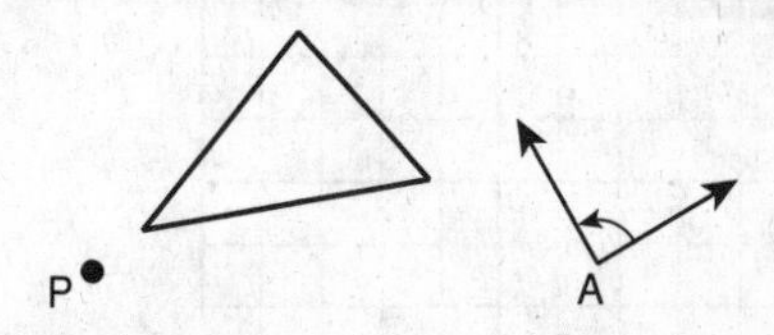

Name ______________________ Date __________ Class __________

LESSON 12-3

Review for Mastery

Rotations continued

Rotations in the Coordinate Plane	
By 90° About the Origin	**By 180° About the Origin**
N(x, y); N′(−y, x); 90°	N(x, y); N′(−x, −y); 180°
$(x, y) \rightarrow (-y, x)$	$(x, y) \rightarrow (-x, -y)$

Rotate △*MNP* with vertices *M*(1, 1), *N*(2, 4), and *P*(4, 3) by 180° about the origin.

The image of (x, y) is $(-x, -y)$.

$M(1, 1) \rightarrow M'(-1, -1)$

$N(2, 4) \rightarrow N'(-2, -4)$

$P(4, 3) \rightarrow P'(-4, -3)$

Graph the preimage and image.

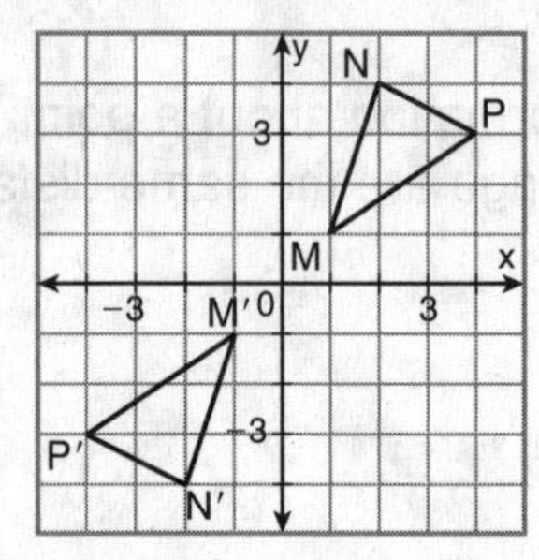

Rotate the figure with the given vertices about the origin using the given angle.

5. $R(0, 0)$, $S(3, 1)$, $T(2, 4)$; 90°

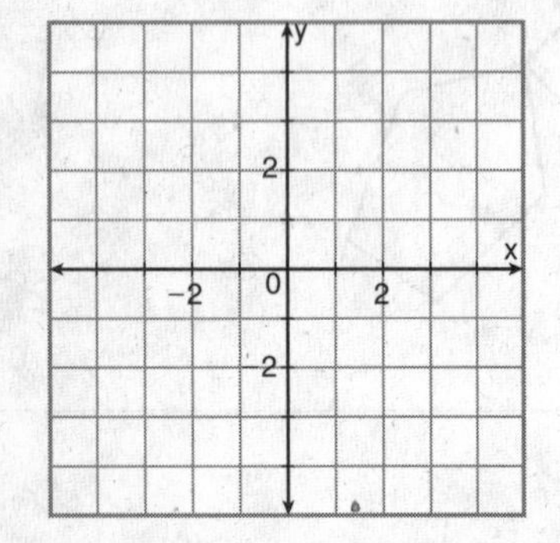

6. $A(0, 0)$, $B(-4, 2)$, $C(-1, 4)$; 180°

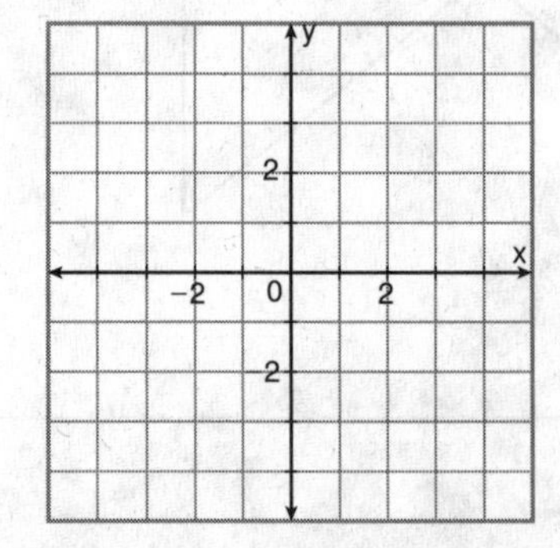

7. $E(0, 3)$, $F(3, 5)$, $G(4, 0)$; 180°

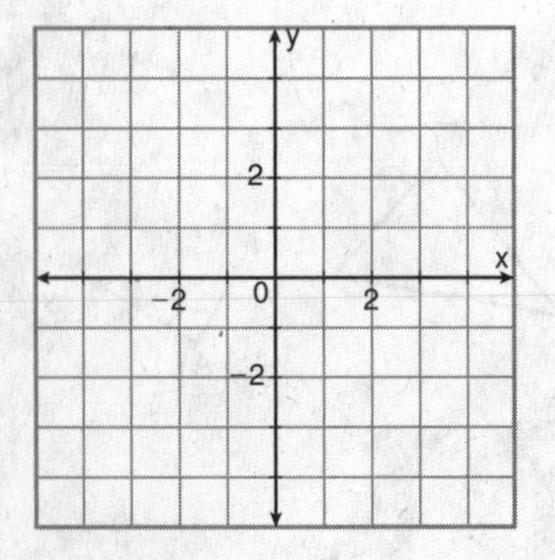

8. $U(1, -1)$, $V(4, -2)$, $W(3, -4)$; 90°

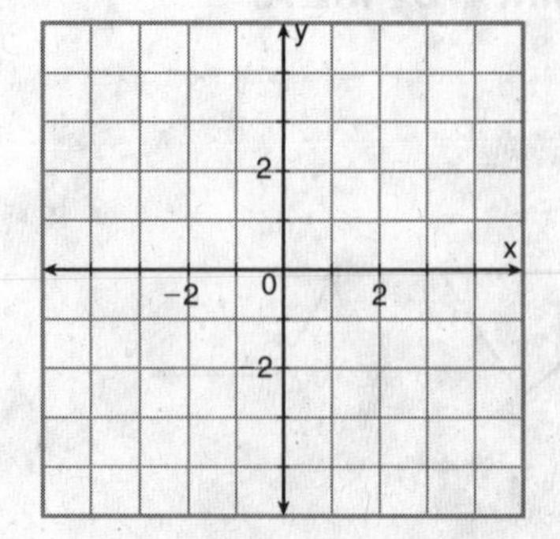

Name ______________________________ Date ______________ Class ______________

LESSON 12-4

Review for Mastery

Compositions of Transformations

A **composition of transformations** is one transformation followed by another. A **glide reflection** is the composition of a translation and a reflection across a line parallel to the vector of the translation.

Reflect △*ABC* across line ℓ along $\vec{v}$ and then translate it parallel to $\vec{v}$.

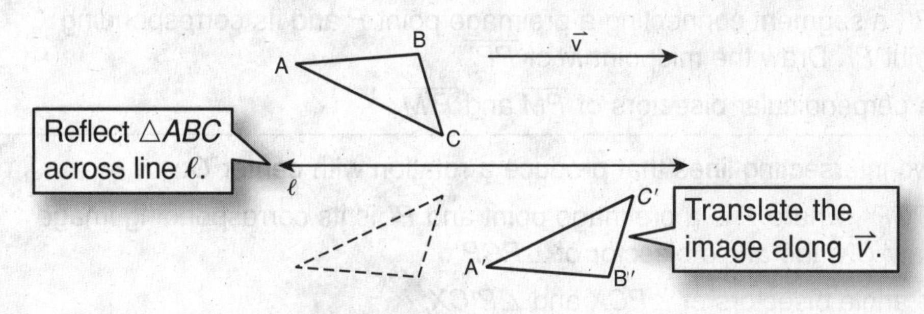

Draw the result of each composition of transformations.

1. Translate △*HJK* along $\vec{v}$ and then reflect it across line *m*.

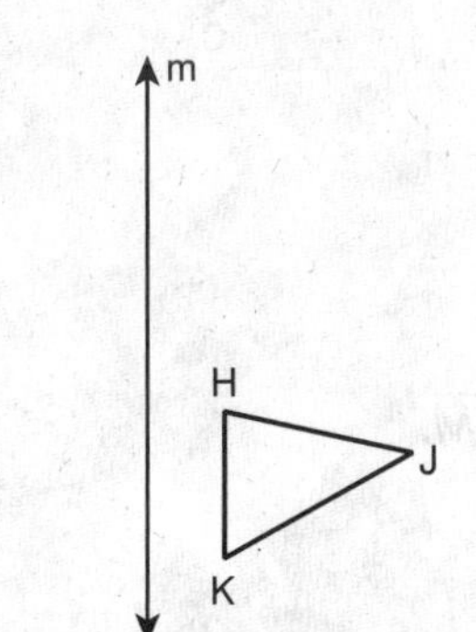

2. Reflect △*DEF* across line *k* and then translate it along $\vec{u}$.

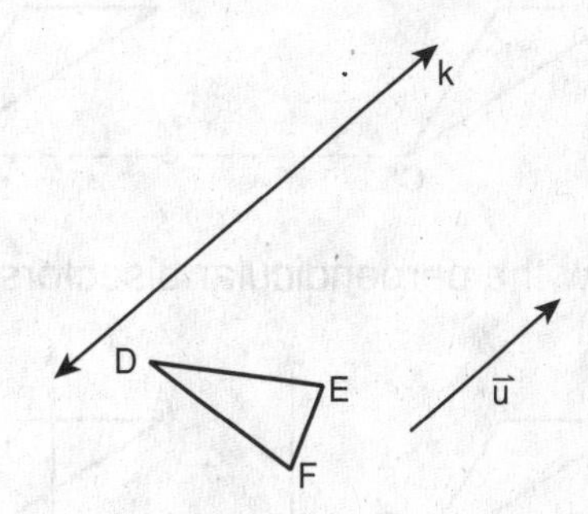

3. △*ABC* has vertices *A*(0, −1), *B*(3, 4), and *C*(3, 1). Rotate △*ABC* 180° about the origin and then reflect it across the *x*-axis.

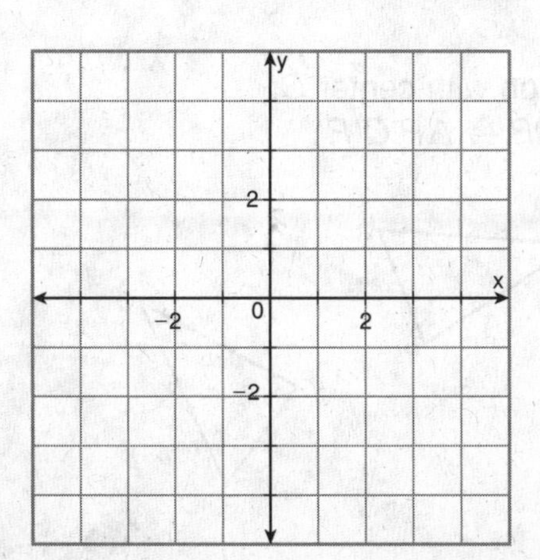

4. △*QRS* has vertices *Q*(2, 1), *R*(4, −2), and *S*(1, −3). Reflect △*QRS* across the *y*-axis and then translate it along the vector ⟨1, 3⟩.

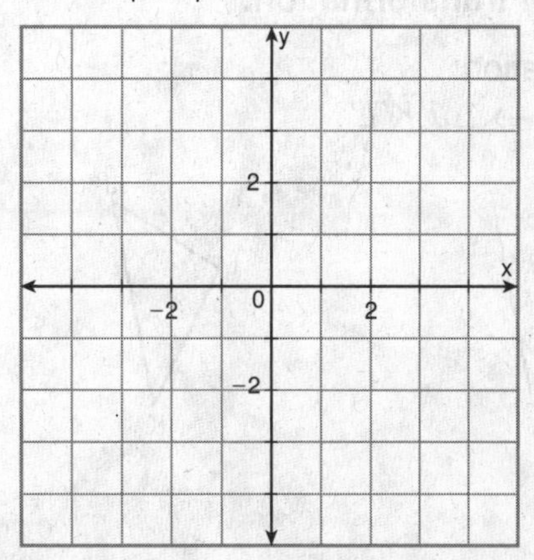

Name ______________________ Date ____________ Class ____________

LESSON 12-4

Review for Mastery

Compositions of Transformations continued

Any translation or rotation is equivalent to a composition of two reflections.

Composition of Two Reflections
To draw two parallel lines of reflection that produce a translation: • Draw $\overline{PP'}$, a segment connecting a preimage point P and its corresponding image point P'. Draw the midpoint M of $\overline{PP'}$. • Draw the perpendicular bisectors of $\overline{PM}$ and $\overline{P'M}$.
To draw two intersecting lines that produce a rotation with center C: • Draw $\angle PCP'$, where P is a preimage point and P' is its corresponding image point. Draw $\overrightarrow{CX}$, the angle bisector of $\angle PCP'$. • Draw the angle bisectors of $\angle PCX$ and $\angle P'CX$.

Copy $\triangle ABC$ and draw two lines of reflection that produce the translation $\triangle ABC \rightarrow \triangle A'B'C'$.

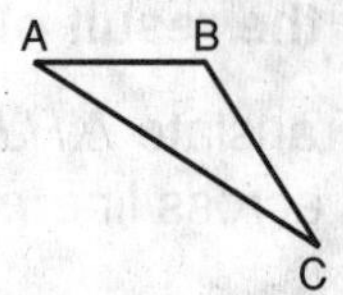

Step 1 Draw $\overline{CC'}$ and the midpoint M of $\overline{CC'}$.

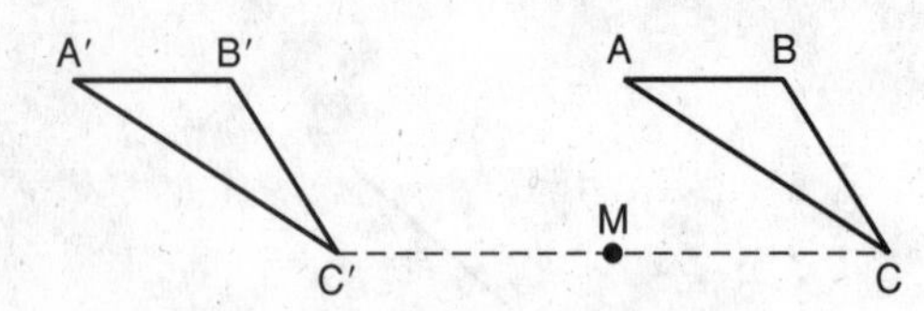

Step 2 Draw the perpendicular bisectors of $\overline{CM}$ and $\overline{C'M}$.

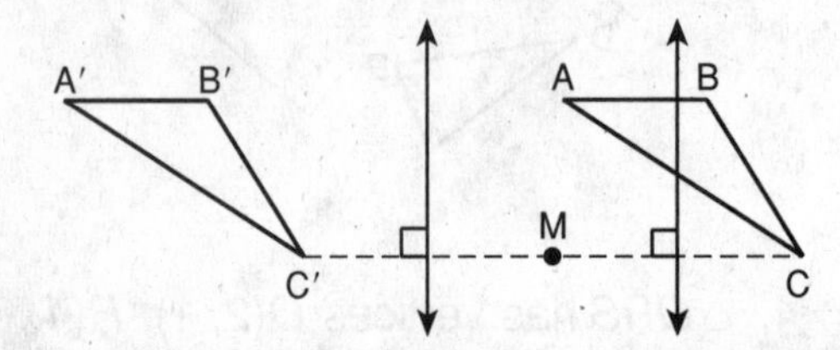

Copy each figure and draw two lines of reflection that produce an equivalent transformation.

5. translation:
$\triangle JKL \rightarrow \triangle J'K'L'$

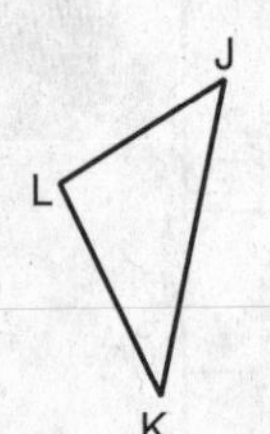

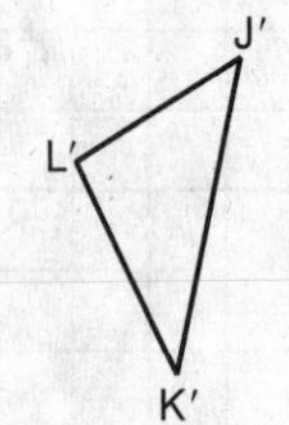

6. rotation with center C:
$\triangle PQR \rightarrow \triangle P'Q'R'$

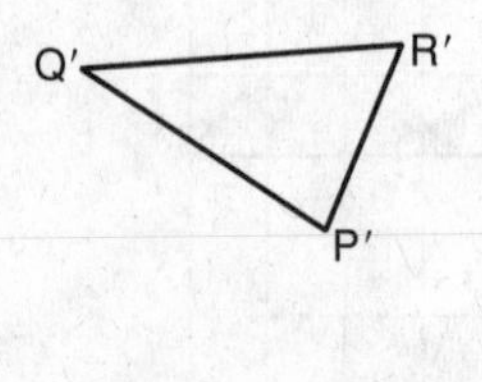

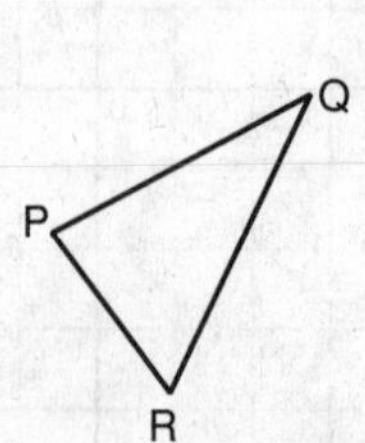

Name ______________________ Date ____________ Class ____________

LESSON 12-5

Review for Mastery

Symmetry

A figure has **symmetry** if there is a transformation of the figure such that the image and preimage are identical. There are two kinds of symmetry.

Line Symmetry	The figure has a **line of symmetry** that divides the figure into two congruent halves. one line of symmetry two lines of symmetry no line symmetry
Rotational Symmetry	When a figure is rotated between 0° and 360°, the resulting figure coincides with the original. • The smallest angle through which the figure is rotated to coincide with itself is called the *angle of rotational symmetry.* • The number of times that you can get an identical figure when repeating the degree of rotation is called the *order* of the rotational symmetry. angle: 180°, order: 2 angle: 120°, order: 3 no rotational symmetry

Tell whether each figure has line symmetry. If so, draw all lines of symmetry.

1.

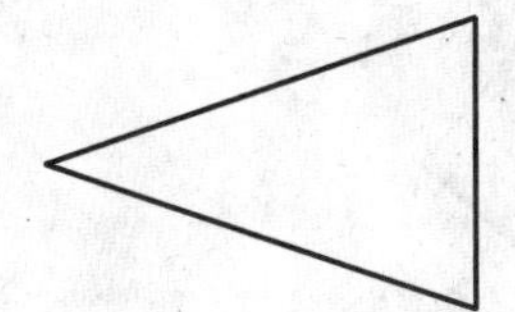

____

2.

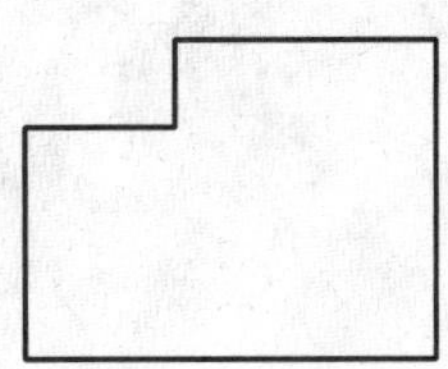

____

Tell whether each figure has rotational symmetry. If so, give the angle of rotational symmetry and the order of the symmetry.

3.

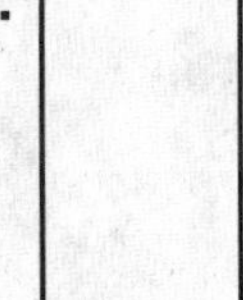

____

4.

____

Name ______________________ Date __________ Class __________

LESSON 12-5

Review for Mastery

Symmetry continued

Three-dimensional figures can also have symmetry.

Symmetry in Three Dimensions	Description	Example
Plane Symmetry	A plane can divide a figure into two congruent halves.	
Symmetry About an Axis	There is a line about which a figure can be rotated so that the image and preimage are identical.	

A cone has both plane symmetry and symmetry about an axis.

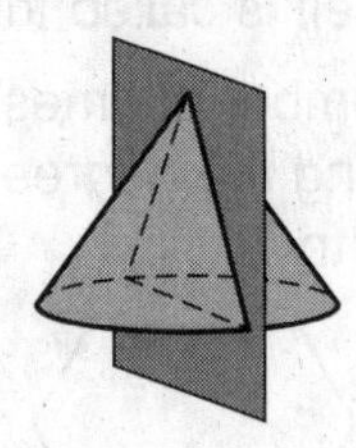

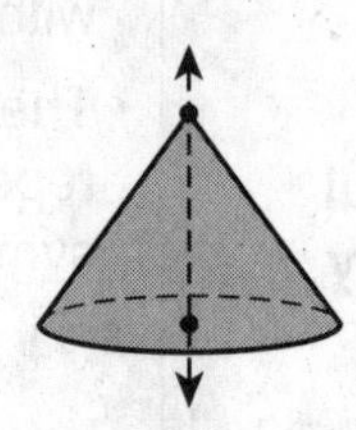

Tell whether each figure has plane symmetry, symmetry about an axis, both, or neither.

5. square pyramid

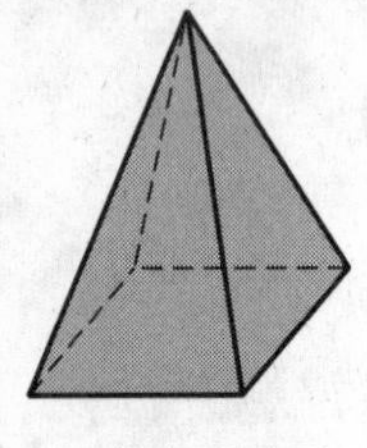

____

6. prism

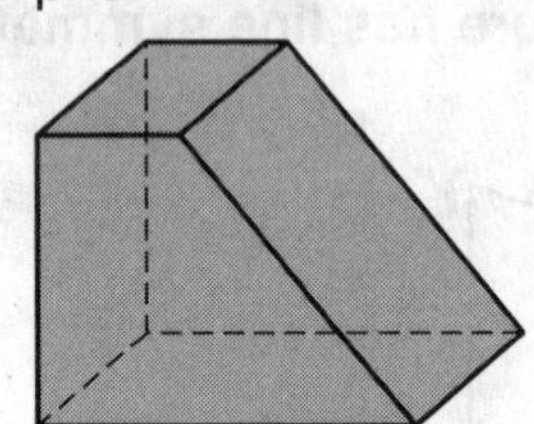

____

7. triangular pyramid

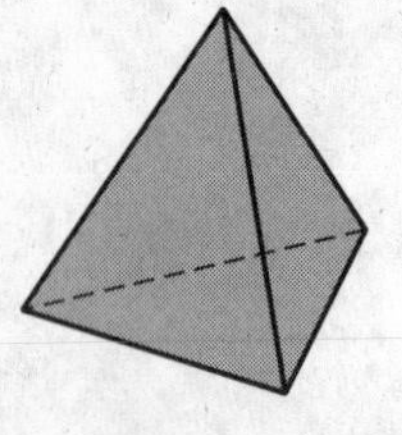

____

8. cylinder

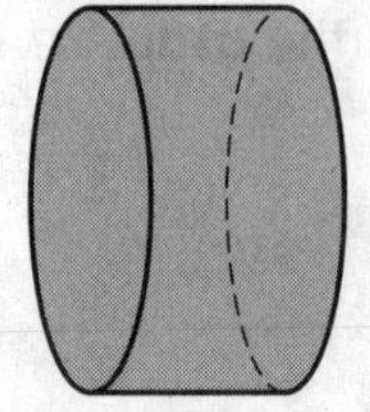

____

Name ______________________________ Date __________ Class __________

LESSON 12-6

Review for Mastery

Tessellations

A pattern has **translation symmetry** if it can be translated along a vector so that the image coincides with the preimage. A pattern with **glide reflection symmetry** coincides with its image after a glide reflection.

Translation Symmetry

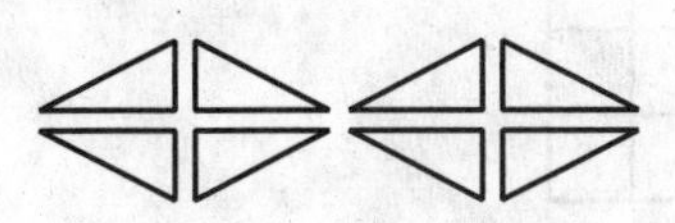

Translation Symmetry and Glide Reflection Symmetry

A **tessellation** is a repeating pattern that completely covers a plane with no gaps or overlaps.

Tessellation

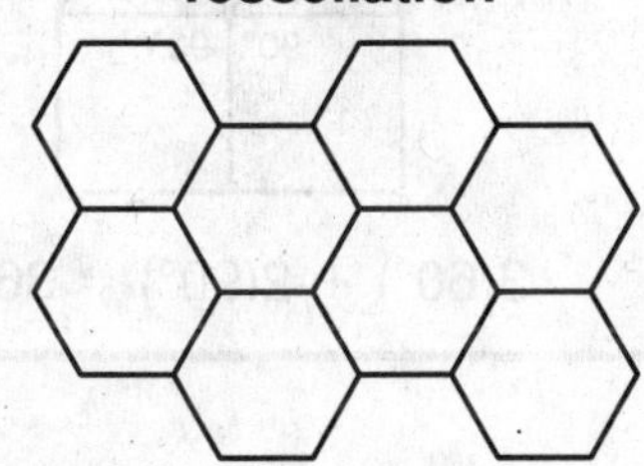

Not a Tessellation

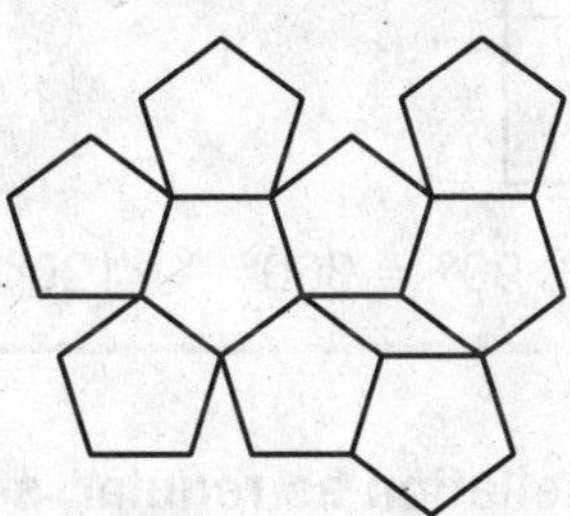

Identify the symmetry in each pattern.

1. 

____

____

2. 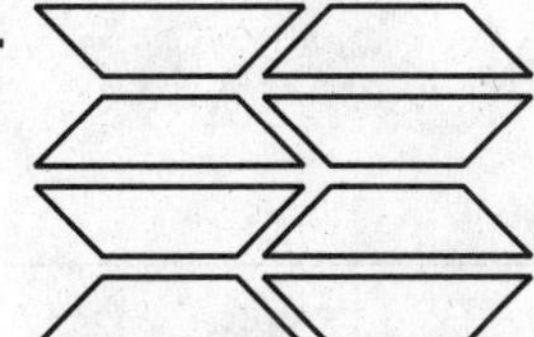

____

____

Copy the given figure and use it to create a tessellation.

3.

4. 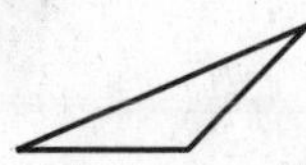

Name ______________________ Date __________ Class __________

LESSON 12-6

Review for Mastery

Tessellations continued

A **regular tessellation** is formed by congruent regular polygons. A **semiregular tessellation** is formed by two or more different regular polygons.

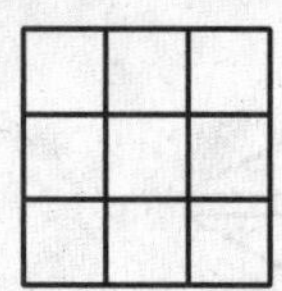

Regular Tessellation

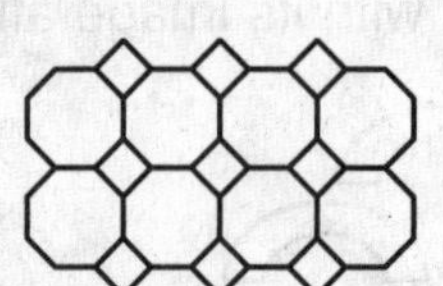

Semiregular Tessellation

In a tessellation, the measures of the angles that meet at each vertex must have a sum of 360°.

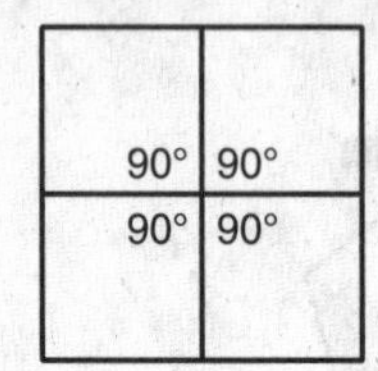

$90° + 90° + 90° + 90° = 360°$

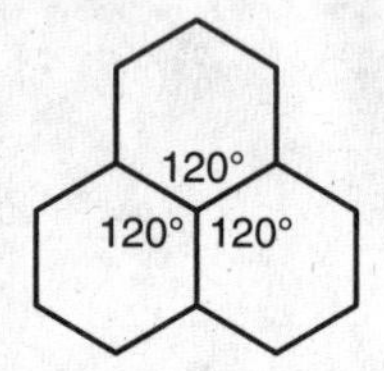

$120° + 120° + 120° = 360°$

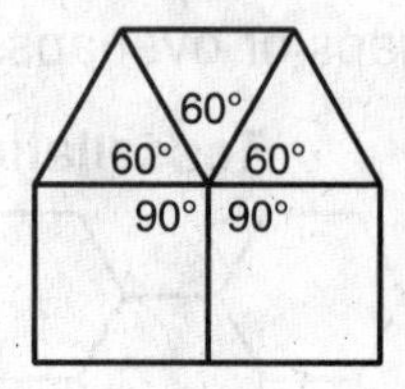

$3(60°) + 2(90°) = 360°$

Classify each tessellation as regular, semiregular, or neither.

5.

____

6.

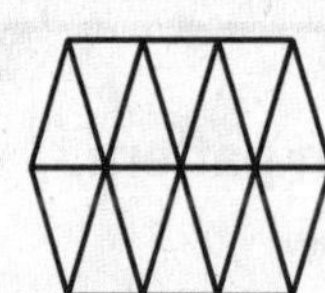

____

Determine whether the given regular polygon(s) can be used to form a tessellation. If so, draw the tessellation.

7.

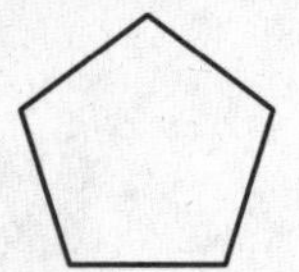

____

8.

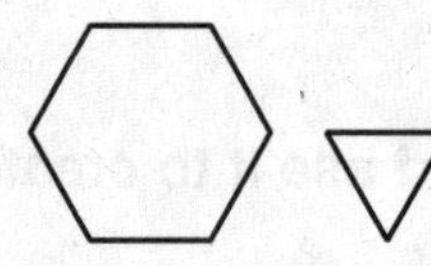

____

Name ______________________ Date __________ Class __________

LESSON 12-7

Review for Mastery

Dilations

A dilation is a transformation that changes the size of a figure but not the shape.

Dilation

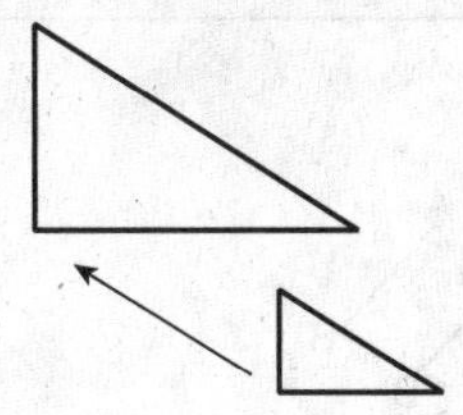

Not a Dilation

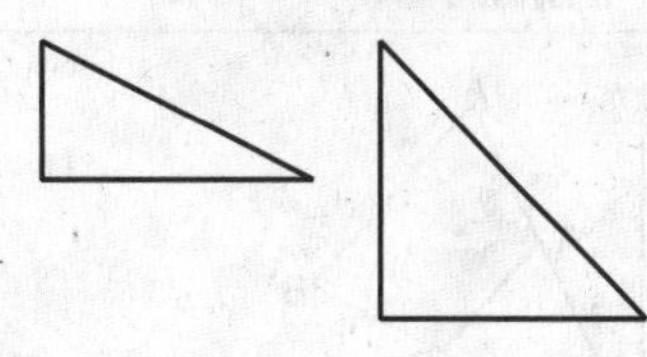

A dilation is a transformation in which the lines connecting every point *A* with its image *A′* all intersect at point *P*, called the **center of dilation**.

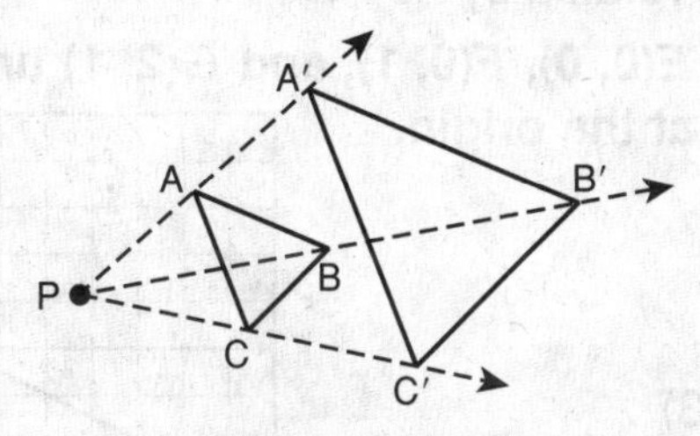

Tell whether each transformation appears to be a dilation.

1.

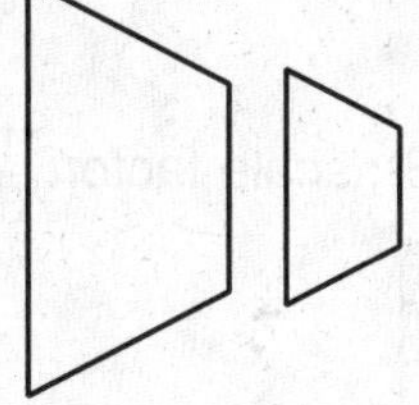

____

2.

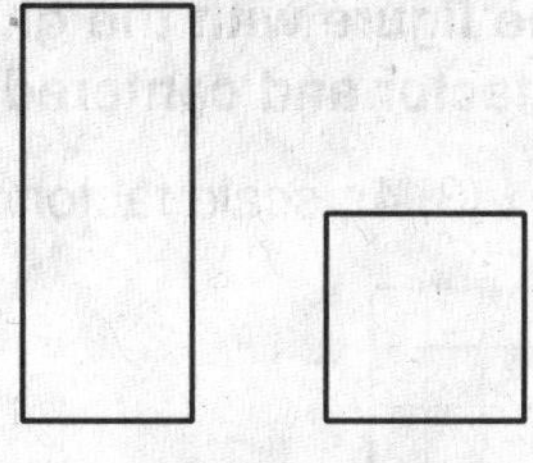

____

Copy teach triangle and center of dilation. Draw the image of the triangle under a dilation with the given scale factor.

3. scale factor: 2

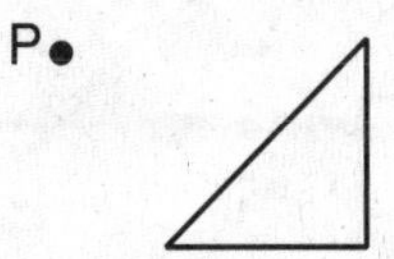

4. scale factor: $\frac{1}{2}$

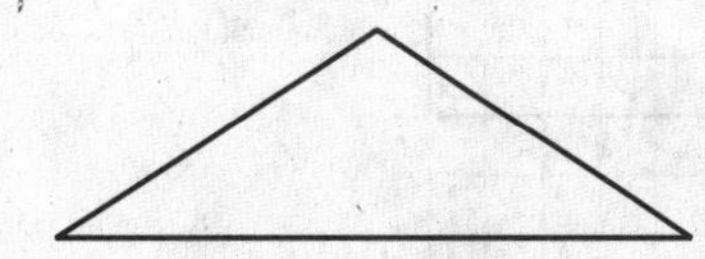

Name _______________ Date _______ Class _______

LESSON 12-7

Review for Mastery

Dilations continued

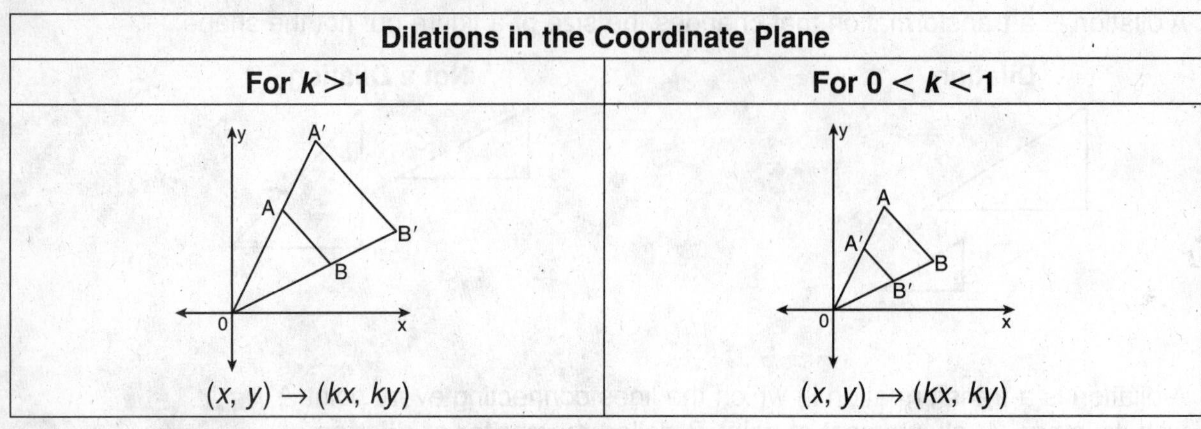

If k has a negative value, the preimage is rotated by 180°.

Draw the image of $\triangle EFG$ with vertices $E(0, 0)$, $F(0, 1)$, and $G(2, 1)$ under a dilation with a scale factor of -3 and centered at the origin.

The image of (x, y) is $(-3x, -3y)$.

$E(0, 0) \rightarrow E'(0(-3), 0(-3)) \rightarrow E'(0, 0)$

$F(0, 1) \rightarrow F'(0(-3), 1(-3)) \rightarrow F'(0, -3)$

$G(2, 1) \rightarrow G'(2(-3), 1(-3)) \rightarrow G'(-6, -3)$

Graph the preimage and image.

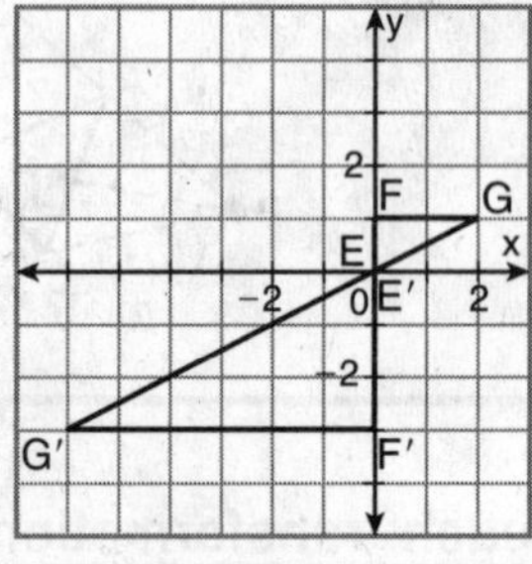

Draw the image of the figure with the given vertices under a dilation with the given scale factor and centered at the origin.

5. $J(0, 0)$, $K(-1, 2)$, $L(3, 4)$; scale factor: 2

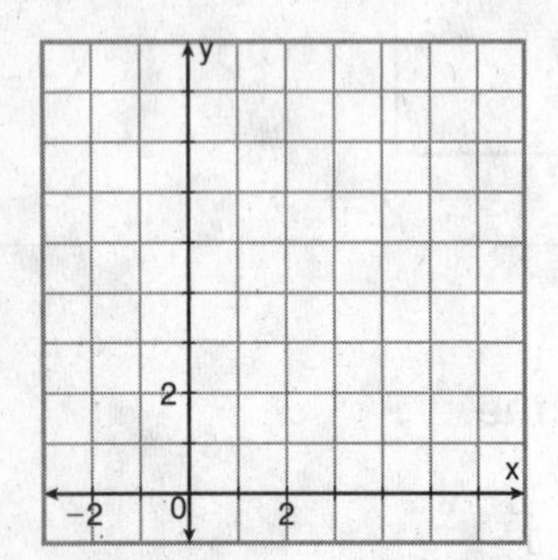

6. $A(0, 0)$, $B(0, 6)$, $C(6, 3)$; scale factor: $\frac{1}{3}$

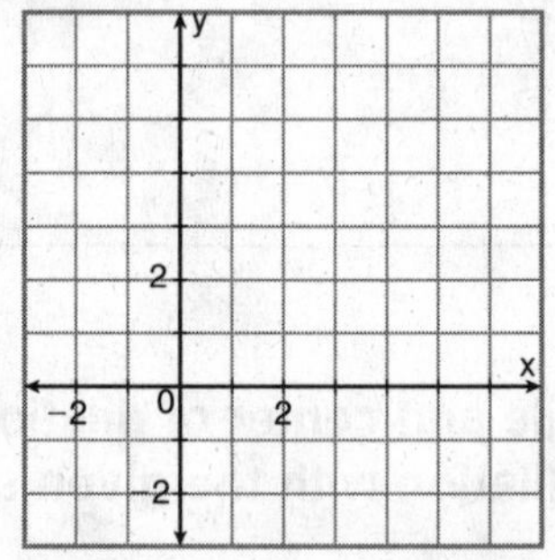

7. $R(1, 0)$, $S(1, -2)$, $T(-1, -2)$; scale factor: -2

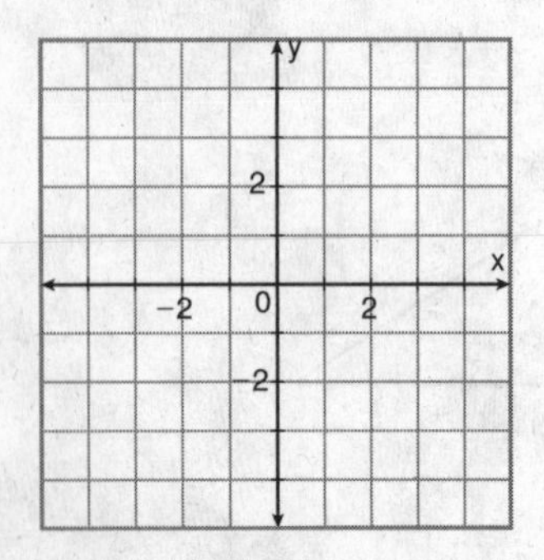

8. $G(2, 0)$, $H(0, 4)$, $I(4, 2)$; scale factor: $-\frac{1}{2}$

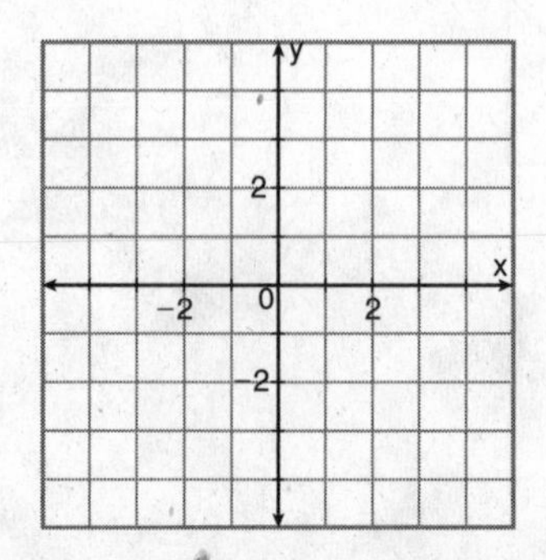